ACID WATERS IN WALES

MONOGRAPHIAE BIOLOGICAE

VOLUME 66

Series Editors

H.J. Dumont and M.J.A. Werger

Acid Waters in Wales

Edited by

R.W. EDWARDS, A.S. GEE

and

J.H. STONER

KLUWER ACADEMIC PUBLISHERS
DORDRECHT / BOSTON / LONDON

Library of Congress Cataloging-in-Publication Data

Acid waters in Wales / edited by R.W. Edwards, A.S. Gee, J.H. Stoner.
p. cm. -- (Monographiae biologicae ; v. 66)
ISBN 0-7923-0493-4
1. Acid rain--Environmental aspects--Wales. 2. Land use--Wales--Management. 3. Water quality management--Wales. I. Edwards, R. W. II. Gee, A. S. (Alun S.) III. Stoner, J. H. (John H.)
IV. Series.
QP1.P37 vol. 66
[QH545.A17]
574 s--dc20
[574.5'222]
89-39548
CIP

Published by Kluwer Academic Publishers,
P.O. Box 17, 3300 AA Dordrecht, The Netherlands

Kluwer Academic Publishers incorporates
the publishing programmes of
D. Reidel, Martinus Nijhoff, Dr W. Junk and MTP Press.

Sold and distributed in the U.S.A. and Canada
by Kluwer Academic Publishers,
101 Philip Drive, Norwell, MA 02061, U.S.A.

In all other countries, sold and distributed
by Kluwer Academic Publishers Group,
P.O. Box 322, 3300 AH Dordrecht, The Netherlands.

printed on acid free paper

Printed in the Netherlands

Llyn Brianne Reservoir from the dam.

The river's hard at it,
Tries and tries to wash and revive
A bedraggle of dirty bones. Primitive, radical
Engine of earth's renewal. A solution
Of all dead ends – an all-out evacuation
To the sea.

Ted Hughes
Poet Laureate

This book is dedicated to Guardians of the Water Environment

Preface

This book brings together research into the process of stream acidification and its impact on Welsh surface waters, carried out over the past decade or so. It is perhaps surprising that not until the 1980's was clear evidence of stream acidification assembled.

In Wales, concerns over pollution had focused water quality sampling principally on the areas of traditional heavy industry and large urban populations served by inadequate sewerage systems and sewage disposal arrangements. Mistakenly, it had been assumed that, with its prevailing westerly winds, Wales would receive precipitation substantially unpolluted by the industrial and urban emissions from Britain and mainland Europe. Assurance of the high quality of Welsh upland streams, the traditional nursery ground of salmonids, was eroded particularly by studies in the vicinity of Llyn Brianne reservoir in the catchment of the River Tywi of Central Wales. These demonstrated a clear correspondence between the biological quality and fisheries of streams in the catchment and aspects of stream chemistry, particularly pH, aluminium and calcium on the one hand, and catchment land-use on the other.

It is salutary to record that the first signals were of an inexplicable failure of the runs of migratory salmonids into the upper catchment, occupied by the Llyn Brianne reservoir and its influent streams, and the failure to restore the fishery by re-stocking with eggs and fry. Only then did the significance of the recent decline in some other upland lake and reservoir fisheries in Wales become apparent.

Subsequent studies have developed along the following paths:

i) analytical and experimental investigations of inputs, pathways and effects of acid precipitation, particularly in Llyn Brianne catchment but also elsewhere (see Chap. 1), taking into account the potentially modifying effects of forms of land-use and management. These studies include the manipulation of land-management practices, particularly in the application of lime products.
ii) a synoptic survey of stream chemistry at over 100 sites throughout Wales, relating water quality to upstream land-use and, wherever possible,

establishing the biological status of these sites.

iii) palaeoecological studies to establish the historical changes in lake ecosystems and explore relationships with land use and atmospheric pollution.

iv) development and application of models which facilitate the prediction of changes in the biological and chemical quality of surface waters with changes in emission patterns of atmospheric pollutants and in land-use and management practices.

All these components seem essential to establishing the extent and severity of damage, in understanding the elements essential for its causation, and in assessing the opportunities for ameliorating damage within an economic framework and within a realistic time-scale of recovery.

These studies have together demonstrated that in a substantial area of upland Wales, particularly the Cambrian Mountains, streams and lakes have either acidified or are likely to acidify over the next century unless there are significant changes in emission and land-use policies. Effective amelioration within a reasonable timescale will only be achieved if proposed reductions in emissions are further renewed downwards, procedures to control land-use and management in sensitive areas are strengthened and liming is encouraged in suitable locations.

Such a wide-ranging programme could not have been attempted without, firstly, the collaboration of many research organisations and individuals contributing complementary expertise (see Chap. 1) and, secondly, the injection of adequate funding for the provision of equipment, costs of land-treatments and the recruitment of contract staff. The major costs of the programme have been borne by the Welsh Water Authority, particularly in providing core staff and extensive analytical services (and from September 1989 by Welsh Region of the National Rivers Authority). We are also particularly grateful to the Department of the Environment and Welsh Office for much of this supplementary funding which will ensure that the programme of catchment studies is maintained at least until 1991.

The Editors wish to thank the many people who helped with the individual studies described here but whose efforts are not always recognised by authorship. The preparation of the book was greatly assisted by Alistair Donald, Bob Hemmings and Sue Price who respectively checked the draft, traced figures and organised references.

Ron Edwards
Alun Gee
John Stoner

Contents

1.	Introduction to Welsh Studies A.S. Gee	1
2.	Chemical and ecological evidence on the acidification of Welsh lakes and rivers S.J. Ormerod and A.S. Gee	11
3.	Recent acidification of upland lakes in North Wales: palaeolimnological evidence S.C. Fritz, A.M. Kreiser, P.G. Appleby and R.W. Battarbee	27
4.	Atmospheric Deposition A.P. Donald, J.H. Stoner, B. Reynolds, F. Oldfield, B. Rippey and J. Natkanski	39
5.	The role of geology and soils in controlling surface water acidity in Wales M. Hornung, S. Le-Grice, N. Brown and D. Norris	55
6A.	Recent land use change R. Boon and D. Kay	67
6B.	Acidified Welsh lakes: the significance of land use and management S.T. Patrick and A.C. Stevenson	81
7.	The role of acidity in the ecology of Welsh lakes and streams S.J. Ormerod and K.R. Wade	93
8.	Effects of acidification on fish and fisheries in Wales N.J. Milner and P.V. Varallo	121
9.	Acid deposition – Drinking Water Quality and Health C.B. Buckley and L. Keil	145
10.	Ecotoxicological studies of acidity in Welsh streams N.S. Weatherley, C.P. McCahon, D. Pascoe and S.J. Ormerod	159

11. Dissolved inorganic aluminium in acidic stream and soil waters in Wales — 173
C. Neal, B. Reynolds, P.A. Stevens, M. Hornung and S.J. Brown

12. Determination of aluminium speciation in acid waters — 189
X. Goenaga and D.J.A. Williams

13. Catchment characteristics and basin hydrology: their effects on streamwater acidity — 203
S.C. Bird, R.P.D. Walsh and I.G. Littlewood

14. Water quality changes from input to stream — 223
M. Hornung, B. Reynolds, P.A. Stevens and S. Hughes

15. The influence of land management on stream water chemistry — 241
S.C. Bird, S.J. Brown and E. Vaughan

16. Hydrochemical modelling of acidification in Wales — 255
P.G. Whitehead, T.J. Musgrove and B.J. Cosby

17. Modelling the ecological impact of changing acidity in Welsh streams — 279
S.J. Ormerod, N.S. Weatherley and A.S. Gee

18. Llyn Brianne – Acid deposition modelling — 299
S.E. Metcalfe and R.G. Derwent

19. Amelioration of surface water acidity by catchment management — 311
M. Hornung, S.J. Brown and A. Ranson

Conversion table (equivalents and weights) — 329

List of authors' addresses — 331

Index — 333

Opposite page:

A. Checking recording equipment and automatic samplers in catchment 7. Note stream gauge and wind turbo-generator.

B. Biological stream sampling below recording and gauging station in catchment 10.

C. Spraying lime over forest in catchment 2.

D. Felling trees to clear stream banks in catchment 3.

A

B

C

D

1. Introduction to Welsh studies

A.S. GEE

It is now over a century since acid precipitation was first studied in Britain[11] but it is only in recent years that it has been recognised to have widespread impact. As a result of a doubling of the consumption of fossil fuels in Europe between 1940 and 1970, there has been a large increase in the emissions of oxides of sulphur and nitrogen. Together with changes in emmission strategy, this had led to an increase in the amount of acidity deposited in areas remote from major sources. Where soils and rocks offer little buffering capacity, this deposition has resulted in the acidification of surface waters and major ecological changes.[15]

The majority of published information on surface waters acidity is of North American or Scandinavian origin.[3, 9] More recently however, the problem of acidification has been identified in several areas of upland Britain.[2, 4, 13, 15] These and other studies[7] have shown that the water quality of streams, rivers and lakes is determined by the quality and quantity of atmospheric deposition, the ability of soils and rocks to neutralise acids deposited or generated within catchments and differences in land use. Acidification is thought to have resulted in ecological changes including elimination of many plant and invertebrate taxa, the loss of salmonid fisheries and a reduction in the density and distribution of higher animals such as certain amphibians, birds and mammals.

There is also some concern about the direct effects on man through the impact on potable supplies.

Large areas of Wales are particularly vulnerable to acidification as the rocks of the uplands are slow weathering and the soils have little or no acid neutralising capacity; the soft water streams and lakes have been particularly severely affected. This process of acidification has been exacerbated by land use changes – especially conifer afforestation, which alters the hydrological regime and increases the loading of pollutants on the system.

Studies of acid waters in Wales are now sufficiently advanced to permit a synthesis of ideas on mechanisms and to allow an assessment of the regional impact. This book is largely based on work undertaken by the Welsh Water Authority (WWA), Institutes of Terrestrial Ecology (ITE) and Hydrology (IH) and the Universities of London and Wales and has been structured so as

R.W. Edwards et al. (eds) Acid Waters in Wales. 1–9.

to cover the following broad subject areas: historical perspective including palaeoecological evidence, atmospheric deposition, processes within vegetation and soils, the effects of land use, biological consequences, ameliorative measures, the modelling of the interactions and the prediction of the effects of different strategies for reducing acidity.

Due to the frequent reference in this book to studies by the IH, ITE and WWA, a brief description is given in this introductory chapter of the work done by these organisations at Plynlimon, Beddgelert, and Llyn Brianne respectively, the principal research areas. Particular emphasis is placed on the Llyn Brianne study and the work that led to it because of its importance to understanding the processes and pathways of acidification and the ways in which the problem may be reduced in Wales.

Principal Welsh study sites

Beddgelert

The ITE has been studying mature conifer plantations and the impact of clearfelling and whole tree harvesting on site fertility, soil solution and stream water chemistry at Beddgelert forest since 1982. The experimental site is situated in the upper Glaslyn catchment, in the mountains of Snowdonia, some 12 km from the coast (Fig. 1.1) and receives about 2800mm of rainfall annually. The geology consists of Ordovician slates with dolerite intrusions overlain by soils dominated by ferric stagnopodzols. The forest was planted mainly between 1931 and 1936 and is dominated by Sitka spruce (*Picea sitchensis)*.

A 'small-catchment' approach has been used to determine the fluxes and concentrations of major solutes through the mature spruce plantations, and to gauge the effects of clearfelling. In addition, plot scale studies have investigated the effects of 'conventional' versus 'whole tree' harvesting on soil and soil water chemistry, vegetation establishment and long-term site nutrient depletion.

Whilst the studies have been primarily aimed at determining the effects of the clearfelling which took place in 1983–84, they have provided extensive and valuable information on the processes of acidification and the transport and exchange of pollutants through the vegetation and soils.[6, 12]

Plynlimon

Work by the ITE at Plynlimon in the upper Wye and Severn catchments (Fig. 1.1) began in 1979, to study the release and transfer of elements within soils and from soils to freshwaters and the processes controlling these transfers. The impact of upland management and land use change by agricultural improvement and plantation forestry on element transfer and release has also

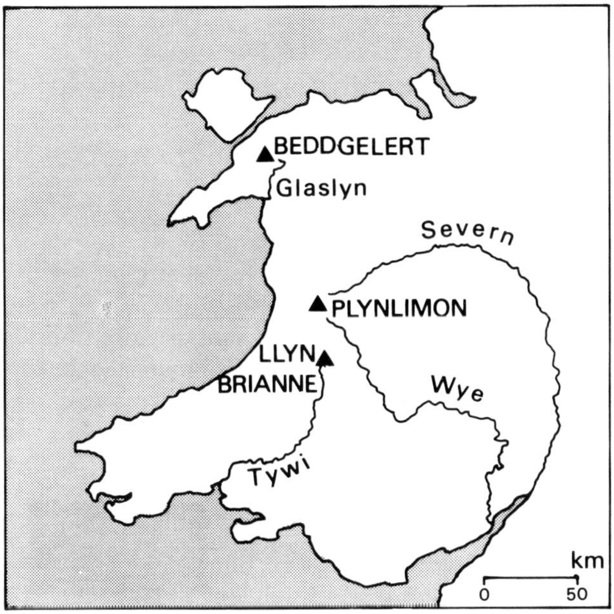

Fig. 1.1. Locations of principal Welsh study sites at Beddgelert, Plynlimon and Llyn Brianne.

been investigated. A 'small-catchment' approach was adopted for the work and within each catchment, plot based studies were undertaken on the dominant soil-vegetation complex. At each catchment site, the changes in water chemistry from precipitation input to drainage water output have been determined, encompassing measurements of the chemistry of bulk precipitation, throughfall and stemflow, soil waters from the main soil horizons and stream water. The work by ITE was set against a background of ongoing hydrological and fluvial geomorphological studies conducted by the IH.[1, 8] More recently, IH have undertaken studies of trace element hydrogeochemistry in the forest catchments of the upper Severn.[7]

The study sites are located 24km from the mid-Wales coast and receive about 2500mm of rainfall annually. In common with most of upland Wales, the geology is principally Ordovician and Silurian shales, mudstones and grits with occasional and local veins of calcite. The soils are mostly peats and stagnopodzols with some acid brown earths and stagnogleys.

The vegetation of the upper Wye comprises typical moorland communities, dominated by *Nardus-Festuca* grassland; *Eriophorum*, *Calluna* and *Vaccinium* occur on the hilltops and *Molinia* and *Juncus* in the valleys. Some parts of the catchment have been improved by means of liming, reseeding and the addition of fertilisers. In contrast, the upper Severn catchment is extensively afforested with Sitka and Norway spruces planted in 1937–64 after ploughing and digging of drainage ditches. One of the forested sub-catchments (Afon Hore) was clearfelled in 1986–87 in order to investigate the effects on the catchment water balance, sediment losses, soil and stream water chemistry. As in the Beddgelert study, measurements of the quantity and quality of water passing through the terrestrial systems have provided information on the pathways and processes of acidification.[5, 10]

Llyn Brianne

Llyn Brianne is a reservoir constructed in 1969 to regulate the flow of the River Tywi in West Wales. Studies by the WWA at Llyn Brianne date back to 1981 when the effects of water quality and land use on a fishery protection scheme were initiated. More recently, several organizations contributing to this book have extended studies to the numerous small catchments whose streams drain into the reservoir and its principal tributaries, the upper Tywi and Camddwr (Fig. 1.2).

The study sites are some 33km from the sea (Fig. 1.1), receiving about 1800mm of rainfall annually. The bedrock is very similar to that of Plynlimon, consisting of Ordovician and Silurian shales, grits and mudstones. Hill top soils are typically oligo-morphic peats with ferric stagnopodzols on upper slopes and brown podzolic soils on the lower slopes. Subcatchments of the Camddwr are unimproved moorland whilst most subcatchments of the upper Tywi were afforested in the late 1950's with Sitka spruce and Lodgepole Pine (*Pinus contorta*).

As indicated above, WWA's studies began in 1981 with an investigation of the survival of young salmonid fish in streams draining into Llyn Brianne. The very poor survival rate of these fish was attributed to the low pH and calcium and high aluminium concentrations of the streams, particularly those draining conifer afforested catchments.

Thirteen stream sites were monitored regularly allowing a classification on the basis of water hardness (as a measure of their acid neutralising capacity). Of the three primary classes two were further subdivided according to the presence or absence of conifers. Streams in each category of the classification had characteristic pH and aluminium concentrations which were reflected in the biological status (Table 1.1).

At the time of completion of this phase of the work there was little information on the acidity of atmospheric deposition[14] and the extent of the impact of stream acidification, although contemporary studies in North West Wales were demonstrating that lake fishery decline could also be attributed to acidification. Consequently, the WWA established a 15 month Regional Monitoring Programme in October 1983. A grid of 41 bulk deposition collectors was established throughout Wales to monitor rainfall quality weekly. At three additional sites rainfall was monitored daily. One hundred and twenty streams were also sampled chemically at weekly intervals and the fishery and biological status determined. Details of the studies are given in later chapters of this book.

From these investigations it became evident that more intensive and

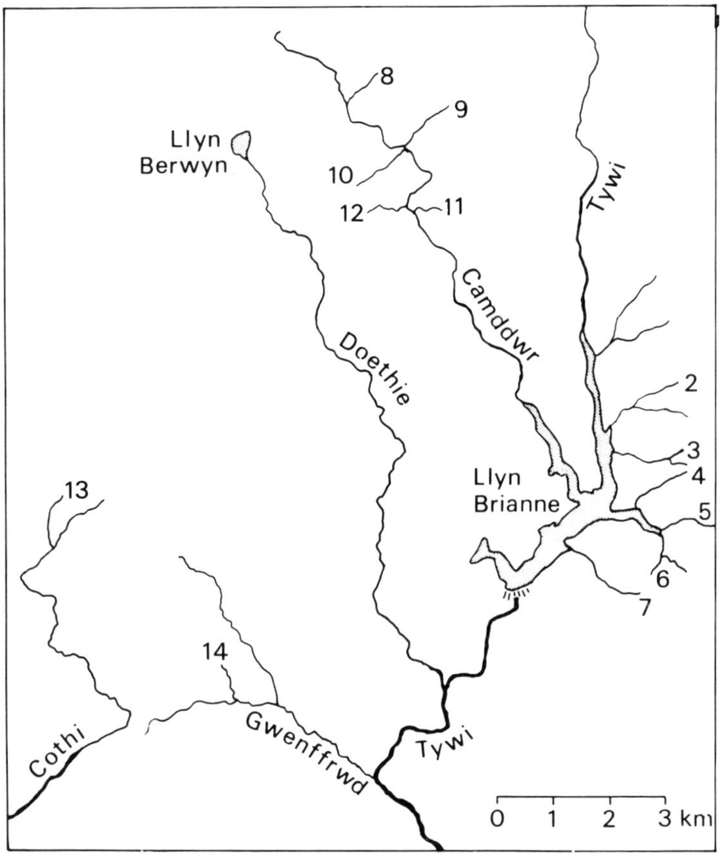

Fig. 1.2. Land management study catchments at Llyn Brianne.

detailed studies at a more limited range of sites were required to unravel the complex relationships between precipitation, geology, soils, vegetation, stream quality and aquatic ecology. In 1984 the Department of the Environment and Welsh Office provided resources to undertake such work and to carry out experimental land treatments within catchments at Llyn Brianne.

Particular emphasis is placed on the development of mathematical models which allow the prediction of the effects of changes in land-use or deposition on both stream quality and ecology. This programme, continuing until 1991, is coordinated by the Welsh Region of the National Rivers Authority (formerly Welsh Water Authority) and involves the following participants: IH and ITE; University Colleges of Aberystwyth, Bangor, Cardiff, Lampeter and Swansea; Forestry Commission and Economic Forestry Group Ltd; Welsh Office.

The catchment studies, which are referred to extensively in most chapters of this book, involve the comprehensive monitoring of deposition, meteorology, flows and quality of water through vegetation and in soils and streams.

Table 1.2. Llyn Brianne study catchments – description and treatments.

Site Number		Stream classifications (see table 1.1)	Land use	Treatment description	Treatment date	Catchment area (ha)	Treatment area (ha) and % of catchment
As used in this book	As in previous publication						
1	LI1	III B	Closed canopy conifer forest	Reference.	–	253	–
2	LI2	III B	Closed canopy conifer forest	Bankside clearance (30m width) liming at 10 tonnes ha^{-1}	May/Aug Sept 1986	105	6 (6%)
3	LI3	II B	Closed canopy conifer forest	Bankside clearance (20m width)	1983	64	2 (3%)
				Liming of water source areas in upper catchment with pelletised limestone	Summer 1988		32 (50%)
4	LI4	II B	Closed canopy conifer forest	Liming of water source area in stream headwaters (30 tonnes ha^{-1})	Sept 1987	33	4 (12%)

5	LI6	I	Unimproved unacidic moorland	Reference	–	68	–
6	LI7	I	Unimproved unacidic moorland	Reference and acid pulse experiments	–	73	–
7	LI8	II B	Open canopy conifer forest	Reference	–	66	–
8	CI2	III A	Unimproved acidic moorland	Water source area liming at 15 tonnes ha^{-1}	Summer 1988	59	4 (7%)
9	CI3	III A	Unimproved acidic moorland	Standard agricultural improvement; ploughing, liming (10 tonnes ha^{-1}), fertilizer addition and reseeding	May/June 1986	84	20 (24%)
10	CI4	III A	Unimproved acidic moorland	Contour ploughing in accordance with current forestry practice	Sept/Oct 1986	49	8 (16%)
11	CI5	III A	Unimproved acidic moorland	Surface liming at 9 tonnes ha^{-1}	Sept 1987	34	34 (100%)
12	CI6	III A	Unimproved acidic moorland	Reference	–	72	–
13	UC4	III A	Unimproved acidic moorland	Ploughing and planting with conifers	Spring 1989	237	50 (21%)
14	GI1	III A	Acid oak woodland	Reference	–	18	–

Table 1.1. Classification of streams in relation to water hardness, pH and aluminium concentration and their fishery status.

Class	Mean hardness* (mgl^{-1}, $CaCO_3$)	Land use	Mean pH	Mean soluble aluminium concentration (u equiv.1^{-1})	Fishery status
I	>10	Moorland	>6	<15	Abundant
IIA	8–10	Moorland	5.5–6.0	<15	Abundant
IIB	8–10	Conifer	5.0–5.5	25–38	Absent
IIIA	6–8	Moorland	5.0–5.5	15–20	Poor
IIIB	6–8	Conifer	4.5–5.0	28–50	Absent

* Due to exceptionally unseasonal weather in 1985 and 1986, the water hardness of some streams was reduced below the limit shown in Table 1.1. However the values shown are thought to be typical of average conditions.

Five catchments, remaining unchanged, are used as references with measurements being taken of existing conditions. An additional stream is used for studies of artificially induced acid episodes. In eight other catchments, after a period of base-line monitoring, various land management techniques have been applied: these are summarised in Table 1.2.

Throughout this book reference will be made to these study catchments and their land-use treatments. For ease of reference the catchments have been numbered 1 to 14 (Table 1.2 and Fig. 1.2). However, most other publications on the Llyn Brianne study use a different site numbering system which is included in Table 1.2 in order to assist the reader who wishes to consult these research papers. To further facilitate the comparison of the water quality and ecology of these streams with the earlier Welsh studies, Table 1.2 also includes their designation within the classfication shown in Table 1.1.

References

1. Arkell, B., Leeks, G., Newson, M.D. and Oldfield, F. (1983). Trapping and tracing: some recent observations of supply and transport of coarse sediment from upland Wales, *Publ. Int. Ass. Sediment.*, 6: 107–119.
2. Fritz, S.C., Stevenson, A.C., Patrick, S.T., Appleby, P.G., Oldfied, F., Ripey, B., Barley, J. and Battarbee, R.W. (1986). Paleoecological evaluation of the recent acidification of Welsh lakes. 1. Llyn Hir, Dyfed. Department of Geography, University College London, Working Paper No. 16.
3. Haines, T.A. (1981). Acidic precipitation and its consequences for aquatic ecosystems. A review. *Trans. Am. Fish. Soc.*, 110: 669–707.
4. Harriman, R. and Morrison, B.R.S. (1981). Forestry, fisheries and acid rain in Scotland. *Scott. For.*, 36: 89–95.
5. Hornung, M., Stevens, P.A. and Reynolds, B. (1986). The impact of pasture improvement on the soil solution chemistry of some stagnopodzols in mid-Wales. *Soil Use and Manage.*, 2: 18–26.
6. Hornung, M., Stevens, P.A. and Reynolds, B. (1987). The effects of forestry on soils, soil water and surface water chemistry. In: *Environmental Aspects of Plantation Forestry in Wales*, pp. 25–36, Ed. J.E.G. Good, Institute of Terrestrial Ecology, Grange-over-Sands.
7. Neal, C., Smith, C.J., Walls, J. and Dunn, C.S. (1986). Major, minor and trace element mobility in the acidic upland forested catchment of the River Severn, mid-Wales, *Q.J. Geol. Soc.*, 143: 635–648.
8. Newson, M.D. (1979). The results of ten years experimental study on Plynlimon, mid-Wales, and their importance for the water industry. *J. Water Eng, Scientists*, 33: 321–333.
9. Overrein, L.N., Seip, H.M. and Tollan, A. (1980). Acid precipitation-effects on forest and fish. Final report of the SNSF-project 1972–1980, 2nd edn. SNSF Project, Oslo-As.
10. Reynolds, B., Hornung, M. and Stevens, P.A. (1984). Factors influencing bulk precipitation chemistry at an upland site in mid Wales. Int. J. *Environ. Stud.*, 22: 195–205.
11. Smith, R.A. (1872). Air and rain: the beginnings of a chemical climatology. London.
12. Stevens, P.A. (1987). Throughfall chemistry beneath Sitka spruce of four ages in Beddgelert Forest, North Wales, UK, *Plant and Soil*, 101: 291–294.
13. Stoner, J.H., Gee, A.S. and Wade, K.R. (1984). The effects of acidification on the ecology of streams in the upper Tywi catchment in West Wales. *Environmental Pollution* (Series A), 35: 125–157.
14. UKRGAR (1983). Acid deposition in the United Kingdom. United Kingdom Review Group on Acid Rain. Warren Spring Laboratory, Stevenage.
15. UKAWRG (1986). Acidity in United Kingdom Fresh Waters. United Kingdom Acid Waters Review Group Interim Report, Department of the Environment, April, 1986.

2. Chemical and ecological evidence on the acidification of Welsh lakes and rivers

S.J. ORMEROD and A.S. GEE

Introduction

Over the last decade, the acidification of surface waters has been a major focus for political and scientific activity. Effects on the chemical, ecological, economic and aesthetic status of lakes, rivers and wetlands have all been suggested, and have serious implications (Chaps. 7, 8 & 9).

There are many water bodies which today have low pH, low alkalinity and elevated metal concentrations[1] indicating that acidification has occurred at some stage in the past. However, both natural[4, 22] and man-induced phenomena[2, 5, 9] could be responsible for such change, and the key question is not 'has acidification occurred?', but 'when, how much and by what processes?' In many cases we rely on inferential techniques such as modelling (Chap. 16) or palaeoecology (Chap. 3) to provide the answers. These inferential techniques give important indications on the causes and timing of acidification, and long-term modelling and diatom stratigraphy are both techniques which have now been applied in Wales. They are mutually corroborative in suggesting recent acidification. On the one hand, pH records inferred from the diatom remains in seven Welsh lakes show a pH reduction of 0.5–1.8 units from initial values at the beginning of the 19th century of pH 6–6.6. Pronounced chemical change in these systems occurred from 1850–1940 onwards, and can only be explained adequately by acid deposition (Chap. 3). On the other hand, models linking deposition quality to runoff chemistry show that historical trends in acid deposition can explain the pattern reconstructed from diatom remains using simple chemical processes in sensitive catchments (Chap. 16). Both types of investigation also show that forest can exacerbate the chemical influence of acid deposition.

In few instances, however, have trends in acidification been described directly using historical data on chemical and ecological change.[7, 11, 29] This chapter reviews such historical and contemporary evidence in Wales by pursuing four lines:

1. Spatial comparisons between streams draining conifer and moorland
2. Historical data on water quality

R.W. Edwards et al. (eds) Acid Waters in Wales. 11–25.

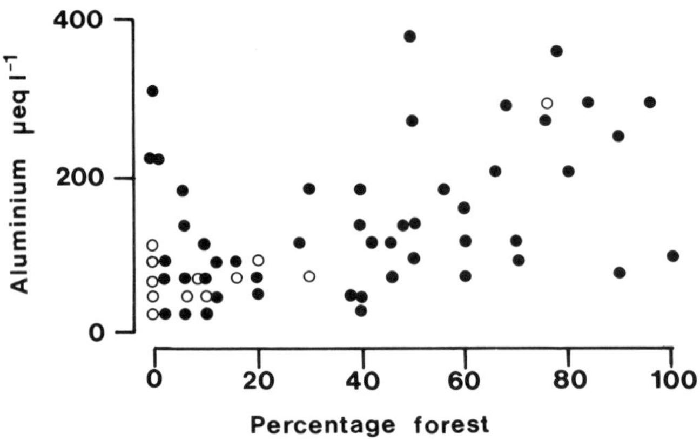

Fig. 2.1. Aluminium concentrations in relation to catchment forest cover at sites from the Welsh Water Authority's regional survey of acid waters (●, 1 point; ○ > 1 point).

3. Historical data on ecology
4. Historical data on fisheries

1. Spatial comparisons between streams draining conifer and moorland

Data from several 'acid-sensitive' regions of the UK[3, 9, 26] have now revealed that streams draining plantations of exotic conifers are more acid, and contain higher concentrations of aluminium, than streams draining moorland. The phenomenon is particularly well noted at Llyn Brianne (Chap. 15), at Plynlimon (Chaps. 14 & 15), and throughout those parts of the Welsh region vulnerable to acidification; data from WWA's regional survey revealed a striking correlation between catchment forest cover and aluminium concentration (Fig. 2.1).

The contrast between forest and moorland streams persists even where they are adjacent, and drain similar soil types (Chap. 15). This would indicate that only forest presence can explain the observed difference in water chemistry, and hence that the increase in aluminium concentrations and fall in pH have arisen since planting. The alternative explanation, that forests in acid sensitive areas were planted predominantly alongside streams which already had increased aluminium can probably be discounted: process studies in afforested catchments have shown elevated sulphate concentrations in throughfall and stemflow at afforested sites, which in turn mobilize aluminium in forest soils (Chap. 14). Thus, because most conifer forest in Wales has been planted since 1950 (Fig. 2.2), some degree of recent change in pH and aluminium concentrations has often been inferred. However, this logic does not necessarily eliminate the role of acid deposition in preferentially influencing streams draining forest at some stage since planting. Indeed, major accelera-

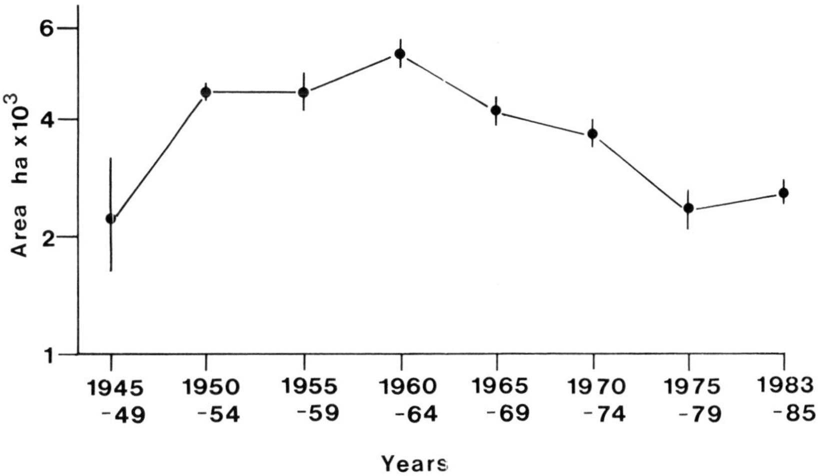

Fig. 2.2. Annual rates of forest planting and restocking in Wales. The values are five-year means ± S.D. (from Ormerod and Edwards 1985).

tion in acidification in some unafforested systems (e.g. post 1960 in Llyn Llagi, post 1940 in Llyn Hir, post 1945 in Llyn Gynon; see Chap. 3) occurred contemporaneously with the major increase in Welsh forest cover. Such a pattern would suggest that sensitive Welsh catchments were responding to acid deposition at about the same time that forest cover was being extended. Nevertheless, in view of the markedly elevated aluminium and H^+ concentration in forest streams, exacerbation of acidification seems almost certain to have occurred in association with increased conifer cover over the last five decades.

2. Historical data on water quality

One way to examine the relative roles of forest, acid deposition or other factors in influencing stream chemistry would be through the examination of historical data on water chemistry. For example, Haines[7] and Howells[11] reviewed several data-sets from Scandinavia and North America which had been used for similar purposes. When such data are carefully collected from the same site and using the same methods, they provide valuable indication on the rate and causes of change.[29] However, many problems arise during attempts to assess trends in acidity from chemical data. Warren et al.[29] could find few historical runs of pH for any site in the UK which were not subject to methodological changes, possible errors in measurement, relocation of the site of measurement, gaps in the run, and confounding changes. These features are often absent only from recent data sets (i.e. post 1970), which could post-date major changes due to acid deposition. Over such short time

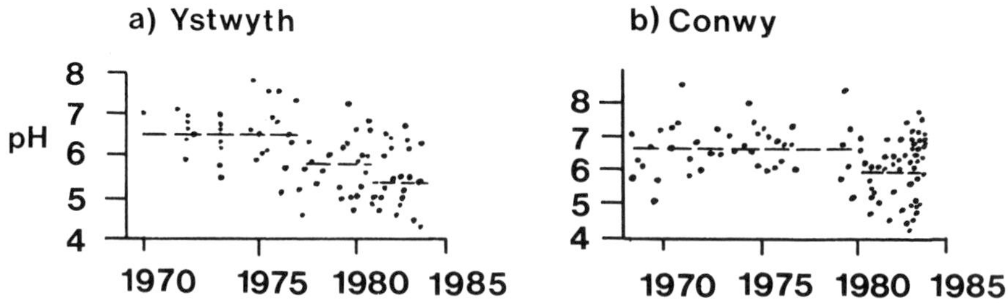

Fig. 2.3. pH values from the upper Ystwyth and Conwy between 1970 and 1984. The lines show the changing mean values according to the CUSUM analysis performed by Warren et al. (1986).

scales, it is often difficult to establish clear trends in pH that can be separated from flow-related, seasonal or annual variability.[23] A further, often overriding, problem is that the upland areas most sensitive to acidification have traditionally been considered 'unpolluted' and hence not in need of intensive monitoring. Consequently, historical data are usually lacking. Jones[16, 17] provides some of the few published pH records for Welsh rivers, indicating by colorimetric determination that the upper Rheidol and Tywi had respective pH values of 5.8–6.4 and 6.6 in the late 1940's. These rivers now have lower pH, but the limited seasonal extent of sampling in the earlier studies (for example only July values were available for the Tywi) makes comparison with present values difficult.

Amongst the 25 data sets from acid-sensitive regions which were assessed by Warren et al.,[29] six showed evidence of recent acidification according to the 'cusum' time-series analysis. Two of these were from Wales, with the Conwy showing a decline in mean pH from 6.6 to 5.8 (1969–1984), and the upper Ystwyth a decline from 6.5 to 5.5 over the same time period (Fig. 2.3).

A further 6 data sets from acid-sensitive areas of Wales were not considered suitable by Warren et al.[29] for the cusum technique because of gaps in the data-set, and because over half the samples from the streams had been taken in the final year of the run, 1984. The summer of this year was notably dry, and most pH values were unusually high. Ormerod & Edwards,[18] therefore, re-analyzed data from these six sites using a different statistical technique, confining their assessment only to winter values (i.e. October–March inclusive) which were liable to be lowest and most damaging biologically. They also pooled data separately from the afforested (Tarenig, Bidno and Irfon) and moorland catchments (upper Wye, Gwesyn), which in general had similar chemistry (Table 2.1). Overall, mean pH measured during the winter at the afforested sites fell by 1.7 pH units between 1964–70 and 1983–85, and at the moorland sites by 0.7 units (Fig. 2.4). Although the latter change might be consistent with the trend in moorland locations detected by palaeoecological investigations (Chap. 3), it was not sufficient to discount measurement error.[12] The pH reduction at the afforested sites, however, could not be

Table 2.1. Chemical data from streams in the upper catchment of the River Wye for which historical pH data were analysed by Ormerod and Edwards (1985). The values are annual means for 1984 (n = 7–45).

Site	Total hardness (mg l^{-1} $CaCO_3$)	pH	% of catchment afforested
Tarenig	8.5	5.5	36
Wye at Tarenig confluence	9.7	6.4	0
Bidno	8.7	5.9	24
Irfon, Abergwesyn	7.5	5.6	48
Irfon, Dol-y-Coed	8.4	5.5	51
Gwesyn	8.9	6.1	0

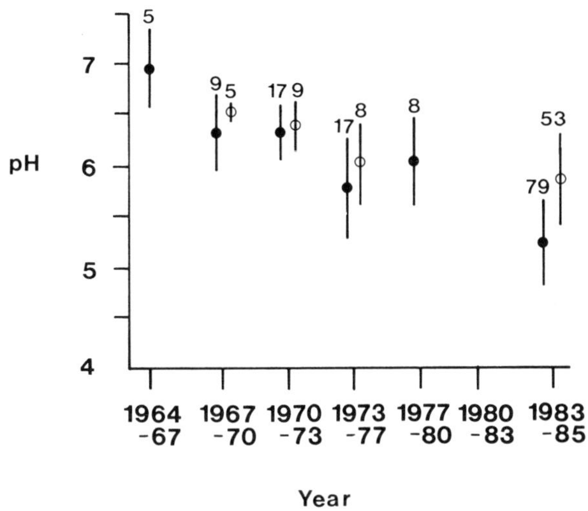

Fig. 2.4. Change in winter measures of pH at three sites on streams draining forest catchments (●), and two draining moorland (○). The values are means, with standard deviation headed by sample size (after Ormerod and Edwards 1985).

accounted for in this way. Moreover, the trend in measured pH in the forest streams occurred following major afforestation on their catchments (Fig. 2.5), and began around the time when tree growth would be in progress (1970 onwards). This period is thought to be an important phase in the acidification process under conifer trees (Chap. 14). Furthermore, the trend at the afforested sites was consistent with the pH fall detected by Warren et al.[29] in the Ystwyth, a stream not only adjacent to those assessed by Ormerod & Edwards,[18] but also draining similar soil types and with recent afforestation in its catchment. Such pH reductions could clearly be representative because of the low pH apparent today in several afforested systems.

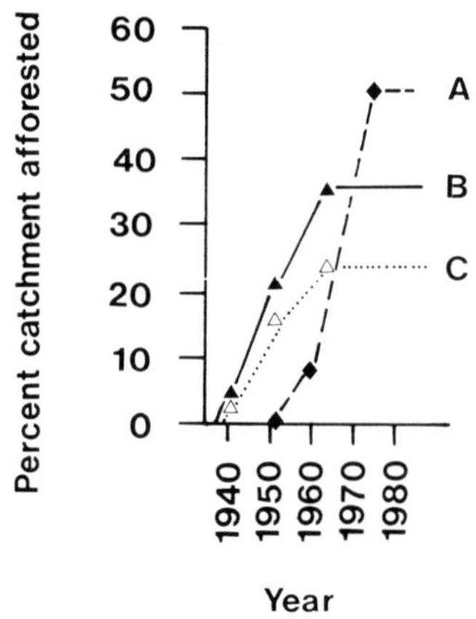

Fig. 2.5. Cumulative areas of afforestation on the upper catchments of the River Irfon (A), Tarenig (B) and Bidno (C). After Ormerod and Edwards (1985).

Although not dependent on historical data, one parameter often used in assessments of historical changes in acidity is the 'acidification index' of Henriksen.[10] In simple terms, the index represents the difference between alkalinity in the present day, and that prior to acidification. The pre-acidification alkalinity is inferred from concentrations of calcium and magnesium (sometimes with sodium and potassium) now present in surface waters, whilst the apparent loss of alkalinity is assumed to result from the deposition of strong acids. Warren et al.[30] have recently derived the index from 94 sites throughout the UK which had clear, oligotrophic waters, and pH either always or occasionally below 5.6. Overall, the UK data showed a strong correlation between the acidification index and mean 'excess' (i.e. non-marine) sulphate concentration. Amongst the twelve Welsh sites, nine had positive index values showing evidence of acidification, and included five streams in the upper Severn, one in the upper Tywi, and three lakes (Fig. 2.6). Water bodies in afforested catchments showed particularly strong acidification according to the index values, and again these were represented in Wales, most notably at the experimental sites at Plynlimon.

Many caveats apply to the Henriksen index. For example, increased calcium and magnesium concentrations sometimes occur in runoff as a response to acid deposition, and pre-acidification alkalinities can be overestimated as a result. Wright & Henriksen (1983) recognized this limitation in the model, and proposed a correction factor, yet to be applied to British surface waters. Recent comparisons between the acidification index and degree of pH change shown by diatom stratigraphy indicate that the index may have overestimated the degree of acidification (C. Neal, pers. comm), a feature which would occur if acid deposition had caused increased concentrations of calcium and magnesium.

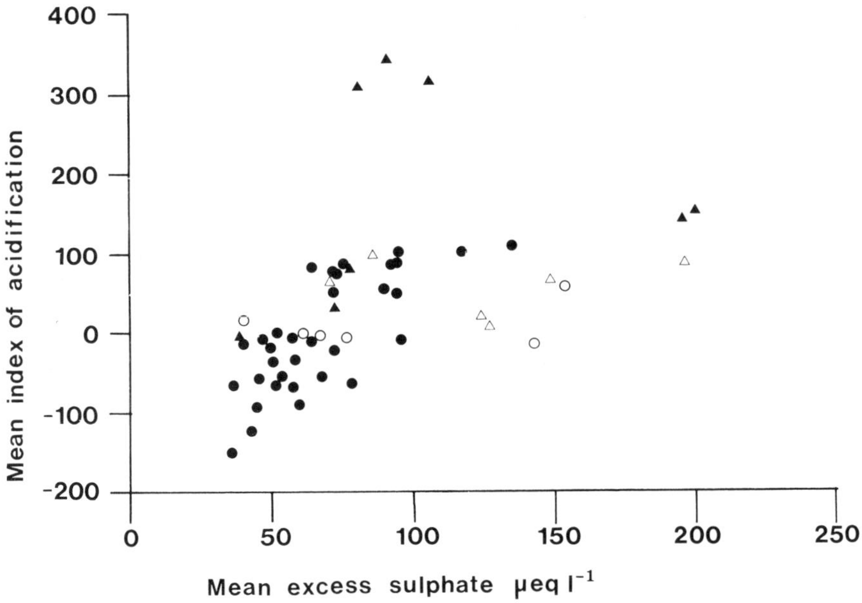

Fig. 2.6. The acidification index of Henriksen (1979) plotted against mean concentration of excess sulphate for moorland (●○) and forest streams (▲△) throughout the U.K. The Welsh streams are shown by open symbols (after Warren et al. 1988).

3. Historical data on ecology

Pronounced relationships between the water quality of rivers and their ecology have led many authors to suggest that biology can be a reliable and often accurate indicator of chemical status.[14] Advantages include the direct relevance of using biological systems in assessing chemical changes whose impacts are also biological. Additionally, ecological systems have an important indicator role in responding to both chronic and episodic conditions, a feature particularly relevant with respect to acid streams. Chapters 7 and 10, and Ormerod & Tyler[19] discuss the impact of acidification on several plants and animals which, because of their sensitivity or tolerance to acid conditions, might be useful indicators, particularly if historical changes in their distribution could be used in assessing trends.[32]

Many of the problems relating to historical chemical data also apply to ecological data, however. Usually, data are simply unavailable, other than through chance. Where data are available, recording methods may change, or impacts other than acidification could be responsible for apparent ecological differences. For example, in the River Tywi, several species of invertebrates (notably mayfly nymphs, elminthid beetle larvae and molluscs) were present in the upper reaches when the river was surveyed in the late 1940's by Jones.[17] Amongst these, *Ecdyonurus*, *Ephemerella*, *Baetis rhodani*, *Elmis aenea*

(= *Elmis maugei*) and *Ancylus fluviatilis* were not recorded when the river was resurveyed in 1985 (F.H. Jones, pers. comm). The moss, *Fontinalis antipyretica* was also recorded by Jones[17] as 'abundant', though it was absent from this site in 1988, when the liverwort *Scapania* was common (N. Stringer & R.A. Jenkins, pers. comm). From the known sensitivity of these species to low pH,[21, 28] such changes would be consistent with increased acidity. The upper reaches of the Tywi catchment have been predominantly afforested since the 1950's and 1960's, with consequent effects on acidity and aluminium concentrations. However, the construction of Llyn Brianne dam, in 1969, and the onset of river regulation, might also have altered physico-chemical conditions to the detriment of some species.[24] Causal effects by acidification are thus difficult to establish and the argument becomes circular: ecological data are required to assess chemical trends, but are unable to do so unless chemical data are also available which indicate why biological change has occurred. This problem applies widely in indicator systems which depend on 'sensitive' taxa disappearing, because it is often difficult (if not impossible) to eliminate all other possible causes.

One useful function of recorded ecological change is where it occurs in conjunction with available chemical data. In such cases, the two could be corroborative in indicating a trend. One such example in Wales concerns the river Irfon, in which a reduction in winter pH has been recorded between the 1960's and present day (see above). Ecological changes along the river involve the dipper, a species of river bird specially adapted to feed wholly on fish and benthic invertebrates, and whose density along rivers correlates strongly with stream pH and aluminium concentrations (Chap. 7). It now seems likely that food availability in acid streams limits their breeding performance.[19] Dipper populations along the upper 15 km of the River Irfon were surveyed in the mid 1950's by the Welsh ornithologist Col. H. Morrey-Salmon (pers. comm). Around 9–10 pairs per 10km of river were recorded, but this population had fallen by 1982 to only 1–2 pairs per 10 km,[20] a change clearly consistent with the reduction in measured pH. Other populations along adjacent circumneutral streams, also surveyed by Morrey-Salmon, had remained stable over the same period.

4. Historical data on fisheries

Several fish species are clearly sensitive to low pH, and there are many instances from north America and Europe in which populations have declined under circumstances where only acidification can explain the change.[7, 11] Information on such changing status arises from anecdotal accounts, catch statistics and fish population assessments, with reliability increasing in this order. Amongst these sources in Wales, there are some anecdotal accounts which describe past conditions in fisheries,[8] but, as with ecological data, it is often difficult to identify the factors responsible for change in the absence of other physico-chemical data. The gathering of scientific data for stock assessment has only occurred relatively recently, and no data are available from which to assess any long-term trends which might have followed acidification. Quantifiable assessments of changing status are, therefore, confined to catch statistics.

As with ecological data and anecdotal information, however, it may be difficult to identify all possible influences on changing fish catches. Not only do a range of environmental perturbations influence fish populations (and hence potential catch), but also catch records are dependent on other aspects of a fishery such as fishing effort and catch declaration. Nevertheless, there are some Welsh cases where definitive links have been identified between catch decline and low pH. For example Stoner & Gee[25] described two groups of adjacent lakes in mid-Wales which are located on similar soil types, but they differ in their surrounding land-use and chemical status. Craig-y-Pistyll drains moorland where its neighbour, Syfydrin, drains forest. Similarly, Rhosgoch drains moorland where Blaenmelindwr and Pendam drain forest. In comparison with the moorland lakes, the forest lakes have low pH ($< \sim$ pH 5) and elevated aluminium concentrations (> 0.2 mg l^{-1}), and in each case has shown either a decline or failure of brown trout *Salmo trutta* catches from the late 1960's or early 1970's onwards (Fig. 2.7). The moorland lakes, by contrast, have shown no such catch decline. Recent parallel studies have confirmed impaired fish survival in the afforested systems.[25]

Catches from river systems require cautious interpretation. Catch statistics for Wales are collected routinely by NRA, although declarations on rod catches are made by anglers only on a voluntary basis. Around 60% of licence holders now submit information annually, a value which represents considerable improvement on previous years. Catch returns from estuarial netsmen are much greater, and nearer to 100%, although there are instances of marked discrepancy between declared catches and those examined by bailiffs.[31] Catches will also depend on features such as time expended and efficiency of fishing gear, but it is only since 1984 that information has been requested on fishing effort. Licence sales offer some indication of the number of fishermen operating in the past, and on the River Dee, for example, the 400 licences sold in 1952 had increased to 1600 in 1984. However, changes over recent decades in mobility, economic status and leisure time clearly mean that the fishing effort expended by licensees may also have changed. Licence sales are not, therefore, a simple effort index. A further confounding influence on the catch of migratory salmonids is that there has been a trend over the last thirty years towards fish ascending rivers later in the year, probably resulting in some fish running outside the angling season (P.V. Varallo pers. comm.). Whilst the relationship between population abundance and catch is unclear, such late running is liable to remove some salmonids from the pool of catchable fish.[15]

Notwithstanding these cautions, there is some evidence of changing salmonid catches in Wales, and most concern has been expressed over rod-catches

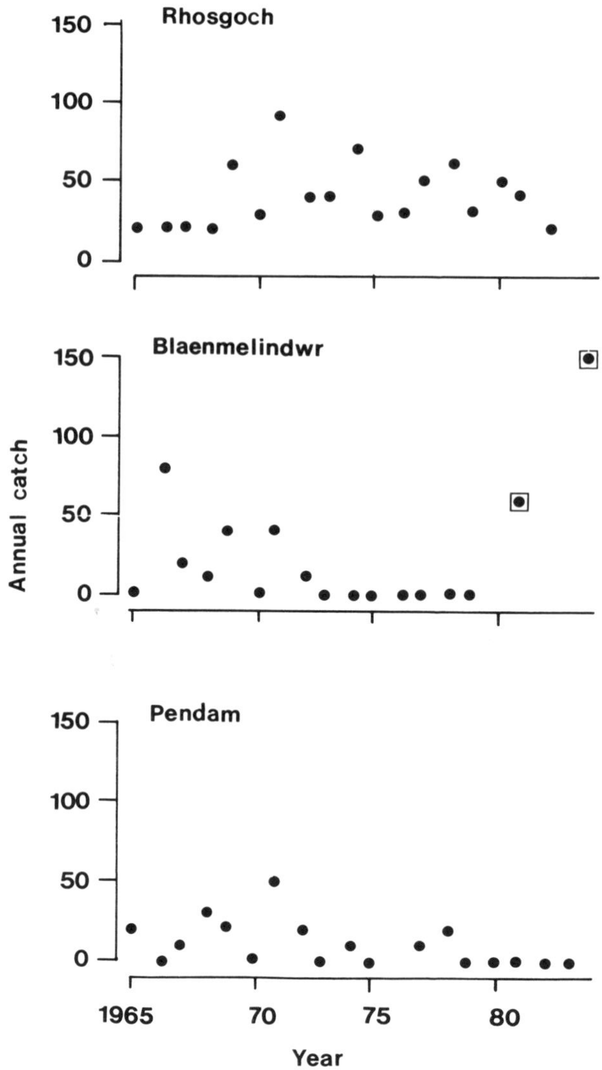

Fig. 2.7. Changes in the annual catch of brown trout in five Welsh lakes. The lakes Syfydrin, Blaenmelindwr and Pendam drain forest, whereas Rhosgoch and Craig-y-Pistyll drain moorland. The encircled values are for brook char *Salvelinus fontinalis*.

of salmon *Salmo salar*. Around 8,000 are currently caught annually, with most taken in the Wye (~ 4000), Tywi (~ 1,000) and Teifi (~ 800). Declared rod-catches of this species generally increased between the decade 1956–65 to 1966–75, which might be expected from the increasing effort over this period. Similar patterns were shown in Scotland.[6] By contrast, catches in all but two of the 22 major Welsh rivers declined between 1966–75 and 1976–85, by 2–68% (expressed as a change in annual means). The two exceptions were the

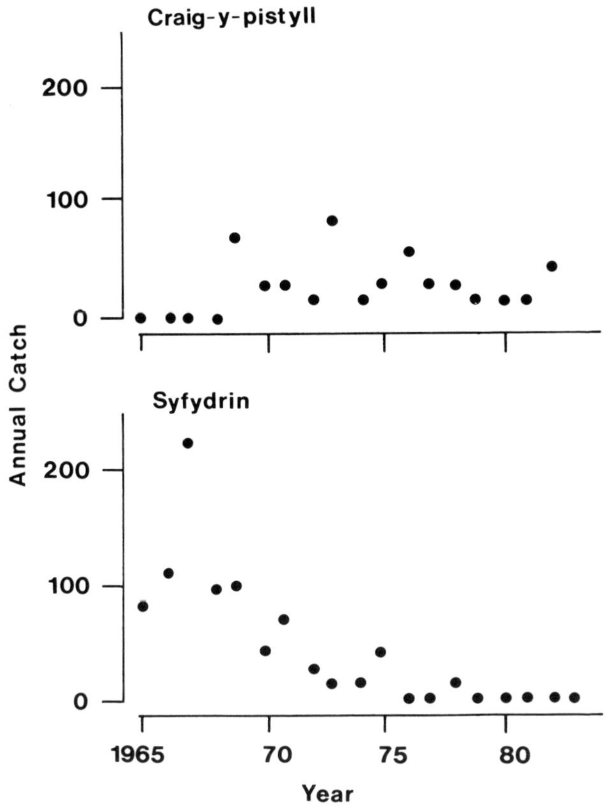

Afon Nevern, which only produces a declared catch of 2–73 fish annually (decadal means 17–24), and the Teifi. Interestingly, there is a significant correlation ($r_s = -0.605$, $P < 0.01$) between the degree of decline and the area in each catchment which gives rise to acid runoff at all flows (Chap. 5) (Fig. 2.8a). Catches declined most markedly where over 20% of river catchments were occupied by this 'land' category. Whilst such a correlation analysis requires considerable caution in view of the possibility of spurious effects, there is no obvious reason to believe that the confounding features which affect catch statistics have operated preferentially in acid sensitive catchments. Indeed, some other factors which are liable to have influenced salmonid populations can probably be excluded. Agricultural pollution incidents, for example, although increasing in number over the last decade, occur predominantly in areas of dairy farming in Dyfed and south east Wales where acid sensitivity is low.[13] Impacts by the Greenland, Irish and Faeroes sea fisheries, or by ulcerative dermal necrosis (U.D.N.), are also unlikely to have removed more fish from some Welsh rivers than from others. One conclusion might, therefore, be that the observed pattern of decline is a consequence of acidification. An anomaly, however, is that the same pattern of catch decline has not been shown by net returns. Like rod catches, net catches predominantly declined over the last two decades (Fig. 2.8c), but there is no significant correlation with the degree of acid sensitivity in the catchment, even after removing one outlying value (that for the Dwyryd; $r_s = -0.370$, N.S.). It could be argued that the rod fishery is more closely influenced by conditions in the river system than the net fishery: around 20% of salmon moving into estuaries eventually ascend other rivers,[27] and such fish possibly create an enlarged pool which could be caught by nets. Moreover, the combined decline in rod and net catches correlates more strongly with acid sensitivity than does either parameter alone ($r_s = -0.68$, $P < 0.001$). The pattern is also consistent with the significant decline in salmon catches which has occurred predominantly in Scottish fishery districts with over 20% forest cover.[6] These correlations clearly warrant closer investigation over how acid sensitivity and afforestation might have influenced salmon catches. Possibilities include the loss or impairment of spawning habitat, the development of acidic conditions which act as barriers to salmon in parts of some catchments, and the perception by anglers of worsening conditions which result in reduced fishing effort in some rivers.

Conclusions

Unlike evidence from diatom studies, there are major problems in examining the evidence for acidification on the basis of historical data on chemistry, ecology or fisheries. Most often, such data are scarce or collected only on an '*ad hoc*' basis which precludes a general assessment of long-term chemical or biological change at the Welsh regional level. Where data are available, difficulties of interpretation arise from changes in methods, study site or frequency of measurement. The causes of change are also difficult to identify since none of the available data have been collected for this purpose. Nevertheless, there are instances where ecological and chemical data are corroborative in indicating increased acidity over recent decades. Pronounced changes appear to have been in water catchments which contain conifer plantations, with evidence accumulated from pH records, acidification indices and declines amongst biological indicators. Such change is consistent with what probably has occurred since afforestation to create present-day differences between some moorland and forest streams. Such changes are also consistent with the acidification processes identified in afforested catchments. A measured decline in pH in moorland streams might also be consistent with acidification in systems independent of forest presence, as indicated by diatom studies, although the change in pH over recent decades was not greater than that expected by measurement error. The fall in the rod catch of Atlantic salmon in streams draining acid sensitive catchments warrants consideration as a possible consequence of acidification. From this analysis, it is not possible to identify separately the roles of acid deposition and afforestation, though the latter is part of the method of identifying acid sensitivity. In

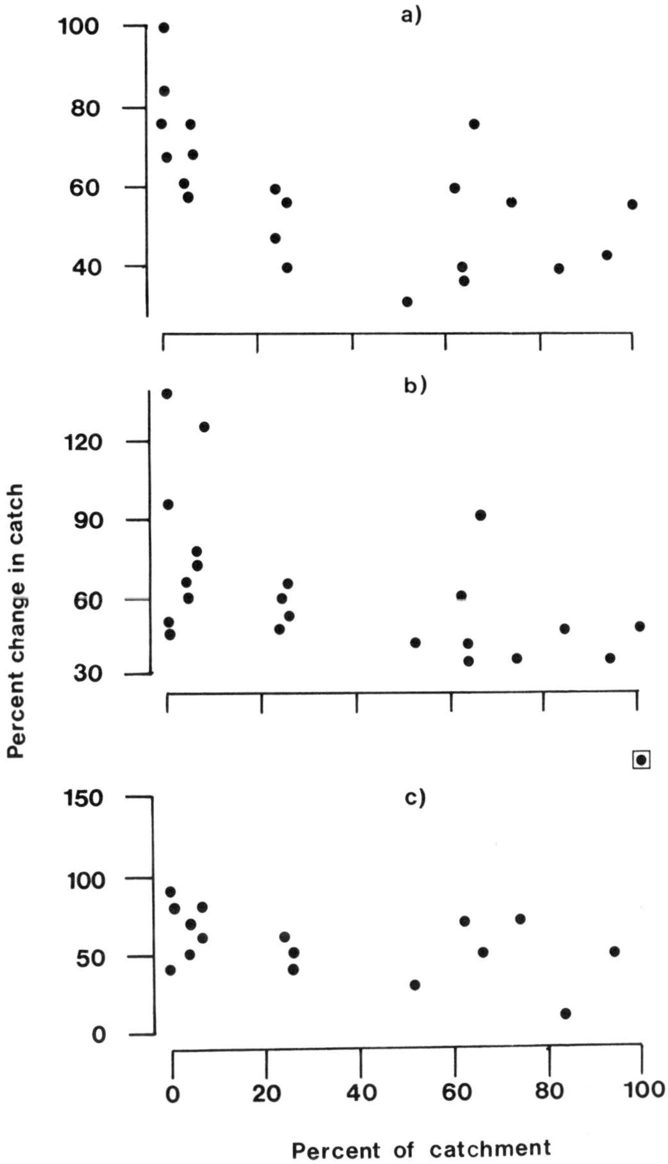

Fig. 2.8. Changes in the catch of salmon in 22 Welsh rivers. The values are mean annual catches in 1976–85 expressed as a percentage of the catch in 1966–75, plotted against the percentage area of each catchment which gives rise to acid waters at all flows. a. Rod and net catches combined ($r_s = -0.68$, $P < 0.001$);. b. rod catches ($r_s = -0.60$, $P < 0.01$); c. net catches only ($r_s = -0.37$, N.S., with one value removed).

Rivers used in assessment of catch statistics: Clwyd (49), Conwy (58), Dau Cleddau (63), Dee (53), Dwyfor (92), Dwyryd (46), Dyfi (60), Dysynni (37), Glaslyn (38), Gwyrfai (39), Llynfi (36), Loughor (98), Maddawch (45), Nevern (123), Rheidol (47), Seiont (41), Taf (75), Teifi (136), Tywi (74), Usk (50), West Cleddau (66), Wye (58). The values in parentheses are the mean catch in the decade 1976–85 expressed as a percentage of the mean catch in 1966–75.

view of similarities with the Scottish data on salmon catch, where afforestation was statistically important, there is a need for further analysis of catch change in catchments of different land use.

References

1. Altshuller, A.P. and Linthurst, R.A. (1984). The Acidic Deposition Phenomenon and its Effects: Critical Assessment & Review Papers. Vol. 2. Effects Sciences. North Carolina State University Acid Deposition Programme, EPA, Washington.
2. Battarbee, R.W. and Charles, D.F. (1986). Diatom based pH reconstruction studies in Europe and North America: a synthesis. *Water, Air & Soil Pollution*, 30: 347–354.
3. Bull, K.R. and Hall, J.R. (1986). Aluminium in the Rivers Esk and Duddon, Cumbria, and their tributaries. *Environmental Pollution (Series B)*, 12: 165–193.
4. Clymo, R.S. (1973). The growth of *Sphagnum*: some effects of environment. *Journal of Ecology*, 61: 849–869.
5. Dillon, P.J., Reid, R.A. and Grosbois, E. de (1987). The rate of acidification in aquatic ecosystems in Ontario, Canada. *Nature*, 329: 45–48.
6. Egglishaw, H., Gardiner, R. and Foster, J. (1986). Salmon catch decline and forestry in Scotland. *Scottish Geographical Magazine*, 102: 57–61.
7. Haines, T.A. (1981). Acidic precipitation and its consequences for aquatic ecosystems: a review. *Transactions of the American Fisheries Society*, 110: 669–707.
8. Hansard, G.A. (1834). *Trout and Salmon Fishing in Wales*. Longman, Rees, Orme, Brown, Green & Longman.
9. Harriman R. and Morrison, B.R.S. (1982). Ecology of streams draining forested and non-forested catchments in an area of central Scotland subject to acid precipitation. *Hydrobiologia*, 88: 251–263.
10. Henriksen, A. (1979). A simple approach for identifying and measuring acidification in fresh water. *Nature*, 278: 542–545.
11. Howells, G.D. (1983). Acid waters – the effects of low pH and associated factors on fisheries. *Advances in Applied Biology* (Ed T. H. Coaker). pp. 143–255. Academic Press, London.
12. Howells, G.D. and Haines, T.A. (1982). Interpretation of aquatic pH trends. *Transactions of the American Fisheries Society*, 111: 779–786.
13. Howells, W.R. and Merriman, R. (1986). Pollution from agriculture in the area of the Welsh Water Authority. *Effects of land use on fresh water* (Ed. J.F. de L.G. Solbe), paper 18. Water Research Centre and World Health Organization, Medmenham.
14. Hynes, H.B.N. (1960). *The biology of polluted waters*. Liverpool University Press.
15. Jenkins, D. and Shearer, W.M. (1986). *The status of the Atlantic Salmon in Scotland*. Institute for Terrestrial Ecology Symposium Number 15, Abbots Ripton.
16. Jones, J.R.E. (1949). An ecological study of the River Rheidol, North Cardiganshire, Wales. *Journal of Animal Ecoloy*, 18: 67–88.
17. Jones, J.R.E. (1951). An ecological study of the River Towy. *Journal of Animal Ecology*, 20: 68–86.
18. Ormerod, S.J. and Edwards, R.W. (1985). Stream acidity in some areas of Wales in relation to historical trends in afforestation and the usage of agricultural limestone. *Journal of Environmental Management*, 20: 189–197.
19. Ormerod, S.J. and Tyler, S.J. (1987). Dippers *Cinclus cinclus* and Grey Wagtails *Motacilla cinerea* as indicators of stream acidity in upland Wales. *The Value of Birds*, (Eds. A.W. Diamond and F.L. Filion) pp. 191–208. International Council for Bird Preservation, Cambridge.
20. Ormerod, S.J., Tyler, S.J. and Lewis, J.M.S. (1985). Is the breeding distribution of dippers influenced by stream acidity? *Bird Study*, 32: 33–39.

21. Ormerod, S.J., Wade, K.R. and Gee, A.S. (1987). Macro-floral assemblages in upland Welsh streams in relation to acidity and their importance to invertebrates. *Freshwater Biology*, 18: 545–558.
22. Pennington, W. (1984). Long-term natural acidification in Cumbria: evidence from post-glacial lake sediment. *Fifty-second annual report of the Freshwater Biological Association*, Freshwater Biological Association, Ambleside.
23. Schindler, D.W. (1987). Detecting ecosystem response to anthropogenic stress. *Canadian Journal of Fisheries and Aquatic Sciences*, 44: (Supplement 1), 6–25.
24. Scullion, J. and Sinton, A. (1983). Effects of artificial freshets on substratum composition, benthic invertebrate fauna and invertebrate drift in two impounded rivers in mid Wales. *Hydrobiologia*, 107: 261–269.
25. Stoner, J.H. and Gee, A.S. (1985). Effects of forestry on water quality and fish in Welsh rivers and lakes. *Journal of the Institute of Water Engineers and Scientists*, 39: 27–45.
26. Stoner, J.H., Wade, K.R. and Gee, A.S. (1984). The effects of acidification on the ecology of streams in the upper Tywi catchment in west Wales. *Environmental Pollution (Series A)*, 36: 125–157.
27. Varallo, P.V. (1988). Report on M.A.F.F./W.A.A. fisheries technical liaison committee workshop on acoustic and radio tracking of adult migratory salmonids. *Report of the Welsh Water Authority, No. FTLC 88/8*, Brecon.
28. Wade, K.R., Ormerod, S.J. and Gee, A.S. (1988). Classification and ordination of macroinvertebrate assemblages to predict stream acidity in upland Wales. *Hydrobiologia*, 171: 59–78.
29. Warren, S.C., Bache, B.W., Edmunds, W.M., Egglishaw, H.J., Gee, A.S., Hornung, M., Howells, G.D., Jordan, C., Leeming, J.B., Pugh, K.B., Sutcliffe, D.W., Wells, D.E., Cape, J.N., Ellis, J.C. and Hunt, D.T.E. (1986). Acidity in United Kingdom Fresh Waters. United Kingdom Acid Waters Review Group Interim Report. Department of the Environment, London.
30. Warren, S.C., Alexander, G.C., Bache, B.W., Battarbee, R.W., Crawshaw, D.H., Edmunds, W.M., Egglishaw, H.J., Gee, A.S., Hildrew, A.G., Hornung, M., Howells, G.D., Hunt, D.T.E., Pugh, K.B., Wells, D.E., Whitehead, P.G., Cape, J.N., Harriman, R., Neal, C., and Ormerod, S.J. (1988). United Kingdom Acid Waters Review Group Final Report. Department of the Environment, London.
31. Welsh Water (1986). *Fisheries and conservation review*, 1983–85. Welsh Water Authority, Brecon.
32. Wium-Anderson, S. (1987). The development of the acid lake Gribso in Denmark after 1950. *Reversibility of Acidification*, (Ed. H. Barth) pp. 76–84. Elsevier Applied Science, London.
33. Wright, R.F. and Henriksen, A. (1983). Restoration of Norwegian lakes by reduction in sulphur deposition. *Nature*, 405: 422–424.

3. Recent acidification of upland lakes in North Wales: Palaeolimnological evidence

S.C. FRITZ, A.M. KREISER, P.G. APPLEBY and R.W. BATTARBEE

Introduction

Acidification of lakes, streams, and forests by atmospheric pollution is of widespread environmental concern, but in most cases recent acidification is difficult to demonstrate conclusively because historical records are lacking (Chap. 2). In the case of lakes, the absence of historical data can be overcome by palaeolimnological analyses, which allow the reconstruction of past environments from fossil remains in lake sediments.[4, 6]

The primary tool in the palaeolimnological reconstruction of lake pH history is diatom analysis. Diatoms are a widely distributed class of algae, which preserve well in lake sediments due to their siliceous cell wall. Diatoms are sensitive indicators of various limnological conditions, and the distribution of many taxa is strongly related to lakewater pH or factors correlated with pH. Indices and predictive equations based on this relationship allow the quantitative reconstruction of past lakewater pH from diatom remains in lake sediments.[3]

Lake acidification has been attributed to four primary causes: 1) increased deposition of atmospherically transported acids and sulphates produced by fossil-fuel combustion, 2) natural acidification caused by long-term changes in the soil and vegetation, 3) increased inputs of catchment-derived acids resulting from land-use and land-management changes, such as a reduction in the frequency of grazing and burning[13] and 4) coniferous afforestation, leading to increases in catchment-derived acids due to changes in soils and drainage, plus the capture and concentration of atmospherically derived acids.[10]

Palaeoecological analyses in Galloway, Scotland have demonstrated recent acidification of both non-afforested and afforested lakes on granite bedrock.[9] These data clearly suggest that acid deposition is the major cause of lake acidification in the non-afforested catchments, with pH declines of 0.5 to 1.2 units since 1850. Acidification of two afforested catchments (L. Dee and L. Grannoch) precedes afforestation, indicating that pH decline at these sites is also a result of acid deposition, not primarily of ploughing and forest growth.

The upland areas of mid- and north Wales are dominated by lower Palaeozoic base-deficient bedrock and contain numerous lakes susceptible to

R.W. Edwards et al. (eds) Acid Waters in Wales. 27–37.

acid deposition. Our research aims to evaluate the magnitude, timing, and causes of recent lake acidification in both moorland and afforested catchments in the Welsh uplands. We use diatom-based reconstruction techniques and ^{210}Pb dating to evaluate the evidence for recent acidification of lakes and carbonaceous particles, magnetic, and geochemical analyses to trace the depositional history of atmospheric pollutants at each site. Documentary records of catchment history and pollen and geochemical analyses of the lake sediments are used to evaluate whether any observed acidification can be correlated with changes in grazing pressure or burning in moorland catchments or with ploughing and forest growth in afforested lake catchments.

To date, we have completed analyses from one afforested and six non-afforested lakes in mid- and north Wales (Fig. 3.1). Llyn Hir and Llyn Gynon are on the central Welsh plateau, a broad plateau of Silurian shales and mudstones. To the north topography is more rugged and the bedrock variable. Llyn y Bi, Cwm Mynach, Eiddew Bach, and Dulyn are in the Rhinogs hills, dominated by Cambrian coarse-grained greywackes of the Harlech grits group. Llyn Llagi, on the southern edge of Snowdonia, is underlain by Ordovician slates and shales of the Glanarfon series.[1] In this chapter we summarise the results from these seven non-granitic sites, focusing on Llyn y Bi with its moorland catchment and Llyn cwm Mynach with its afforested catchment as case studies to exemplify our approach.

CASE STUDY 1: Llyn y Bi – a lake with a moorland catchment

Llyn y Bi is a relatively shallow corrie lake fed by groundwater and surface flows and drained to the east by a small stream. The catchment is largely undisturbed upland moorland used as rough grazing for sheep (Fig. 3.2a). Morphometric characteristics of the lake and its catchment and water-chemistry data are given in Table 3.1. The contemporary fishery status of Llyn y Bi is uncertain. It formerly contained large trout and arctic char[7, 14], but a Welsh Water Authority survey in 1985 (N. Milner, pers. comm.) and several local communicants suggest that the lake is now fishless.

A 65-cm sediment core was obtained from Llyn y Bi. The analyses discussed here concern the uppermost 20 cm of the core. The ^{210}Pb inventory of the core is 5.55 pCi cm^{-2} and represents a mean ^{210}Pb flux of 0.17 pCi cm^{-2} yr^{-1}. This is comparable with several adjacent sites (Appleby, unpublished data) but is lower than the expected value of 0.50 pCi cm^{-2} yr^{-1}. Table 3.2a shows the ^{210}Pb chronologies for Llyn y Bi given by the CRS dating model.[2] A well-defined ^{137}Cs peak at 1.25 cm is consistent with the CRS model.

Sediments below 5.5 cm fall outside the range of ^{210}Pb dating. If we assume a constant sediment-accumulation rate below this depth, the 20-cm depth dates from the mid-fifteenth century.

Diatom taxa can be grouped based on their pH tolerances. Figure 3.3a shows the stratigraphic distribution of individuals in each of Hustedt's pH

Fig. 3.1. Location of study sites.

categories[3] for the Llyn y Bi sediments. Below 5 cm the flora is dominated by attached circumneutral and acidophilous taxa including *Achnanthes minutissima* Kütz., *Brachysira vitrea* (Grun.)Ross, *Fragilaria virescens* var. *exiqua* Grun., and *Eunotia incisa* W. Smith *ex* Greg.. The absence of planktonic species suggests that Llyn y Bi has been moderately acidic throughout the period represented by these sediments. The pH reconstruction (Fig. 3.3a) using multiple regression of pH groups[8] indicates a stable pH of 5.9–6.1 below 5 cm.

The first floristic changes in the Llyn y Bi sediments occur at 5 cm with the

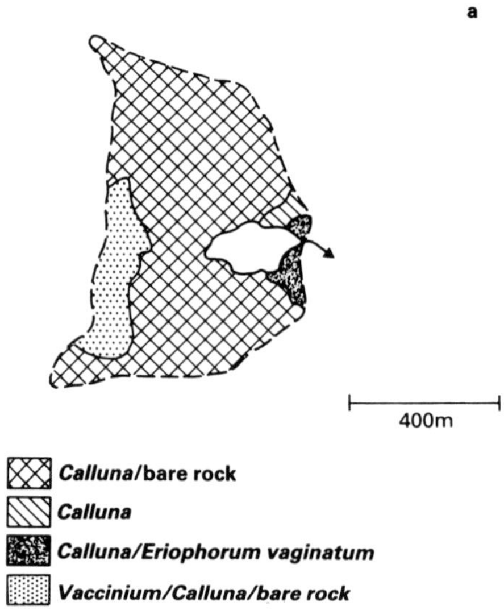

Fig. 3.2. Vegetation maps of a) Llyn y Bi and b) Llyn cwm Mynach.

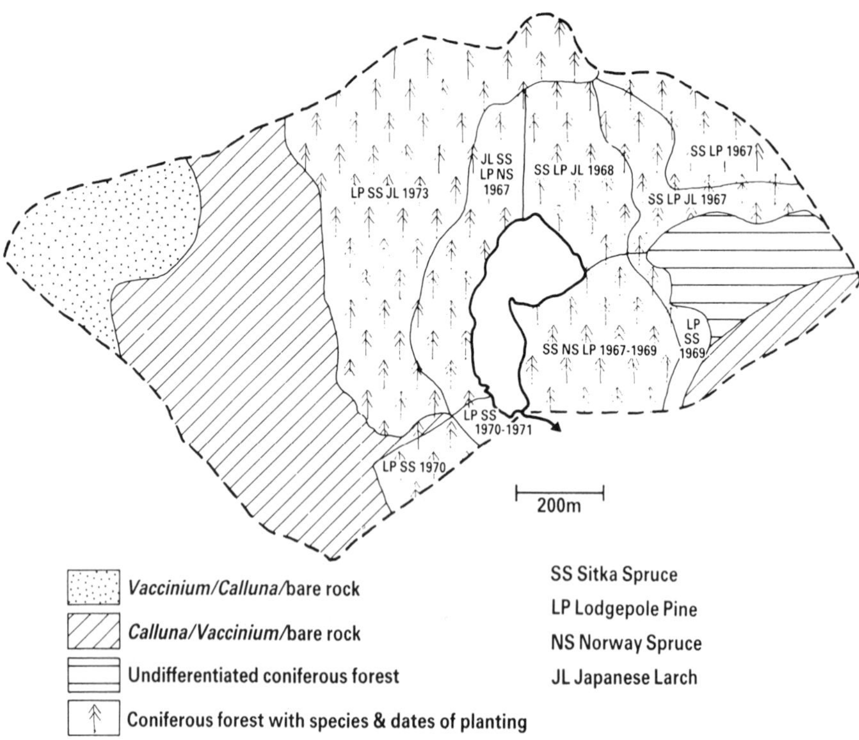

Table 3.1. Catchment and lake characteristics of Llyn y Bi and Llyn cwm Mynach.

Water chemistry data are the mean of eight measurements (1984–1986) courtesy of Welsh Water.

	L. y Bi	L. cwm Mynach
Elevation (m)	445	285
Maximum relief (m)	120	80
Catchment area (ha)	47.7	152
% forest cover	0	51
Lake area (ha)	2.7	5.89
maximum depth (m)	3	11
lake volume (m^3)	42600	58500
pH	4.8	5.7
conductivity (uS cm^{-1})	43	46
SO_4^{2-} ueq 1^{-1})	104.37	108.75
Cl^- ueq 1^{-1})	264.78	280.56
Na^+ ueq 1^{-1})	208.69	237.39
K^+ ueq 1^{-1})	23.02	20.72
Ca^{2+} ueq 1^{-1})	53.39	79.34
Mg^{2+} ueq 1^{-1})	64.97	63.73

expansion of the acidobiontic *Navicula cumbriensis* Haworth and the initial decline of the circumneutral *Achnanthes minutissima*. More dramatic floristic changes begin above 4 cm (1903), with increases in the acidobiontic taxa *Tabellaria binalis* (Ehr.) Grun. and *Navicula hoefleri* sensu Ross & Sims, as well as several acidophilous taxa. Circumneutral taxa, including *Brachysira vitrea*, *Cymbella lunata* W. Smith and *Fragilaria virescens* var. *exiqua*, decline concurrently. The progressive expansion of *Tabellaria binalis* and decline of the circumneutral species is similar to the pattern seen in the uppermost sediments of several Galloway lochs[9] and clearly indicate a progressive decline in lake pH. pH reconstruction (Fig. 3.3a) shows a decline of 1.4 units between 4 cm and the sediment surface (1903–1985).

Concentrations of atmospherically derived particles in the Llyn y Bi sediments all increase in the twentieth century (see Chap. 4). Carbonaceous particles and atmospherically derived trace metals such as lead, zinc, and copper increase sharply above 4 cm (1903). These data clearly indicate an increase in the local deposition of atmospherically transported products of fossil-fuel combustion, and the fact that lake acidification accompanies this increase strongly suggests acid deposition as a major cause of the decline in lakewater pH.

The Llyn y Bi catchment is unimproved moorland used for rough grazing, and there is no documentary evidence to suggest any major change in land use over the last few centuries (S. Patrick, unpublished data). Within the parish, however, sheep numbers have declined throughout the twentieth century, and within the Llyn y Bi catchment there has been a broad decrease in sheep numbers since about 1930, when the National Trust purchased the land. From

Table 3.2a. CRS Model ^{210}Pb chronology for Llyn y Bi core.

Depth	Dry mass	Cumul. unsupp. ^{210}Pb	Date	Chronology Age	Std. error	Sedimentation rate		Std. error
cm	g cm^{-2}	pCi cm^{-2}	AD	Yr		g cm^{-2}	cm yr^{-1}	%
0.00	0.0000	5.55	1985	0				
0.50	0.0329	4.71	1980	5	2	0.0067	0.091	6.2
1.00	0.0711	3.87	1973	12	2	0.0056	0.073	6.1
1.50	0.1102	3.00	1965	20	2	0.0047	0.059	6.3
2.00	0.1499	2.19	1955	30	2	0.0040	0.049	6.8
2.50	0.1897	1.60	1945	40	2	0.0033	0.040	7.4
3.00	0.2320	1.09	1933	52	3	0.0030	0.035	9.0
3.50	0.2769	0.69	1918	67	4	0.0032	0.034	11.9
4.00	0.3264	0.43	1903	82	5	0.0034	0.033	15.7
4.50	0.3805	0.26	1887	98	7	0.0035	0.031	20.5
5.00	0.4382	0.15	1869	116	10	0.0035	0.030	25.2
5.50	0.4996	0.09	1851	134	15	0.0035	0.028	30.0

^{210}Pb flux = 0.17 +/− 0.01 pCi cm^{-2}
90% Equilibrium depth = 3.7 cm or 0.30 g cm^{-2}
99% Equilibrium depth = 5.9 cm or 0.55 g cm^{-2}

Table 3.2b. CRS Model ^{210}Pb chronology for Llyn cwm Mynach

Depth	Dry mass	Cumul. unsupp. ^{210}Pb	Date	Chronology Age	Std. error	Sedimentation rate		Std. error
cm	g cm^{-2}	pCi cm^{-2}	AD	Yr		g cm^{-2}	cm yr^{-1}	%
0.50	0.0287	10.34	1982	3	2	0.0127	0.209	5.5
2.50	0.1507	7.11	1970	15	2	0.0082	0.126	6.0
5.50	0.3579	2.76	1940	45	3	0.0057	0.082	7.5
8.50	0.5697	0.94	1905	80	5	0.0072	0.103	15.2
11.50	0.7808	0.37	1876	109	7	0.0076	0.107	25.4
14.50	0.9939	0.13	1841	144	9	0.0049	0.070	31.9

^{210}Pb flux = 0.35 +/− 0.02 pCi cm^{-2}
90% Equilibrium Depth = 8.3 cm or 0.55 g cm^{-2}
99% Equilibrium Depth = 14.8 cm or 1.02 g cm^{-2}

the mid-nineteenth century to 1930 the catchment was managed for grouse and regularly burned, but since that date the frequency and extent of burning has declined. Thus, although there has been no major land-use change in the Llyn y Bi catchment, management changes have occurred that favour enhanced growth of *Calluna* and other ericaceous shrubs. The mature *Calluna* that presently dominates the catchment is characteristic of low-intensity grazing pressure and the absence of burning.

Rosenqvist[12] has suggested that formation of acid humus from enhanced growth of acid heathland following a decline in the intensity of pastoral agriculture is a primary cause of lake acidification. Because such changes in land management have occurred in the Llyn y Bi catchment over the period of major pH decline in the lake, we are unable to reject this land-use change as proposed by Rosenqvist, as a contributing factor in the acidification of Llyn y Bi. However, since acidification began at an earlier date and since many other lakes in the region with burnt and grazed catchments been have acidified (see below) it is probable that acid deposition has been the dominant influence.

CASE STUDY 2: Llyn cwm Mynach – a lake with a forested catchment

Llyn cwm Mynach lies to the south of Llyn y Bi at an altitude of 285 m above sea level. The lake has an irregular outline and comprises two distinct parts; an 11 m. basin in the northern half and a shallow shelf dominated by aquatic macrophytes to the south. The mean pH of the lake is 5.7. Fifty-one percent of the catchment has been planted with mixed coniferous forest in the period 1967–1973. The remaining unplanted areas consist of either bare rock or moorland vegetation (Fig. 3.2b). Prior to afforestation the catchment was grazed by sheep. An opencast manganese mine, lying partially within the catchment, operated for a short time in the late nineteenth century.

Sediment cores were taken from the deepest part of the lake and one core was dated by the ^{210}Pb method (Table 3.2b). The mean sediment-accumulation rate is calculated at 0.11 cm yr^{-1}. This is slow compared with other lakes with afforested catchments and is probably due to the lack of extensive ploughing at this site. At sites in Galloway, sediment-accumulation rates increased rapidly due to erosion following ploughing.[5]

The results of diatom analysis for the Llyn cwm Mynach core and the pH reconstruction are shown in Fig 3.3b. The diatom assemblages are dominated by circumneutral taxa although two principal circumneutral species, *Achnanthes minutissima* and *Brachysira vitrea* decline from 1870 onward. This decline is accompanied by an increase in the acidophilous *Frustulia rhomboides* var. *saxonica* and acidophilous species within the *Eunotia* genus. The presence of a small proportion of acidobiontic taxa is recorded since the mid-nineteenth century and there has been a slight increase in these diatoms since 1960. The reconstruction of pH from the diatom data shows the maintenance of pH at around pH 6.2 until the 1870's when a slight acidification occurs. An acceleration in the rate of acidification in the 1930's brings the reconstructed pH down to 5.6 in the surface sediments. This agrees well with recent water chemistry measurements. Llyn cwm Mynach has not acidified to the same extent as Llyn y Bi although the trace metal chemistry and carbonaceous particles profiles[11] indicate that the lake has been subjected to acid deposition since the 19th century, in common with other lakes in the region. The lower susceptibility of this site to acidification is largely due to the soil and hydrological characteristics of the Llyn cwm Mynach catchment. The diversity of the conifers planted

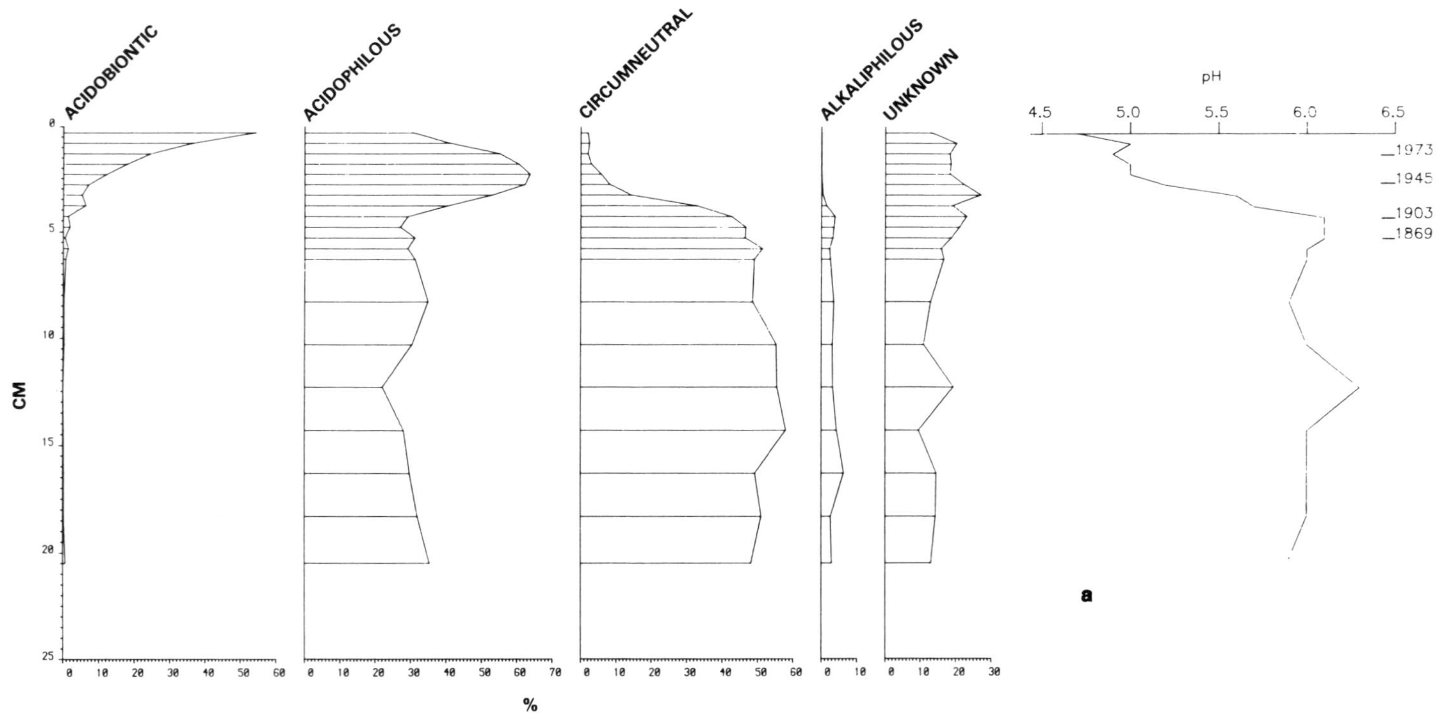

Fig. 3.3. Summary diagram of pH preference categories in a) the Llyn y Bi core and b) the Llyn cwm Mynach core, including pH reconstruction results using multiple regression of pH categories.[8]

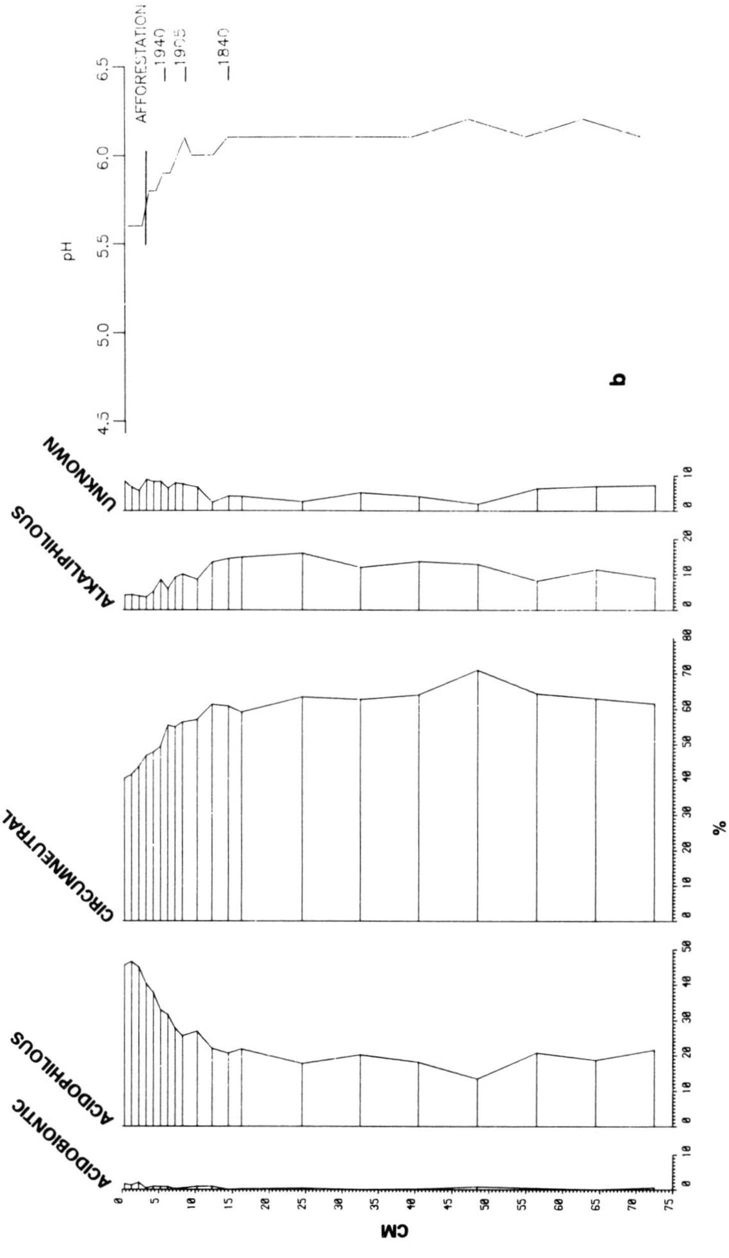
ACIDOBIONTIC
ACIDOPHILOUS
CIRCUMNEUTRAL
ALKALIPHILOUS
UNKNOWN
pH
AFFORESTATION
1940
1840
CM
%
b

Table 3.3. Acidification history of Welsh lakes studied to date.

SITE	GRID REF.	Measured pH	Coring date	Pre-acidification pH	Core top pH	Date of change	pH change	catchment
L. Llagi	SH649483	4.6	1985	6.1	4.6	1850	1.5	moorland
L. Eiddew Bach	SH646345	4.8	1985	6.6	*	–	1.8	moorland
L. y Bi	SH670265	4.6	1985	6.0	4.7	1905	1.3	moorland
L. Dulyn	SH662244	4.6	1985	6.0	4.7	1850	1.3	moorland
L. cwm Mynach	SH678238	5.7	1985	6.2	5.6	1870	0.6	afforested
L. Hir	SN781786	4.5	1984	6.1	4.9	1850	1.2	moorland
L. Gynon	SN800647	5.1	1985	6.0	5.5	1945	0.5	moorland

* The core has time gaps in sediment accumulation. The pH decline is based on an inferred pH from around 1940 and modern measured pH.

and the need for little ground preparation prior to planting is indicative of adequate drainage through the sub-soil allowing acid waters to be neutralised to some extent before passing to the lake. With the increasing amounts of acid deposition after the 1930s this neutralizing capacity was insufficient to prevent further acidification. Although afforestation was not the initial cause of acidification at this site we cannot eliminate the possibility that the forest has subsequently contributed to the acidity of the lake.

Other sites

Similar palaeoecological analyses from other sites in north- and mid-Wales suggest extensive acidification of upland lakes has occurred since 1850 (Table 3.3). The onset of acidification has varied between sites, beginning as early as the mid-nineteenth century at Llyn Dulyn, Hir, and Llagi but not until the mid-1940's in Llyn Eiddew Bach and Gynon. The pattern of acidification also varies greatly, with a rapid decline in pH at some sites and a more gradual decline at others. The magnitude of pH decline ranges from 0.5 to 1.5 units. This variability in the timing, rate, and magnitude of acidification clearly suggests that catchment characteristics play an important role in lake response to increased deposition of acid pollutants.

For the two case-study lakes described here, land-use changes may have had a secondary influence on acidity. However, such changes are absent at other sites and, as in Galloway,[9] the palaeolimnological data clearly indicate acid deposition as the primary cause of acidification at all sites studied in Wales so far.

References

1. Allen, P.M. and Jackson, A.A. (1985). *Geology of the country around Harlech.* HMSO, London.
2. Appleby, P.G. and Oldfield, F. (1978). The calculation of ^{210}Pb dates assuming a constant rate of supply of unsupported ^{210}Pb to the sediment. *Catena*, 5: 1–8.
3. Battarbee, R.W. (1984). Diatom analysis and the acidification of lakes. *Philosophical Transactions of Royal Society of London*, B305: 451–477.
4. Battarbee, R.W. (1986). Diatom analysis. *In* Berglund, B.E. (Ed.). *Handbook of Holocene Palaeoecology and Palaeohydrology.*, Wiley, Chichester.
5. Battarbee, R.W., Appleby, P.G., Odell, K. and Flower, R.J. (1985). ^{210}Pb dating of Scottish lake sediments, afforestation and accelerated soil erosion. *Earth Surface Processes and Landforms.*, 10: 137–142.
6. Charles, D.F. and Norton, S.A. (1986). Palaeolimnological evidence for trends in atmospheric deposition of acids and metals. In *Acid Deposition: Long-term trends.*, Washington, D.C., National Academy of Sciences. pp. 335–431.
7. Cliffe, J.H. (1860). *Notes and recollections of an angler.* London, Hamilton, Adams and Co.
8. Flower, R.J. (1987). The relationship between surface sediment diatom assemblages and pH in 33 Galloway lakes; some regression methods for reconstructing pH and their application to sediment cores. *Hydrobiologia*, 143: 93–103.
9. Flower, R.J., Battarbee, R.W. and Appleby, P.G. (1987). The recent palaeolimnology of acid lakes in Galloway, south-west Scotland: diatom analysis, pH trends, and the role of afforestation. *Journal of Ecology*, 75: 797–823.
10. Harriman, R. and Morrison, B.R.S. (1982). Ecology of streams draining forested and non-forested catchments in an area of central Scotland subject to acid precipitation. *Hydrobiologia*, 88: 251–263.
11. Kreiser, A., Patrick, S.T., Stevenson, A.C., Appleby, P.G., Rippey, B., Oldfield, F., Darley, J., Battarbee, R.W., Higgitt, S.R. and Raven, P.J. (1987). Palaeoecological evaluation of the recent acidification of Welsh Lakes: 10. Llyn cwm Mynach, Gwynedd. *Research Paper*, 26, Palaeoecology Research Unit, Dept. of Geography, University College London.
12. Rosenqvist, I.Th. (1978). Alternative sources for the acidification of river water in Norway. *Science of the total environment*, 10: 39–49.
13. Rosenqvist, I.Th. (1981). Importance of acid precipitation and acid soil in freshwater lake chemistry. *Vann*, 4: 402–409.
14. Ward, F. (1931). *The Lakes of Wales.* London, Jenkins.

4. Atmospheric deposition

A.P. DONALD, J.H. STONER, B. REYNOLDS, F. OLDFIELD,
B. RIPPEY and J. NATKANSKI

Introduction

A primary consideration in accounting for the occurrence of acid waters must be to characterise and quantify inputs from the atmosphere. From an examination of spatial patterns in the quality of deposition, much can be deduced about possible sources of the various ionic constituents. An understanding of short and long term trends will assist in explaining temporal variations in the solute concentrations of streams and lakes.

Although rainfall acidity was first studied in the nineteenth century,[15] it is only in recent years that the phenomenon has been investigated systematically over a widepsread area. The available reviews of U.K. deposition quality[18, 19] show that the patterns of acidity and related anion concentrations in most of northern and eastern Britain are fairly well described, with mean pH varying from about 5.0 in north-west Scotland to around 4.2 in the East Midlands. Wales has hitherto received less attention, and whilst separate investigations in West and Mid Wales indicated a mean pH of 4.5–4.6,[9, 14, 16] a survey in the industrial valleys of South Wales yielded the high figure of pH 6.0.[11] Clarification as to the quality of atmospheric deposition in Wales was therefore required, and so in 1984 an intensive survey of 44 sites was undertaken to identify spatial and seasonal variations throughout the Principality.

The data on atmospheric deposition considered in the present chapter derive principally from that survey. The distribution and likely sources of the major anthropogenically derived ions have been considered in detail elsewhere,[4] although the main findings are also summarised here. In addition, the deposition of dissolved metals is considered, and seasonal and annual variations in deposition quality are discussed. A longer term perspective is provided by considering the record of atmospheric deposition contained in lake sediments.

R.W. Edwards et al. (eds) Acid Waters in Wales. 39–53.

Investigating deposition quality

Atmospheric deposition is generally recognised as consisting of three components. In addition to the 'wet' component, in which pollutants are deposited in dissolved form by precipitation, 'dry' deposition in aerosol or gaseous form occurs. In upland areas these two components are supplemented by 'occult' deposition, in which wind-blown cloud and fog droplets are intercepted directly by vegetation.[3] Because of the technical difficulties of measuring these individual components,[18] bulk deposition collectors which measure total deposition are frequently used. All the data discussed below refer to bulk deposition except where indicated.

The network of 44 sites investigated by WWA in the 1984 survey is shown in Fig. 4.1. It is one of the most intensive of any investigated to date, and permits an excellent resolution of spatial trends. Forty-one of the sites were sampled on a weekly basis, whilst three (sites 4, 14, and 23) were sampled on a daily basis (site 14 on weekdays only). Site 23, the only one in south-west Wales, is known from a pilot survey to be representative of the area as a whole (Welsh Water, unpublished data). Details of the bulk collector design and field procedures are reported in Donald and Stoner,[4] and laboratory methods in Stoner et al.[16] This data set is complemented by the longer-term record from the ITE site at Plynlimon, enabling annual variations in deposition quality in upland Wales since 1980 to be described. Details of site location and laboratory methods are given in Reynolds et al.[14]

Although there are few reliable documentary records of long-term trends in acid deposition in Wales, some understanding of the increasing impact of air pollution can be obtained from a study of lake sediments. In addition to trapping diatoms and pollen grains originating in the lake and its catchment (see Chap. 3), lake sediments also trap material derived from industrial emissions to the atmosphere. A range of independent techniques is available to identify and analyse this material. In acidification studies the most useful methods are trace metal analysis, carbonaceous particle analysis[7] and magnetic mineral analysis.[17] The dating of critical horizons in the sediment record and the calculation of sediment accumulation rates is by ^{210}Pb dating.[1]

Sediment cores from nine lakes in Mid and North Wales have been investigated to date using these techniques and associated palaeoecological methods.[6, 12] The data discussed below refer to Llyn Dulyn in North Wales (Fig. 4.1) and are representative of sediment records from the other lakes examined.

Spatial variations

The main influences on the chemistry of deposition in Wales are sea salts and pollutants of industrial or other anthropogenic derivation. Since hydrogen ion concentrations are determined by a balance with the other ions present,

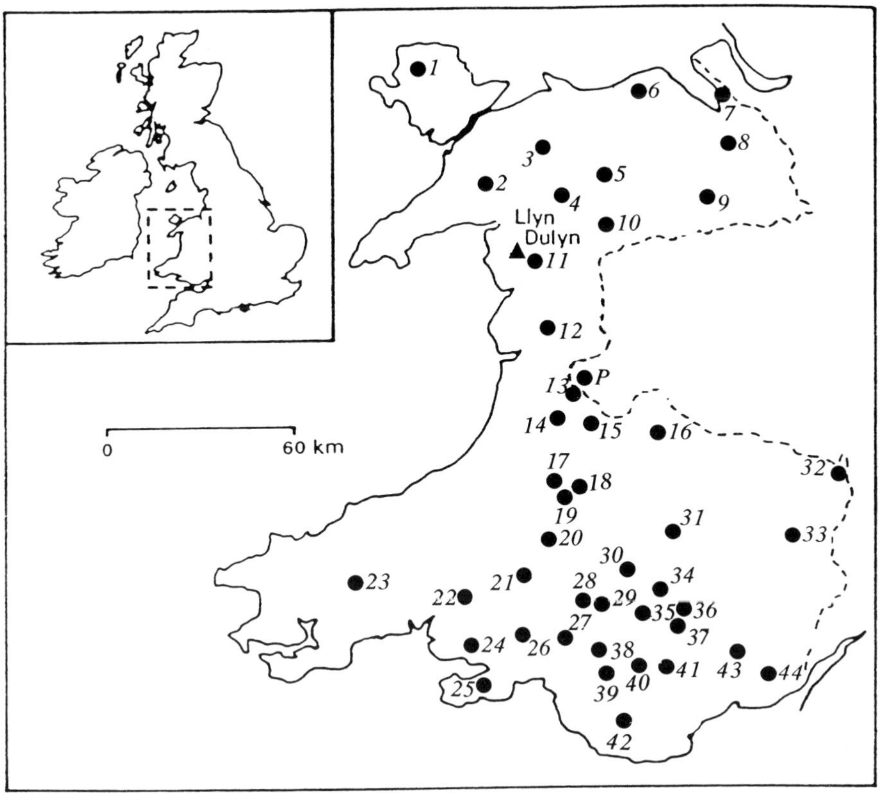

Fig. 4.1. Location of bulk collectors and of Llyn Dulyn. WWA collectors numbered 1–44; ITE Plynlimon collector indicated by P.

discussion of rainfall acidity will be deferred until these other ions have been considered.

Sea salts

In common with all maritime areas, Wales receives significant amounts of sea salts, although the concentrations decline markedly away from the coast. Mean values for sodium exemplify this (Fig. 4.2). The site nearest the coast (site 25) has a mean sodium concentration of 249 μequiv. l^{-1} and a mean conductivity of 70 μs cm^{-1}. Conversely, the site furthest from the sea (site 32) has the lowest mean sodium concentration (67 μequiv. l^{-1}) and low mean conductivity (32 μs cm^{-1}). Other sea salts such as calcium, magnesium, sulphate and chloride are also elevated around the coast (Welsh Water, unpublished data). However these ions also have significant non-marine or 'excess' components, and concentrations do not therefore decline with distance from the coast in the same manner as sodium. Because it is thought to derive almost exclusively from the sea, sodium has been used to calculate values for the 'excess' ions discussed below.[4]

Non-marine ions

Mean concentrations of excess sulphate, nitrate and ammonium broadly reverse the spatial trend of the sea salts (Fig. 4.2), with the lowest values for each ion occurring in western areas and the highest in north-east and south-east Wales. Excess sulphate makes up on average 74% of total sulphate at the sites examined, individual proportions varying from only 54% at site 25 on the coast to 98% at the site furthest inland (site 32). This ion is derived from power station and other major industrial emissions, and the observed spatial variations are clearly related to sources of sulphur dioxide emissions. Published figures for 1983 show that emissions are highest in the South and North-East of Wales, exceeding 10,000 tonnes of sulphur at three separate 20 × 20 km grid squares around the sites of power stations.[19]

Nitrate is also derived from industrial emissions, although the fact that 40% of emitted oxides of nitrogen originate from vehicle exhausts makes nitrate more difficult to relate to specific sources. Nevertheless the association between elevated concentrations and the major centres of population and industry is clear, with mean concentrations rising from around 15 μequiv. l^{-1} in the west to >40 μequiv. l^{-1} in the north-east and south-east. A similar spatial pattern was revealed in a survey of ambient NO_2 concentrations undertaken in 1986 throughout Wales.[8] A further industrial contributor to anionic strength in parts of Wales is excess chloride.[4] This is thought to derive from burning coal containing a high chloride content, and elevated mean concentrations are found in central South Wales where there are known emissions from such coal.

Mean concentrations of ammonium display a marked west-east trend comparable to that of excess sulphate and nitrate. From being <25 μequiv. l^{-1} in parts of West and South Wales, values rise to around 60 μequiv. l^{-1} in the east. The fate of ammonium in the atmosphere during transport is poorly documented by comparison with that of sulphate and nitrate,[2] but likely emission sources are the fertilizer industry and, in south-west Wales, intensive dairy farming. A further non-marine contributor to cationic strength is excess calcium, whose concentrations display similar spatial trends to the other ions discussed here[4] and which may originate both from industrial emissions and from arable agriculture.

The major exception to the west-east trends noted above is site 1 on the Isle of Anglesey, where elevated concentrations occur for all the anthropogenically derived ions. These values are thought to be directly assignable to emissions from an aluminium smelter 12km west of the monitoring site. Corroborative evidence of localised atmospheric pollution on Anglesey is given by a lichen survey of the island.[13]

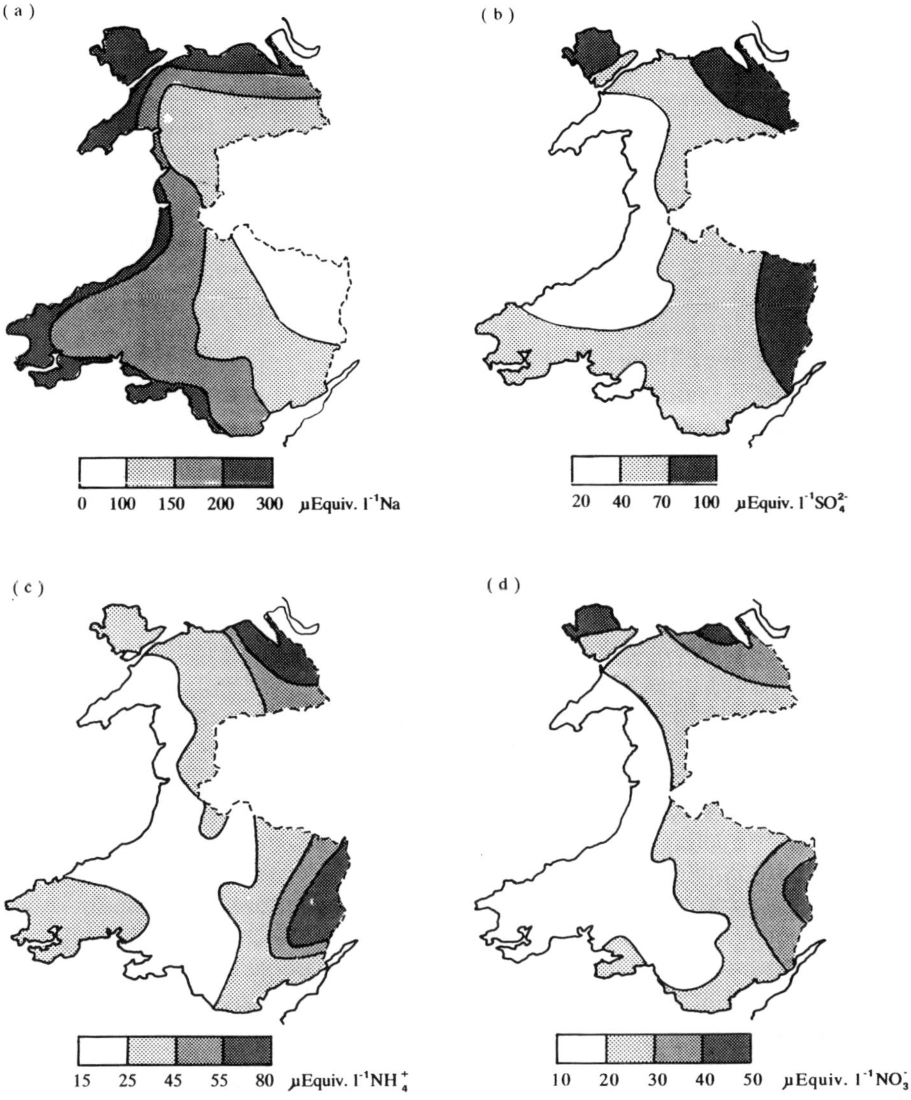

Fig. 4.2. Variations in mean concentrations (μequiv. 1^{-1}) of (a) Na^+ (b) SO_4^{2-} (c) NH_4^+ and (d) NO_3^-.

Acidity of deposition

The frequency distribution of mean pH (Fig. 4.3), calculated from volume weighted mean hydrogen ion concentration, shows that values range from 4.4 to 5.0 with a mode of 4.7. The previously published results from West and Mid Wales[9, 14, 16] are thus broadly confirmed. Page and Swires-Hennessy[11] found that deposition in the South Wales valleys had a mean pH of 6.0,

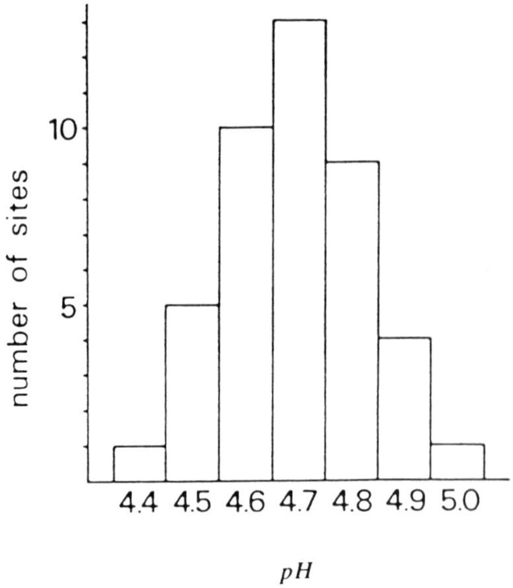

Fig. 4.3. Frequency distribution of mean pH (derived from weighted mean H^+ concentration) for the 44 study sites (after Ref. 4)

although none of the many sites from that area in this study had a mean value >4.8, and in national surveys values in excess of pH 5.0 are restricted to north-west Scotland.[19]

Spatial variations in mean pH are such that values are generally lower in the east of Wales than in the west, although the lowest values are found in the upland areas of Mid and North Wales (Fig. 4.4a). The lack of correlation between mean pH and mean concentrations of the principal anionic constituents, evident from Figs 4.2 and 4.4a and discussed in more detail by Donald and Stoner,[4] contrasts with the situation usually found in areas receiving acidic deposition.[5] The Welsh data can be explained by the high concentration of ammonium (and excess calcium) that occur in those eastern areas where excess sulphate and nitrate concentrations are also elevated. Mean ammonium is closely correlated with both excess sulphate ($r = 0.88$) and nitrate ($r = 0.87$), and the high levels of each ion in eastern areas result in acidity being moderated. The lower mean pH of the upland areas arises from concentrations of ammonium and calcium being substantially lower than further east while sulphate and nitrate are only slightly lower. The influence of occult deposition in the upland areas may be of significance, since fogs and mists can be highly acidic.[3]

Variations in deposited acidity throughout Wales closely reflect the pattern of rainfall, which is in turn controlled by relief. The mean annual deposition of hydrogen ions exceeds 0.6 kg ha^{-1} year in upland North Wales (Fig. 4.4b) and is over 0.4 kg ha^{-1} $year^{-1}$ in upland areas further south. Because of the

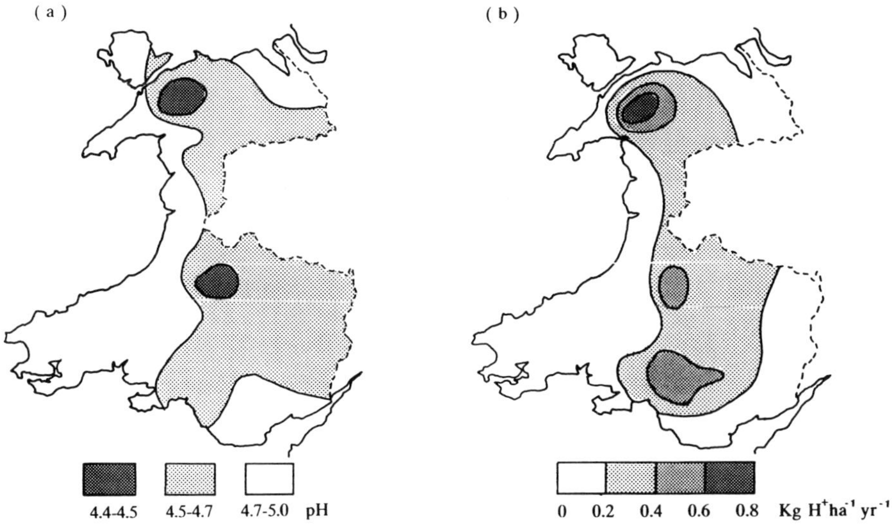

Fig. 4.4. Variations in (a) mean pH and (b) deposited acidity.

relatively high rainfall in the Welsh uplands, deposited acidity is similar to that received in the East Midlands, which has the most acid U.K. deposition (mean pH 4.2) but lower rainfall.[19] The deposition of excess sulphur and nitrate-N in the Welsh uplands at around 10 kg ha^{-1} $year^{-1}$ and 5 kg ha^{-1} $year^{-1}$ respectively are also comparable with East Midlands values, despite concentrations being only about 30% of those found in the latter area. It is likely that sulphuric acid accounts for 60–70% of acidity, with the remainder being accounted for by nitric acid except in parts of south and north-east Wales where hydrochloric acid may account for 5%.[4]

The episodicity of acid deposition and of related anion deposition is particularly marked, with analyses on the three daily sampled sites (4, 14 and 23) showing that 30% of deposition falls on less than 5% of wet days.[4] Acid episodes were recorded in all parts of Wales, with pH readings of <4.0 occurring at all the study sites. The most consistently acidic event recorded throughout Wales occurred over a three day period in May 1984, when pH was on average 3.9. Only five sites had pH >4.0 and the maximum and minimum readings were 4.7 and 3.5 respectively. No less than 10% of the total deposited annual acidity fell over this period, during which winds were moderate north-easterly.

The importance of wind direction as an influence on deposition pH is emphasised by the plots of back trajectories of air mass for days on which non-acid (pH >6.0) and acid (pH <4.2) rainfall was recorded at a minimum of two out of three daily sites (Fig. 4.5). Whereas all the non-acid days can be traced to the Atlantic, reflecting the buffering effect of sea salts, most of the acid events are assignable to air masses originating in the East and South. Even those air masses which originated in the Atlantic yet yielded acidic

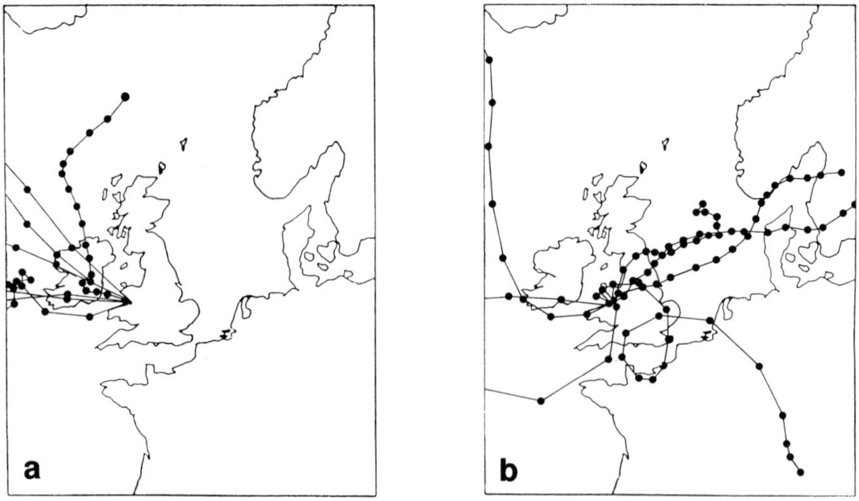

Fig. 4.5. 96 hour back-trajectories of air masses for days on which rainfall pH was (a) >6.0 and (b) <4.2 for at least 2 out of 3 'daily' sites. Each interval represents 6 hours of travel (after Ref. 4).

deposition passed over at least one major industrial source, although the possibility remains that long distance trans-Atlantic pollution may be implicated.

Deposition of metals

Few data concerning the atmospheric deposition of metals have been collected on an intensive scale in the U.K., and the hitherto unpublished information available for Wales is the most comprehensive currently available. Summary information for concentrations and deposition of lead and zinc is presented in Table 4.1. Analyses refer to the 'dissolved', or filterable at 0.45 μm, component.[16] In the case of lead, the 40% of values that were below the detection limit led to the substitution of zero for all such values and overestimation was thus avoided.

Spatial variations for dissolved metals were not as clear as for the other ions of presumed industrial origin. Nevertheless the highest values were found in the north-east and in the south, especially in the case of lead which might be expected to be related to the main centres of population.

Seasonal trends

Seasonal variations in the concentration and deposition of the principal ions are illustrated in Fig. 4.6, which shows data from rural site 12 during 1984.

Table 4.1. Summary statistics of weighted mean filterable lead and zinc from the 44 study sites.

	Pb		Zn	
	concentration (μequiv. l^{-1})	deposition (kg ha^{-1} year)	concentration (μequiv. l^{-1})	deposition (kg ha^{-1} year)
min	0.04	0.04	0.27	0.11
max	0.20	0.20	1.61	0.35
mean	0.18	0.09	0.59	0.20
S.D.	0.06	0.04	0.28	0.06

Values were derived by dividing the monthly means by the annual mean in each case. Although based on only one year's data, masking of the seasonal pattern by episodic events is minimal at site 12 and seasonal variations in ionic concentrations can be clearly seen.

Concentrations of all ions are highest in late winter/early spring, and lowest in late autumn/early winter. The timing of these peaks is thought to be related to a lag effect, with, for example, maximum values of sulphate occurring about three months after the winter maxima of sulphur dioxide emissions due to a variety of meteorological factors.[19] However, there is no equivalent seasonality in deposition (Fig. 4.6b), possibly due to the effects of the 1984 drought.

Annual trends

The data presented above refer to samples from 1984 only. It is therefore important to establish both the degree to which that year was representative in terms of deposition quality, and also the nature of any annual trends. 1984 turned out to have a very dry summer, with an average of 261mm of rainfall in the period March–August or only 52% of the long-term mean. However, total precipitation for the year was not particularly low at 94% of the long-term mean.

Annual trends in deposition quality at Plynlimon during the period 1980–1987 are shown in Fig. 4.7. It can be seen that the 1984 values are broadly representative of the period as a whole, except that mean sulphate concentration (although not deposition) was higher than in other years. The data suggest that both concentration and deposition of hydrogen ion have decreased over the period in question, a finding which is in agreement with data from England and Scotland for the period 1978–1985.[19] However, despite the net reduction in UK emissions of SO_2, concentrations of sulphate at Plynlimon do not display an overall trend, while depositions of sulphate show a less prominent trend than for hydrogen ion. This is consistent with the findings of modelling studies that only a small fraction of deposition in upland Wales is attributable to UK emissions (see Chap. 18). Very little change is

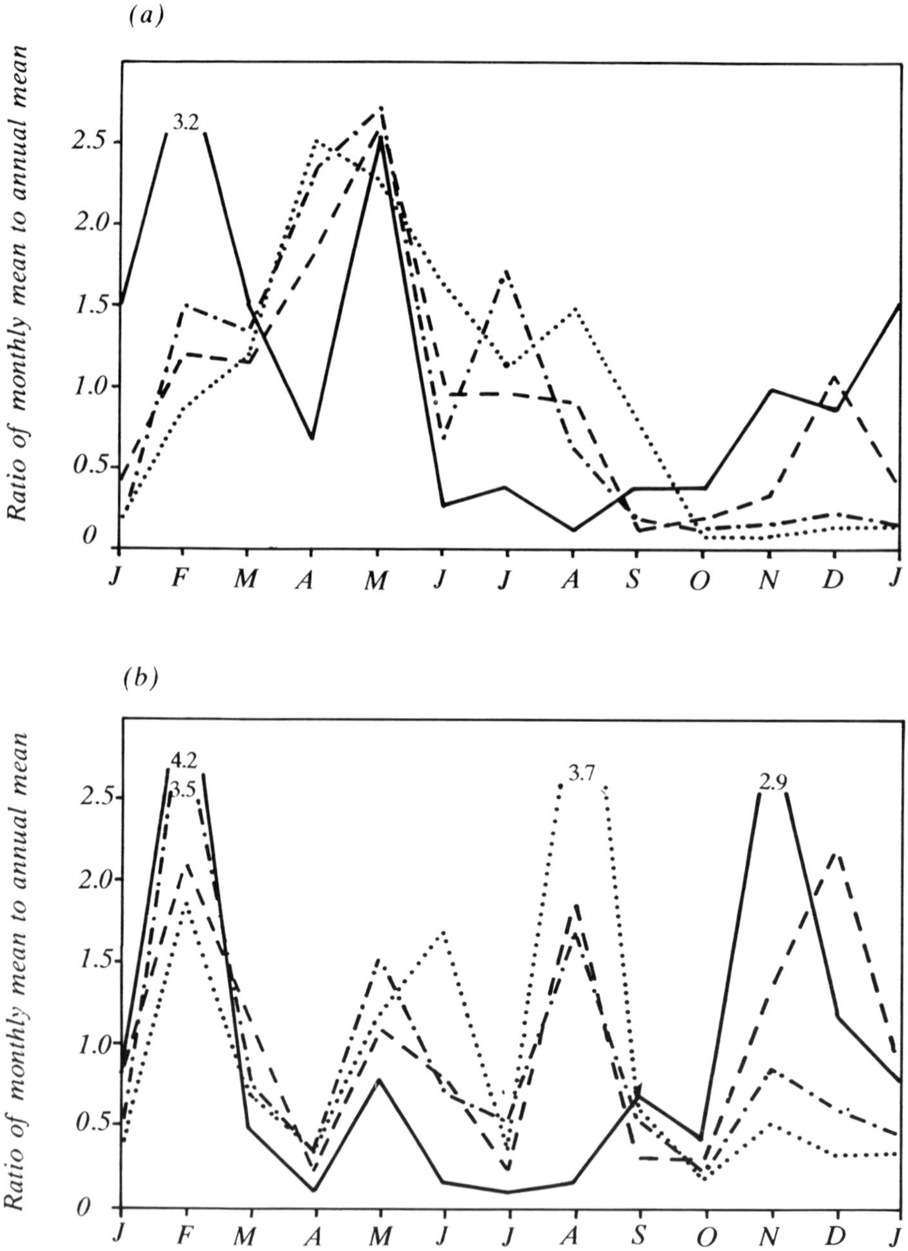

Fig. 4.6. Seasonal variations in (a) concentration and (b) deposition of selected ions at Plynlimon.
—— H^+; - - - SO_4^{2-}; -.-.- NO_3^-; . . . . NH_4^+.

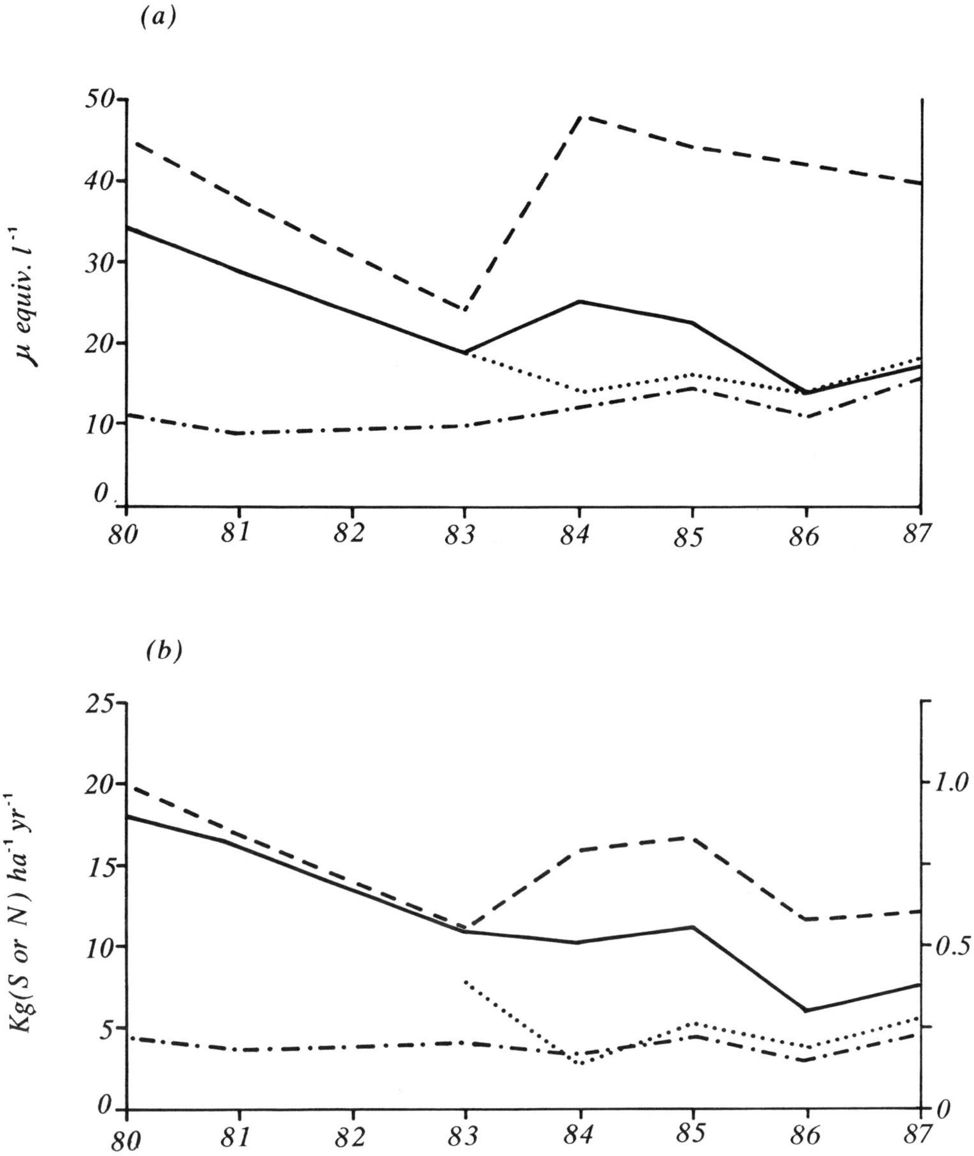

Fig. 4.7. Annual trends (1980–87) in (a) concentration and (b) deposition of selected ions at Plynlimon. —— H^+; – – – SO_4^{2-}; –.–.– NO_3^-; . . . . NH_4^+. No data for 1982; NH_4^+ commences 1983.

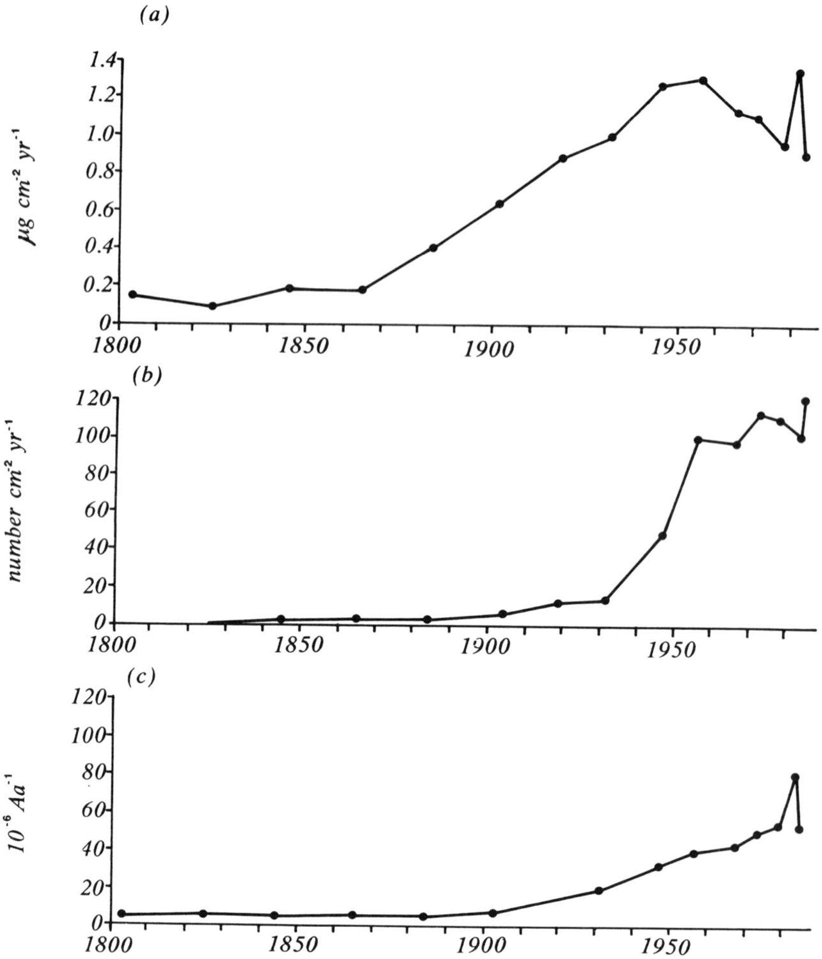

Fig. 4.8. Trends in atmospheric contaminants in Llyn Dulyn sediments. (a) Pb accumulation; (b) carbonaceous particle accumulation; (c) magnetic accumulation.

evident in the annual deposition of ammonium or nitrate over the period in question, although concentrations of the latter ion were about 25% higher in 1987 than in 1980.

Long-term trends

Figure 4.8 shows data from the analysis of a sediment core from Llyn Dulyn. Since there have been no major changes in land use within the catchment since medieval times, these results provide a long-term record of atmospheric deposition in remote areas of upland Wales.

Trace metals

The trace metal record in lake sediments derives partly from the catchment and partly from the industrially polluted atmosphere. The increased atmospheric flux is especially reflected by large increases in recent sediments in the concentration of lead, zinc and cadmium, and other elements. Lead is widely regarded as the most stable element, unlike zinc and cadmium which can be mobilised from sediments in acidified lakes.[10]

Fig. 4.8a shows a lead profile for Llyn Dulyn. Background fluxes prior to about 1850 AD are very low, but in the latter half of the nineteenth century there is a steep increase. Since there has been no metal mining in the catchment of this lake the increase reflects the historic rise in atmospheric contamination following the rapid industrialisation of Wales and England during this period. Values continue to rise during the twentieth century and reach maximum levels in post 1940 sediments.

Spherical carbonaceous particles

Whereas the lead record reflects the general development of industry, spherical carbonaceous particles found in the sediments are a direct record of fossil fuel combustion.[7]

Fig. 4.8b shows the record for Llyn Dulyn. Unlike the trace metal record where there is a background flux from the catchment there are no records of carbonaceous particles prior to about 1850. The increase during the late nineteenth and early twentieth centuries is probably due to local coal-derived material but the major increase in post-1930's sediments is almost certainly due to the introduction and rapid development of fuel oil combustion by industry and for electricity generation by power stations.

Magnetic mineral measurements

A range of industrial processes, including iron and steel manufacture, solid-fuel power generation and non-ferrous metal smelting give rise to particulate emissions with high ferrimagnetic concentrations. These particles are found both in ombrotrophic mires and in lake sediments. Scanning Electron Microscopy shows that they are predominantly spherical and morphologically indistinguishable from power station fly-ash. So far suitable techniques have not been developed to quantify these particles in sediments but their presence and relative concentration can be derived from magnetic measurements of the sediment.[17]

The best available measure of the mainly multidomain magnetite that is carried by fly-ash particles is $SIRM\text{-}IRM_{-20mT}$. Figure 4.8c shows the profile of these measurements converted to magnetic accumulation rates for the Llyn Dulyn core. Before 1900 AD background levels were very low and stable, but in the twentieth century there has been a continuous rise in values reflecting the progressive input of fly-ash from high temperature solid-fuel combustion during this period.

Since there is a close correlation between SO_2 emissions and the use of fossil-fuel derived energy, the trends shown in Fig. 4.8 can be regarded as a general representation of historical trends in acid deposition in this part of Wales. Similar records are found at other sites in North and Central Wales so it can be concluded that sensitive surface waters in this region have been subjected to a progressive increase in acid deposition since the middle to late nineteenth century.

Conclusions

It is clear from the Llyn Dulyn sediment record that the atmospheric effects of industrial activity have been felt even in the remotest parts of rural Wales for more than a century. Whereas the accumulation of lead and of magnetic particles has increased steadily since the late nineteenth century, spherical carbonaceous particles deriving from fossil fuel combustion have shown a particular increase since the 1930s. That these developments have been intimately linked with surface water acidification is clear from the diatom record from Llyn Dulyn and other lakes discussed in Chap. 3.

Atmospheric deposition today is dominated by marine and anthropogenic sources, which together determine the spatial variability of deposition quality. In the west the influence of sea salts results in mean pH being in the range 4.7–5.0. Although sea salts are much less important in the east, considerable neutralisation of the high sulphate and nitrate concentrations is afforded by ammonium and calcium, and pH values are comparable with those of the western seaboard. The upland areas receive the most acid deposition, with mean pH in the range 4.4–4.6.

Although Welsh deposition is not as acidic as that of eastern Britain (pH 4.2) or central Europe (pH 4.0), the relatively high rainfall in Wales results in annual deposited acidity being comparable with that received in more highly polluted areas. Moreover, extremely acid episodes with pH <4.0 occur throughout the region, the most acid being associated with winds from easterly points. Even if the apparent downward trend of recent years in hydrogen ion concentrations continues, the continued occurrence of discrete highly acidic episodes is likely to remain a major source of concern.

References

1. Appleby, P.G. and Oldfield, F. (1978). The calculation of lead-210 dates assuming a constant rate of supply of unsupported ^{210}Pb to the sediments. *Catena*, 7: 1–8.
2. Buckley-Golder, D.H. (1984). Acidity in the Environment. Energy Technology Support Unit, Harwell: HMSO, 181 pp.

3. Dollard, G.J., Unsworth, M.H. and Harvey, M.J. (1983). Pollutant transfer in upland regions by occult precipitation. *Nature*, 302: 241–243.
4. Donald, A.P. and Stoner, J.H. (1989). The quality of atmospheric deposition in Wales. *Arch. Env. Contam. Toxicol.*, 18: 109–119.
5. Fowler, D., Cape, J.N., Leith, I.D., Paterson I.S., Kinnaird, J.W. and Nicholson, I.A. (1982). Rainfall acidity in northern Britain. *Nature*, 297: 283–296.
6. Fritz, S.C., Stevenson, A.C., Patrick, S.T., Appleby, P.G., Oldfield, F., Rippey, B., Darley, J. and Battarbee R.W. (1986). Palaeoecological evaluation of the recent acidification of Welsh lakes. In: Llyn Hir, Dyfed. Dept of Geography, U.C. London.
7. Griffin, J.J. and Goldberg, E.D. (1981). Sphericity as a characteristic of solids from fossil fuel burning in a Lake Michigan sediment. *Geochimica et Cosmochimica Acta*, 45: 763–769.
8. ITE (1988). Institute of Terrestrial Ecology Annual Report for 1987.
9. Martin, A. (1982). A survey of the acidity of rainfall over large areas of Great Britain. *Sci. Tot. Environ.*, 13: 119–130.
10. Norton, S.A. (1986). A review of the chemical record in lake sediments of energy related air pollution and its effects on lakes. *Water, Air and Soil Pollution*, 30: 331–346.
11. Page, R.A. and Swires-Hennessy, (1983). Acid deposition in Wales. *Envir. Health*, 92: 7–9.
12. Patrick, S.T., Fritz, S.C., Stevenson, A.C., Appleby, P., Rippey, B., Oldfield, F., Darley, J., Battarbee R.W., Higgitt, S.R. and Raven, P.J. (1987). Palaeoecological evaluation of the recent acidification of Welsh lakes. VII: Llyn y Bi, Gwynedd. Dept. of Geography, U.C. London.
13. Perkins, D.F., Millar, R.O. and Neep, P. (1980). Accumulation of airborne fluoride by lichens in the vicinity of an aluminum reduction plant. *Envir. Poll.*, A 21: 155–168.
14. Reynolds, B., Hornung, M. and Stevens, P.A. (1984). Factors influencing bulk precipitation chemistry at an upland site in Mid-Wales. *Intern. J. Environmental Studies*, 22: 195–205.
15. Smith, R.A. (1872). Air and rain: the beginnings of a chemical climatology. London.
16. Stoner, J.H., Gee A.S. and Wade, K. (1984). The effects of acidification on the ecology of streams in the upper Tywi catchment in West Wales. *Environ. Poll.*, A 35: 125–157.
17. Thompson, R. and Oldfield, F. (1986). Environmental Magnetism. London.
18. U.K.R.G.A.R. (1983). United Kingdom Review Group on Acid Rain–Acid Deposition in the United Kingdom. Warren Spring Laboratory, Stevenage.
19. U.K.R.G.A.R. (1987). United Kingdom Review Group on Acid Rain–Acid Deposition in the United Kingdom 1981–1985. Warren Spring Laboratory, Stevenage.

5. The role of geology and soils in controlling surface water acidity in Wales

M. HORNUNG, S. LE-GRICE, N. BROWN and D. NORRIS

Introduction

The chemistry, including the acidity, of surface waters is a function of the chemistry of atmospheric inputs; the amount, type and distribution of precipitation; the vegetation; the type and thickness of soils; the catchment hydrology; and the drift and bedrock geology. Land use is also a major influence as it can modify several of the above factors. A comprehensive predictive model of water chemistry, and acidity, or of the sensitivity of surface waters to acidification, would include the above factors. At a regional scale, however, geology and soils are the major factors influencing surface water chemistry and acidity.[4, 9, 11] The following chapter examines the influence of bedrock and soils on surface water acidity in Wales.

The influence of bedrock geology

Several authors have used schemes based solely on bedrock geology to predict the occurrence of naturally acidic waters or waters sensitive to acidification.[5, 9, 11] Solution of carbonate minerals and hydrolysis of silicate minerals are the main processes which assimilate H^+ ions, and hence neutralise acidity, as waters percolate through rocks:

$$CaCO_3 \text{ (calcite)} + H^+ = Ca^{2+} + HCO_3$$

$$2\ KA1Si_3O_8 \text{ (orthoclase)} + 2H^+ + 9H_2O = A1_2Si_2O_5\ (OH)_4 + 2K^+ + 4H_4SiO_4$$

The main properties of rocks controlling the amount of buffering or neutralisation of percolating waters are therefore, the content of carbonates and of weatherable silicate minerals. Contact area and residence time also influence solution-rock interactions and the rate and amount of buffering. The contact area available for solute-rock reactions will be greater in shattered, finely bedded or jointed rocks than in massive rocks. Reactions are also more likely to reach equilibrium, with greater neutralisation, given longer residence

R.W. Edwards et al. (eds) Acid Waters in Wales. 55–66.

times. The dominating influence on water chemistry and acidity is, however, the mineralogy and geochemistry of rocks, particularly, as noted above, the content of carbonates and weatherable silicates. The predictive schemes mentioned above usually classify rocks on the basis of their acid neutralising or buffering capacity. These classifications use known relationships between rock type and the content of carbonates and weatherable silicates.

Norton[11] produced a classification of rocks based on their acid neutralising capacity and used this as a basis for predicting the impact of acidic deposition on surface water acidity in the USA. Kinniburgh and Edmunds[7] developed a similar classification of the buffering capacities of rocks; these classes were then used to predict the sensitivity of UK groundwaters to acidification. The main features of these two classifications have been combined in Table 5.1.

The classification given in Table 5.1 has been applied to the solid geology of Wales, using published maps and data on the lithology and mineralogy of the main stratigraphic units, to produce a map showing distribution of rocks with little, low, moderate or infinite neutralising capacities (Fig 5.1). This indicates that large areas of north, central and south west Wales are underlain by rocks with little or no neutralising capacity. These are areas underlain by Pre-Cambrian schists and metasediments in north west Wales, Lower Palaeozoic slates and mudstones in north central and south west Wales and acid igneous rocks in north west Wales. The younger Carboniferous and Devonian rocks, of north east and south east Wales, have higher buffering capacities. The Carboniferous limestones have very high buffering capacity and the arkosic Pennant sandstone of the South Wales coalfield, and the marls and sandstones to the north of the coalfield have moderate buffering capacities. Thus, on the basis of bedrock geology, naturally acidic, low conductivity waters, sensitive to acidification, are predicted to occur over much of north west, central and south west Wales. In contrast, the surface waters of much of eastern Wales are, on the basis of solid geology, predicted as being moderately to well buffered.

The map (Fig 5.1) of neutralising classes of bedrock varies somewhat from the interpretation of Kinniburgh and Edmunds;[7] they allocated areas to their four classes of buffering capacity mainly on the basis of stratigraphic units. As they pointed out, this approach has limitations due to the facies changes within stratigraphic units. Thus, the Namurian sandstones of northern England tend to be silicious grits with, therefore, little or no buffering capacity, while the Namurian sandstones of south Wales, i.e. the Pennant sandstone, contain arkosic beds and as a result have moderate buffering capacity. In the present study, we have classified the Pennant sandstone as having moderate buffering capacity. Kinniburgh and Edmunds' division between Ordovician rocks (little or no buffering capacity) and Silurian rocks (low buffering capacity) is retained; this is based upon the higher content of slates and grits in the Ordovician rocks.

Table 5.1. The impact of acidic deposition on surface and groundwaters of areas underlain by rocks with different buffering capacities.

Class	Buffering capacity	Rock types	Impact of acidic precipitation on surface waters	Impact of acidic precipitation on groundwaters	Characteristics of first and second order streams
1	Little or no buffering capacity	Granite and acid igneous rocks or metamorphic equivalents; granite gneisses, quartz sandstones and metamorphic equivalents; decalcified sandstones; most metasediments, including slates; non arkosic grits	Widespread impact expected	Most areas susceptible to acidification unless significant thicknesses of glacial drift present	Naturally acidic, low conductivity, poorly buffered
2	Low to medium buffering capacity	Sanstones, shales, conglomerates and metamorphic equivalents; coal measures; intermediate igneous rocks; high-grade metamorphic felsic to intermediate volcanic rocks	Impact restricted to first and second-order streams and small lakes	Many areas could be susceptible	Weakly acidic, low conductivity, poorly buffered
3	Medium to high buffering capacity	Slightly calcareous rocks, eg marlstones; basic and ultrabasic igneous rocks; Mesozoic mudstones; low grade intermediate to mafic volcanic rocks	Impact improbable except for near surface drainage in areas of acid soils	Little general likelihood of acidification	Circum neutral, well buffered
4	Infinite buffering capacity	Limestones, chalk, dolomitic limestones, highly fossiliferous sediments or metamorphic equivalents	No impact	No impact	Alkaline, high conductivity, highly buffered

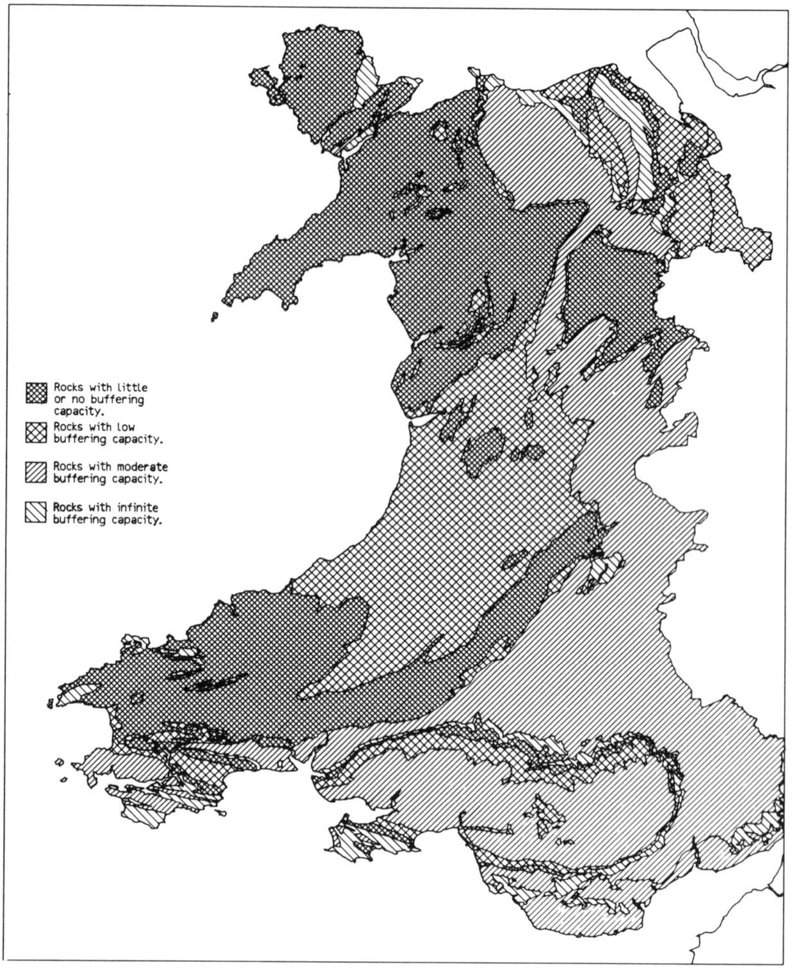

Fig. 5.1. The distribution of rocks with low, medium, high and infinite neutralising capacities in Wales.

The influence of soils

Predictions of the likely occurrence of naturally acidic waters and those sensitive to acidification, based solely on bedrock geology have limitations. Thus, Norton[11] notes the occurrence of lakes in Florida which have become acidified even though they occur over highly calcareous and phosphatic rocks; the lakes are shallow, occuring on acid soils leached of $CaCO_3$ and with little deep groundwater input from the calcareous rocks. Conversely, Norton[11] reports lakes in areas of Maine underlain by granite, which are not acidified as a result of the presence of calcareous tills and marine clays. Also, during periods of high flow a significant part of the water in the stream is not derived from groundwater sources, i.e. from bedrock or drift, but enters the stream as surface flow, throughflow, or return flow.[3] The chemistry of these waters will reflect the chemistry of the soils over, or through, which they have flowed, rather than that of the underlying bedrock.

A more reliable prediction of the likely occurrence of naturally acidic, or of sensitive waters would, therefore, include a consideration of soils. The chemistry of waters held within and moving through soils, including water acidity, is influenced by a variety of processes.[12] The most important of these processes are ion exchange, anion adsorption, solution and hydrolysis of soil minerals, oxidation-reduction processes, and solution, deprotonation and degradation of organic acids. The acid neutralising capacity of soils is largely determined by their content of carbonate and weatherable silicate minerals, cation exchange capacity and base saturation. These properties are dependent on the nature of the parent material, age, and weathering and leaching histories of the soil. Soils with free carbonates, a high content of weatherable silicates or a high base saturation will, in general, give rise to well buffered, circumneutral drainage waters and will neutralise acidic deposition. These soils will have a naturally high pH. Acidic deposition will only be partially buffered, if at all, in acid soils which have low base saturation, no carbonate minerals and a low content of weatherable silicate minerals. Drainage waters from acid, raw peat soils are commonly more acid ($pH < 4.5$) than precipitation, even in unpolluted areas, due to solution of organic acids. Any buffering in the acid mineral soils will be linked to mobilisation of aluminium (see Chaps. 11 & 14).

A number of classifications of soils based on buffering capacity, or neutralising capacity have now been published. Most are related to base saturation and content of carbonates, or to a routinely measured soil parameter which is used as a surrogate for these parameters, eg pH or texture. Thus, Lau and Mainwaring[8] found that soil texture was a good basis for predicting the buffering capacity of soils. Wang and Coote[17] used texture, % clay, cation exchange capacity and pH to determine the sensitivity of soils to acidification by acid deposition: soils could be classified on their system using texture and pH alone. Cation exchange capacity, the presence or absence of carbonates and the occurrence of flooding were used by McFee[10] in his classification of the sensitivity of soils to acid precipitation.

Catt[2] used the base status of subsoil horizons to classify soils on the basis of neutralising capacity (Table 5.2). He then applied the classification to the units of the 1:250,000 soil maps of England and Wales (Soil Survey of England and Wales, 1983) to produce a map showing the distribution of soils with large, moderate and little or no neutralising capacity.

A broadly similar approach has been used here but has been simplified to give two classes of soils: acid soils with little or no neutralising capacity and non-acid to weakly acidic soils with moderate to large neutralising capacity. The map produced by Catt[2] shows very small areas of soils with large

Table 5.2. Soil buffering classes; terminology based on the soil classification of Avery (1980).

Soils with little or no neutralising capacity	Holocene podzols; deeply weathered paleo-argillic soils, thin rankers on non-calcareous Paleozoic rocks; brown earths on silicious gravels; gley soils on non-calcareous or pyritic clays and shales; raw bog, basin and blanket peat soils
Soils with moderate neutralising capacity	Non-calcareous pelosols; brown earths on base-rich materials; gleys soils on non-calcareous clay; fen peats
Soils with large neutralising capacities	Little weathered soils on Holocene marine clays and calcareous sands; rendzinas; calcareous pelosols; brown calcareous earths; calcareous gleys formed during the Holocene or calcareous sediments

neutralising capacity, mainly calcareous dune sands. Figure 5.2 shows the distribution of the two classes of soils in Wales and was derived from the 1:250,000 Soil Map of Wales (Soil Survey of England and Wales 1983). The soil associations used on the Soil Map of Wales were allocated to the two classes on the basis of the known chemical characteristics, particularly pH and base saturation, of the dominant soil series in the soil associations based mapping units. The soil associations classified as being acid, with low neutralising capacity, are dominated by podzols, brown podzolic soils, cambic stagnohumic gleysoils, acid rankers and raw, acid peats. These soils dominate much of upland Wales (Fig. 5.2); podzolic soils and podzols do, however, occur at lower altitudes on base poor soil forming materials.

The predominance of acid soils in the upland areas is the result of the interaction of the base poor soil forming materials, derived from the acid igneous rocks, slates, mudstones and grits, and the humid, cool climate.[14] It has also been suggested that acidification was accelerated and the development of peat and peaty horizons initiated in some areas as a result of the clearance of natural woodland by man. Studies of soil formation in upland Wales[15] show, however, that acid soils have dominated for many hundreds of years, and in most parts for thousands of years.

The combined influence of soils and geology

The maps of the occurrence of 'acid' and 'non-acid' soils and of rocks with low, moderate, high and infinite buffering capacity have been combined to show the occurrence of a number of soil and rock combinations with differing buffering capacities in Wales (Fig. 5.3).

These soil-rock combinations are used to predict the likely occurrence of acidic, sensitive waters, and possible changes in stream acidity with flow. In making these predictions it is assumed that acidic waters will not occur in areas with 'non-acid' soils with moderate to high buffering capacities, what-

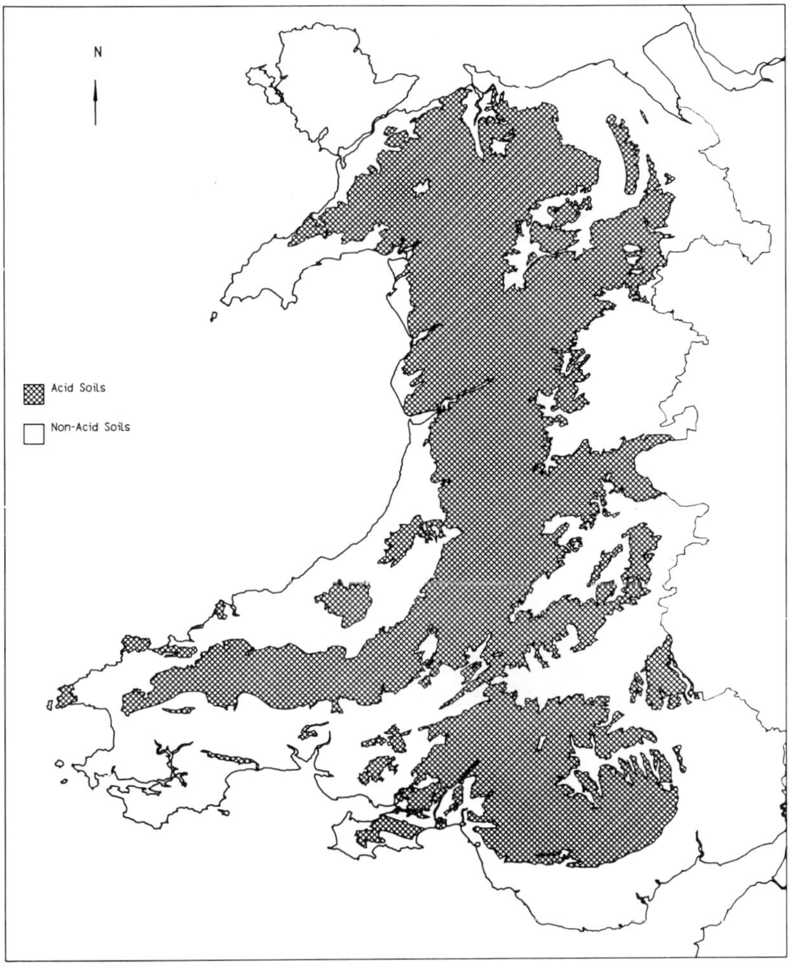

Fig. 5.2. The distribution of acid soils with little or no neutralising capacity and weakly acid to non-acid soils with moderate to high neutralising capacity in Wales.

ever the underlying bedrock. Thus, it is predicted that acid waters will not occur in those areas of north west and west Wales underlain by rocks with little or no buffering capacity. Waters entering bedrock from the non acid soils will already be buffered and are unlikely to be increased in acidity as a result of rock-solute interactions.

The predicted changes in streamwater acidity with flow are rather tentative but are based upon hydrochemical data held by the ITE and WWA. They envisage the changes in drainage water chemistry which might be expected in 1st and 2nd order streams when there is a change from groundwater domination in the stream to a predominance of soil derived water. The influence of this changeover will clearly vary with soil chemistry.

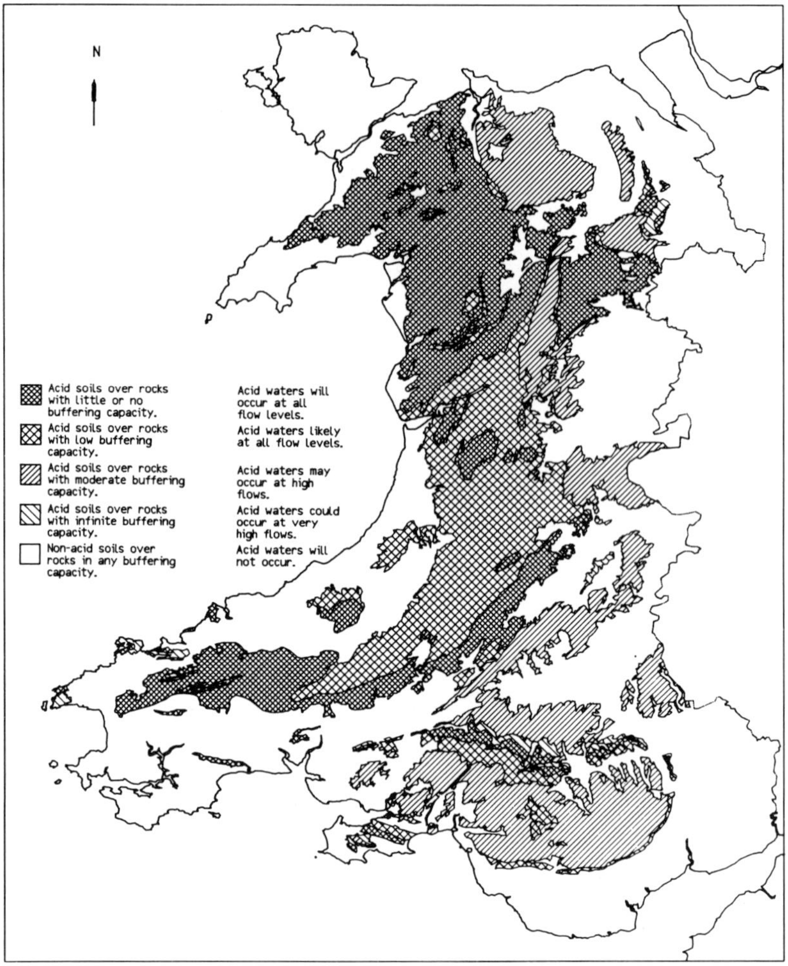

Fig. 5.3. The likely occurrence of acid, sensitive waters in Wales predicted from data on soils and geology.

The influence of agricultural land use

The classification of soils as 'acid' or 'non-acid' relies on known relationships between soil subgroup and soil chemistry. If land management has altered the soil chemistry significantly, the relationship will break down. In large areas of the Welsh uplands brown podzolic soils, stagnopodzols, and in some areas peats, have been limed as part of pasture improvement programmes. This can transform the soil chemistry and soil-water interactions,[6] changing these acid soils into soils with a pH of about 6.0 and moderate to high neutralising capacity. Reliable maps of the extent of pasture improvement are not, however, available. Land use data from agricultural returns, on a parish basis, were therefore used as a surrogate. If the dominant agricultural land use in a group of parishes was recorded as arable or permanent grass, it was assumed that lime would be applied, if necessary, to maintain a suitable pH and that the soils could be regarded as non-acid, with moderate to high neutralising capacity. Agricultural land use maps (Atlas of Wales, 1981), were used to identify areas of Wales in which more than 80% of the agricultural land was arable or permanent grass. Comparison of these maps with Fig. 5.2 allowed us to identify areas where liming had probably changed naturally acid soils into soils with a moderate to high neutralising capacity. Such areas were then reclassified and the distribution map of combined soils and bedrock units revised accordingly (Fig. 5.4). Where parishes with arable and/or permanent grass as the dominant (> 80%) agricultural land use overlayed 'acid soils', the soils were reclassified as 'non-acid'. The modified soil map shows that large areas of south west Wales are now classed as 'non-acid'. When this modified map is compared with Fig. 5.3 it can be seen that large areas of south west Wales are removed from classes 2 and 3 (Fig. 5.3, Table 5.3) and placed into class 5 (Fig. 5.4, Table 5.3).

Testing the predictions

The predicted distribution of acidic, low conductivity waters from Fig. 5.4 has been tested using the results from a regional survey of water quality by WWA. A figure of 15 mg l^{-1} has been used by the WWA to define highly sensitive surface waters. Almost all these sites fall within classes 1 and 2 on Fig. 5.4. The predicted variations in acidity with flow have not been fully tested to date.

Problems of scale and resolution

Figure 5.4 provides a regional prediction and overview. Considerable variations in streamwater acidity may occur, however, as a result of small scale variations in soils and bedrock geology which cannot be shown at the scales of the maps used here. Thus, Reynolds et al.[13] have shown that baseflow waters in areas of mid Wales underlain by slates and mudstones have a pH > 6.0 and significant levels of bicarbonate. The chemistry of these waters reflects the presence of small calcite veins within the slate and mudstone bedrock. These veins cannot be shown even on a large scale geological map. In north Wales, small dolerite dykes within slate country rock produce a similar buffering of groundwaters and baseflow. Again, the dykes are not shown on the 1:500,000 scale map used in the present exercise. In the Llyn Brianne area of central Wales, Stoner, Gee and Wade[16] subdivided a series of streams on the basis of hardness. Streams with apparently similar lower Palaeozoic bedrock, soils

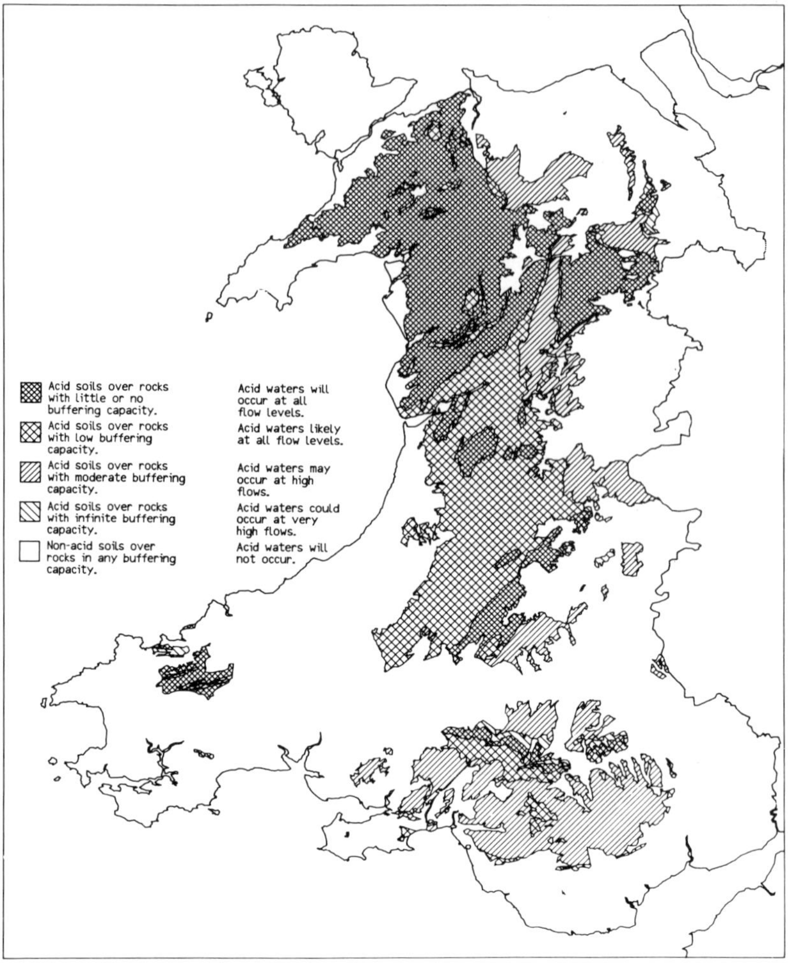

Fig. 5.4. The likely occurrence of acid-sensitive waters in Wales predicted from data on agricultural land use, soils and geology.

and vegetation had significantly different hardness. The streams with the higher hardness seem to drain catchments containing thin fossiliferous or arkosic bands which buffer groundwater and baseflow. Detailed geological and soil data are therefore needed before predictions and interpretations of streamwater chemistry can be made at the individual catchment scale.

Table 5.3. The likely occurrence of acid waters in Wales predicted from soils and geology.

	Soil/rock combination	Occurrence of acid waters
Class 1	Acid soils over rocks with little or no buffering capacity	Acid waters will occur at all flow levels
Class 2	Acid soils over rocks with low buffering capacity	Acid waters likely at all flow levels
Class 3	Acid soils over rocks with moderate buffering capacity	Acid waters may occur at high flow levels
Class 4	Acid soils over rocks with infinite buffering capacity	Acid waters could occur at very high flows
Class 5	Non-acid soils over any rock type	Acid waters will not occur.

References

1. Avery, B.W. (1980). Soil classification for England and Wales. (Higher categories). *Soil Survey Technical Monograph no. 14.* Harpenden, UK: Soil Survey of England and Wales.
2. Catt, J.A. (1985). Natural soil acidity. *Soil Use and Management*, 1: 8–10.
3. Cresser, M.S. and Edwards, A.C. (1987). Acidification of freshwaters. Cambridge, UK: Cambridge University press. 136pp.
4. Drever, J.I. (1982). The geochemistry of natural waters. Englewood Cliffs, USA; Prentice Hall.
5. Galloway, J.N. and Cowling, E.D. (1978). The effects of precipitation on aquatic and terrestrial ecosystems: a proposed precipitation chemistry network. *Journ. Air Pollut. Contr. Assoc.*, 28: 229.
6. Hornung, M., Stevens, P.A. and Reynolds, B. (1986). The impact of pasture improvement on the soil solution chemistry of some stagnopodzols in mid-Wales. *Soil Use & Management*, 2: 18–26.
7. Kinniburgh, D.G. and Edmunds, W.M. (1986). The susceptibility of UK groundwaters to acid deposition. *Hydrogeol. Rep. Br. Geol. Surv.*, No 86/3. Wallingford, UK: British Geological Survey.
8. Lau, W.M. and Mainwaring, S.J. (1985). The determination of soil sensitivity to acid deposition. *Water, Air & Soil Pollution*, 25: 451–464.
9. Likens, G.E., Wright, R.F., Galloway, J.N. and Butler, T.J. (1979). Acid rain. *Scientific American*, 241: 42–51.
10. McFee, W.W. (1980). Sensitivity of soil regions to long-term acid precipitation. In: *Atmospheric sulphur deposition: environmental impact and health effects*, (edited by D.S. Shriner, C.R. Richmond & S.E. Lindberg), 405. Ann Arbor, USA; Ann. Arbor, Science.
11. Norton, S.A. (1980). Geologic factors controlling the sensitivity of aquatic ecosystems to acid precipitation. In: *Atmospheric sulphur deposition*, (edited by D.S. Shriner, C.R. Richmond & S.E. Lindberg). Michigan, USA: Ann Arbor Science, 521–531.
12. Reuss, J.O. and Johnson, D.W. (1986). Acid deposition and the acidification of soils and waters. New York: Springer-Verlag.
13. Reynolds, B., Neal, C. Hornung, M. and Stevens, P.A. (1986). Baseflow buffering of streamwater acidity in five mid-Wales catchment. *Journal of Hydrology*, 87: 167–185.
14. Rudeforth, C.C., Hartnup, R., Lea, J.W., Thompson, T.R.E. and Wright, P.S. (1984). Soils and their use in Wales. *Soil Survey of England and Wales Bulletin no. 11.* Harpenden: Soil Survey of England and Wales.

15. Smith, R.T. and Taylor, J.A. (1969). The post-glacial development of soils and vegetation in northern Cardiganshire. *Trans. Inst. Brit. Geogrs.*, 48: 75–96.
16. Stoner, J.H., Gee, A.S. and Wade, K.R. (1984). The effects of acidification on the ecology of streams in the Upper Tywi catchment in west Wales. *Environ. Poll. A.*, 35: 125–127.
17. Wang, C. and Coote, D.R. (1981). Sensitivity classification of agricultural land to long term acid precipitation in eastern Canada. (*L.R.R.I. contribution 98*). Agriculture Canada Research Board.

6A. Recent land use change

R. BOON and D. KAY

Introduction

Maps defining the susceptibility of a region to acidic precipitation have traditionally concentrated on distributions of soil, drift and geology (see Chap. 5). Each of these parameters is constant over medium timescales.

Land use changes have been proposed as a causative factor in acidification. *Long term* land use changes such as Holocene forest clearances and thc onsct of peat formation have produced acidic, organic soils over much of the British uplands.[9, 36] More recent *short term* land use changes, such as softwood afforestation, have been associated with increased acidity in upland streams and lakes[40] However, afforestation is not the only land use change operative in the Welsh uplands. Land improvement and moorland reclamation are significant features of recent land use change in upland Wales.[30] The effects of this type of land use change on acidified surface waters are not well researched when compared to the extensive literature on the effects of afforestation.

The *land use hypothesis* in the acidification debate has centred on the direct effects of land use change on water quality.[33] Significant research attention has been devoted to the separation of land use and anthropogenic deposition factors in explaining surface waters acidification.[23] Much less attention has been devoted to defining the degree to which land use change can alter the susceptibility of a region to existing acidic precipitation. This is surprising in view of the dynamic nature of land use.[31] Furthermore, managed land use change offers one of the very few control strategies for the amelioration of upland acidification.[42] Present land use change is significantly affected by government and agency policies on agricultural support and afforestation.[8, 38, 41] This management structure is imperfect and fragmented. However, it offers some potential for land use planning in the uplands to reduce acidification.

Recent land use change

Much of the acid susceptible area defined by Hornung and Le Grice[22] is found in upland areas which are suitable only for sheep and cattle rearing or for

R.W. Edwards et al. (eds) Acid Waters in Wales. 67–80.

afforestation. Agriculture and afforestation are economically marginal in this location and without financial support both activities would decline in the harsh environment of upland Wales. Such a decline would have significant social and political impacts in the rural areas and hence a system of subsidies has been developed with the aim of supporting these economically marginal land uses.

Agriculture

The importance of support for upland agriculture in the postwar period was recognised in the *Hill Farming Act* of 1946 and the *Livestock Rearing Act* of 1951. These provided headage payments for stock and financial support for equipment and land improvement.[25] Such payments were designed to compensate farmers for the natural disadvantages of farming in the uplands and this mechanism achieved many of its social aims in maintaining a farming community in the remote upland areas. It was into this structure of existing legislation that the Less Favoured Areas (L.F.A.) Directive (75/268/EEC) was integrated.[12, 13, 38] The L.F.A. Directive was intended to benefit *mountain* areas or regions where the physical environment produces conditions of low agricultural productivity which could induce economic decline and depopulation (Articles 3.4 and 3.5) (Fig. 6A.1). This European legislation extended and broadened the previous policy to encompass the social goals of maintaining rural populations as well as the conservation objectives outlined in Article 1. The exact boundary of the L.F.A. is considered confidential but, within Wales, lists of community councils wholly or partly within the L.F.A. are published by the Welsh Office Agriculture Department.[43, 44]

The main instrument of the L.F.A. policy is through improvement grants and headage payments or Hill Livestock Compensatory Allowances (H.L.C.A.'s). Two aspects of the H.L.C.A.'s have been critisised by Wathern et al.[38] First, the allowances are equal in all the L.F.A.'s regardless of the severity of environmental constraint. This policy runs counter to the instructions contained in Article 7 of Directive 75/268/EEC which suggest that assistance should match the degree of constraint. Second, the maximum allowable stocking density of 6 ewes ha^{-1} represents a 2–3 fold overstocking of sensitive upland environments which can lead to problems of peat erosion and heavy winter stock losses.

The effects of the L.F.A. Directive on agriculture in a 130,483 ha area of mid-Wales have been documented by Parry and Sinclair.[31] Fig. 6A.2 shows the trends in land management observed in this area of L.F.A. between 1949 and 1983. These data illustrate the growth of coniferous woodland during recent years which has replaced rough pasture and moorland. The change of land use from moorland to conifer plantation is rarely reversed.[38] Reversion of improved grazing land to rough pasture is more common and the relation between moorland and rough pasture is a two way dynamic process. In the total study period (1948–83), primary afforestation accounted for about

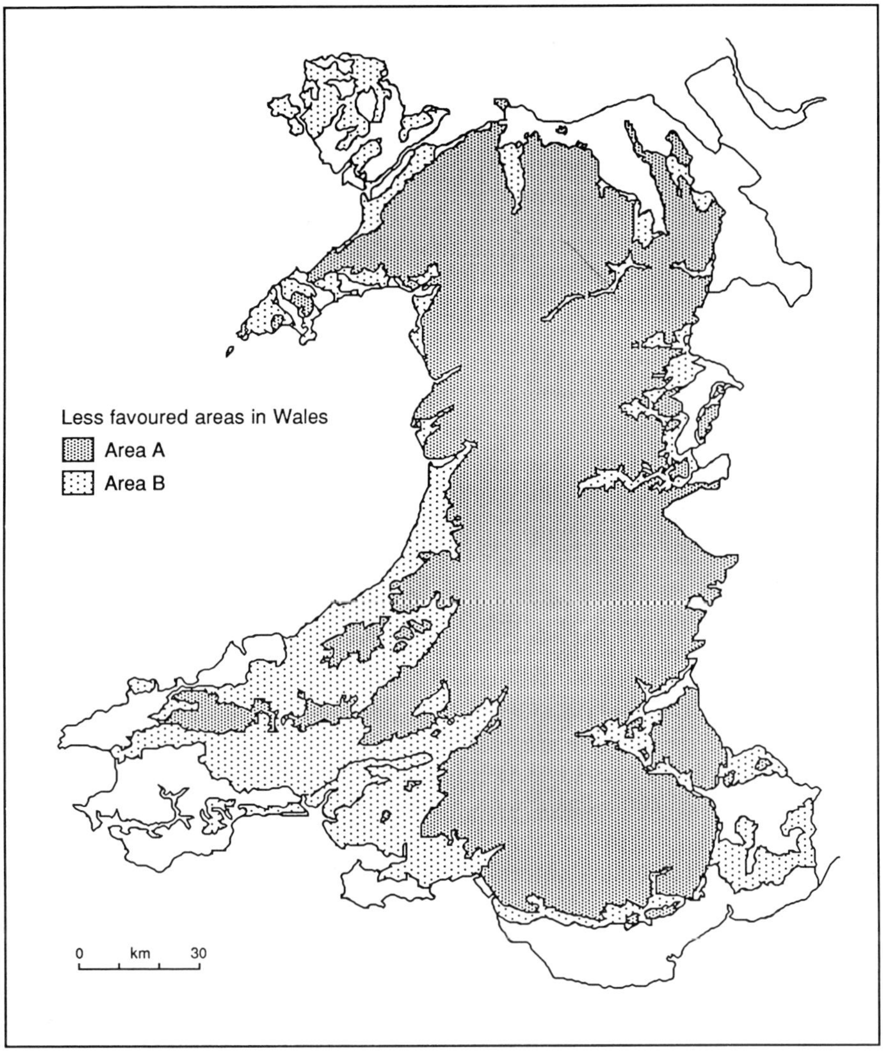

Fig. 6A.1. The Welsh less favoured areas (Sources 13 and 43).

23,000 ha. or 67% of the moorland loss. Moorland improvement and reclamation made up almost 30%, or 9,726 ha.

This pattern is reversed in the most recent period examined of 1978–1983 when 75% of the moorland loss is attributable to reclamation with only 22% due to afforestation. The annual reclamation rate in this period is 863 ha yr^{-1} (i.e. 0.67% of the total study area) compared with an afforestation rate of 255 ha $yr.^{-1}$ The mid-Wales study suggests a pattern of increasing sheep densities leading to land improvements supported by the recent influx of European

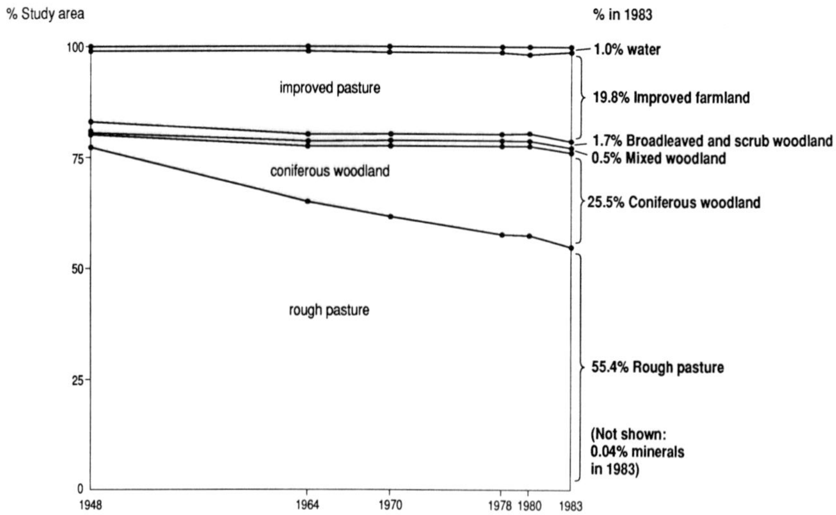

Fig. 6A.2. Land management trends in the mid-Wales L.F.A. between 1949 and 1983 (Source 31).

funding. The extent to which this cycle is ecologically or financially sustainable in upland Wales has not been defined.

A very similar pattern of increased agricultural intensity is seen in the area of the Clwyd L.F.A. in north Wales where Wathern et al.[39] have charted the sheep stocking patterns from 1977 to 1983. Here, stocking rates are approaching 6 ewes ha^{-1} and a switch from low rate breeds to the higher rated '*hill*' breeds is clearly evident within an overall 15% increase in the Clwyd L.F.A. (Fig. 6A.3). Total H.L.C.A. payments increased by 75% (£1.65m to £2.9m) over this period and capital grants increased 8.5 fold from £24K to £209K.

The evident increase of sheepmeat production from the Welsh L.F.A. suggests a successful policy of price support under Directive 75/268/EEC. Within the Welsh L.F.A.'s, rapid moorland reclamation is the most dynamic aspect of land use change at the present time. This change is dependent upon continued high levels of E.C. support.

Afforestation

Zehetmayr[47] has charted the main influences on forest management from the period 1880–1990. The net result was a rapid increase in forest planting in the immediate post-war periods. The total U.K. new plantings peaked at around 40,000 ha yr^{-1} in the early 1960's. More recently, new planting has declined in importance. Since 1980 the '*private*' sector has increased at the expense of the Forestry Commission (F.C.) and Scottish planting has dwarfed the activities in England and Wales (Fig. 6A.4). Today national (U.K.) new plantings total around 23,000 ha $yr.^{-1}$ Wales accounts for about 3% of this figure.

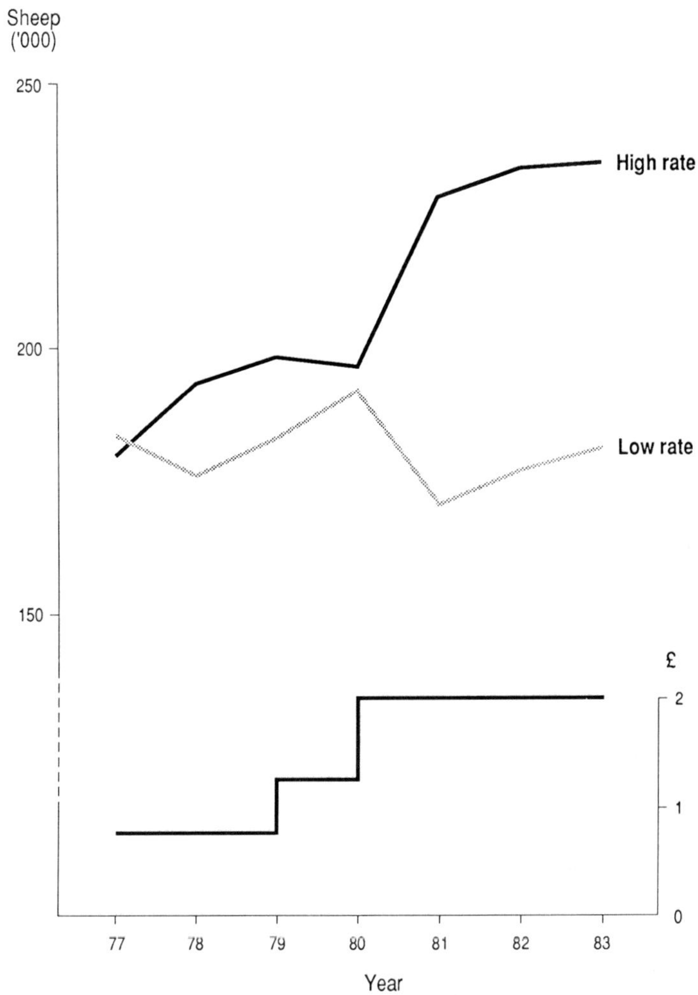

Fig. 6A.3. Sheep numbers in the Clwyd L.F.A. from 1977 to 1983. Note the switch to designated '*hill*' breeds (Source 39).

Total forest cover in the U.K. has doubled from 1.106 Mha to 2.207 Mha in the period 1924 to 1986.[27] Over this period new planting has averaged some 18,133 ha yr^{-1} and the F.C. has certainly achieved the objectives defined at its creation in 1919.[10] Widely quoted projections made in C.A.S.[8] and F.C.[17] envisaged an additional 2.0 Mha and 1.8 Mha of new planting respectively. These projections were looked upon with some concern by the water industry which feared the water loss implications and the possible water quality effects of catchment afforestation.[7, 34] Whilst new planting has now declined from the high levels of the 1960's, many suggest that the planting objectives outlined in C.A.S.[9] and F.C.[17] will be achieved in the early part of the next century.[28]

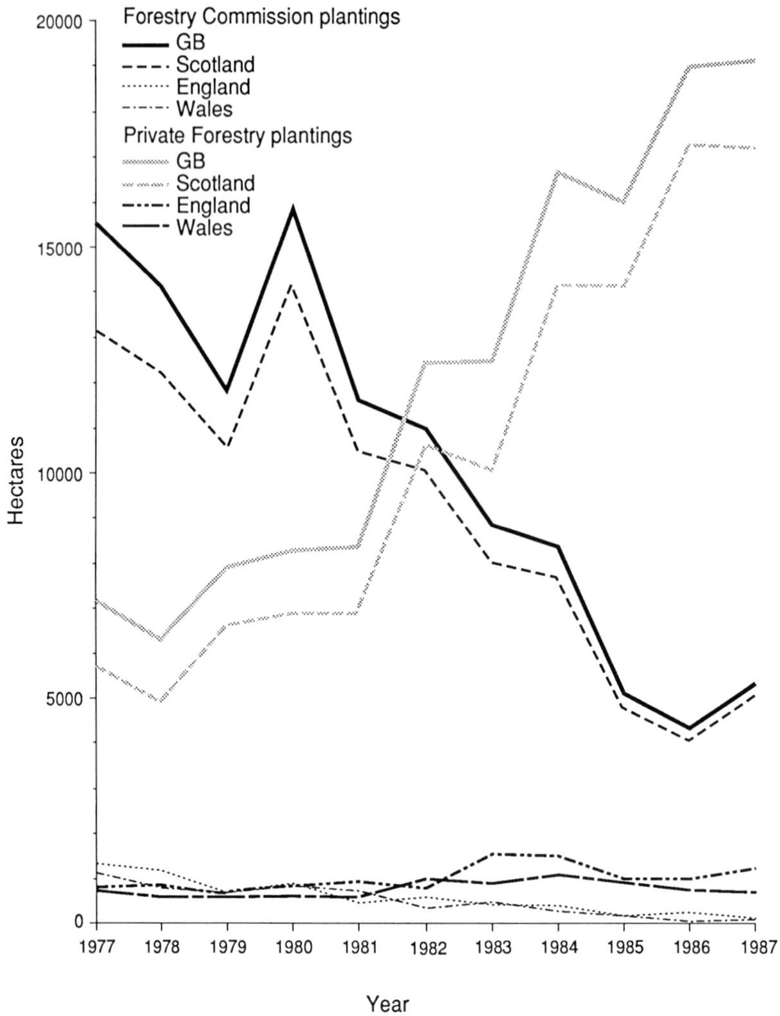

Fig. 6A.4. Trends in United Kingdom afforestation (Source 18).

Approximately 11% of the Welsh land area is afforested. Ten percent (203,000 ha) is commercially productive with 7% F.C. land and 3% privately owned. The average timber yield is 11 $m^3 ha^{-1} yr^{-1}$.[5] C.A.S.[8] estimated that an additional 430,000 ha were technically afforestable. However, the recent afforestation rates outlined in Fig. 6A.4 indicate little possibility that the available area will be utilised within the next century.

The current balance of land use is maintained by a complex system of financial support for agriculture and afforestation. Government policy is to expand production from both sectors.[38, 46] The net result of these conflicting policies can be seen in the Clwyd and mid-Wales case studies where sheep production has shown a recent intensification and conifer forest growth has almost halted. If this balance were to be upset by a reduction in the E.C. support for sheepmeat production or a more advantageous tax regime for the owners of upland conifer plantation, then a return to the longer term trends of increasing afforestation might be seen.

An economic evaluation

An objective economic comparison of afforestation and agriculture is very difficult because the expenditure and income streams from the two activities are so different. Agriculture involves a regular annual income whilst afforestation produces irregular and delayed income. All attempts at economic comparison must therefore employ a method based on discounted cash flow. For activities with very long gestation periods, the assumptions made at the outset regarding discount rates are central to any economic evaluation.[29] The time period of projection (50+ years) makes other important aspects of the economic analysis little more than guesswork. For example, future income from either timber sales or agricultural produce 50 years hence is very difficult to predict.[8] Other, less tangible, future benefits, such as employment or maintaining the social fabric of rural areas, are even more difficult to express financially.[21] A third difficulty derives from the complex system of state support through grants, tax relief and price maintenance which underpins both activities. There is considerable evidence that certain elements such as conservation, price support and employment are in direct conflict.[25] It is unlikely therefore that present support structures will continue in the longer term.

Tomkins[37] compared the subsidy costs of 500 acres (=200 ha) of upland afforestation with a similar moorland area. The forestry subsidies derive from planting grants and tax relief whilst the agricultural subsidies comprise H.L.C.A. payments and sheep annual premiums. Over the first six years of the forest stand the total subsidy cost would be £140,000. The farmer would receive £14,400 over the same period. This comparison clearly indicates the greater cost to the public purse of afforestation in the early years following planting. However, neither land use achieves the 5% rate of return expected of public investment.[20] In his review of the comparative economics of afforestation and agricultural improvement, Bowers[4] concludes that

> The social rates of return on both activities are unacceptably low: investment in them i.e. is not a sensible use of the nation's capital.

Green and Dyfri-Jones[20] have completed the most recent economic comparison between farming and afforestation in Wales using three case study farms to evaluate the comparative economics of farming against afforestation. Their rationale for this approach was to examine the economics of the decision process at the farm level. Institutional decisions take account of wider social issues which may not be relevant to the individual landowner. These authors conclude that

> Undiscounted, unsubsidised net returns from agriculture and forestry in selected hill and upland areas of Wales appeared similar. Therefore assuming similar relative efficiencies of the farmer/landowner in agriculture or forestry, there is probably no significant difference between the two activities in terms of absolute income levels under the prevailing economic conditions.

It is clear from all these studies that '*objective*' economic comparisons between upland agriculture and afforestation are very difficult. There is a lack of agreed information despite several studies since the 1972 Treasury investigation. The proponents and opposers of increased afforestation have both used the economic argument to support their positions. The dependence of both afforestation and upland agriculture on the long term fiscal climate suggests that regional scale forecasts of future trends should be treated with great caution.

Implications for surface water quality

The recent land reclamation and the more '*medium term*' afforestation described above have conflicting effects on upland water quality.

Upland reclamation requires soil pH adjustment through liming to improve nutrient availability to the sward.[1] Initial '*remedial*' liming involves approximately 5–10 t ha^{-1} of ground limestone,[42, 45] Ormerod and Edwards[29] have suggested that standard lime application rates should be sufficient to maintain stream Ca^{2+} status above the 2mg l^{-1} concentration which is critical for healthy salmonid fisheries.[35] However, they also suggest that total Welsh lime applications have decreased since the withdrawal of the liming subsidy in 1976. Despite this overall reduction the impacts on upland '*acid susceptible*' areas is difficult to assess. Boon and Kay[3] have studied liming practice in the Upper Twyi catchment covering some 22,100 ha which includes the Llyn Brianne study area. They could identify little net change in total weight of liming materials applied to the catchment between 1950 and 1986, although there is some evidence of a slight downward trend in the neutralising capacity of the materials applied because of the switch away from burnt lime (Fig. 6A.5). Certainly, this limited case study, does not suggest that the '*national*' downward trend in liming could be responsible for significant streamwater acidification in this susceptible upland catchment in mid-Wales. Patrick's comments,[32] that national lime use trends may be poor indicators of upland application rates in acid susceptible areas, are reinforced by the results of the Twyi study.[3] Nevertheless, the switch to upland '*improvement*' subsidies and away from '*liming*' grants involves the application of high nitrate fertilizers which may induce an undefined degree of acidification by the oxidation of

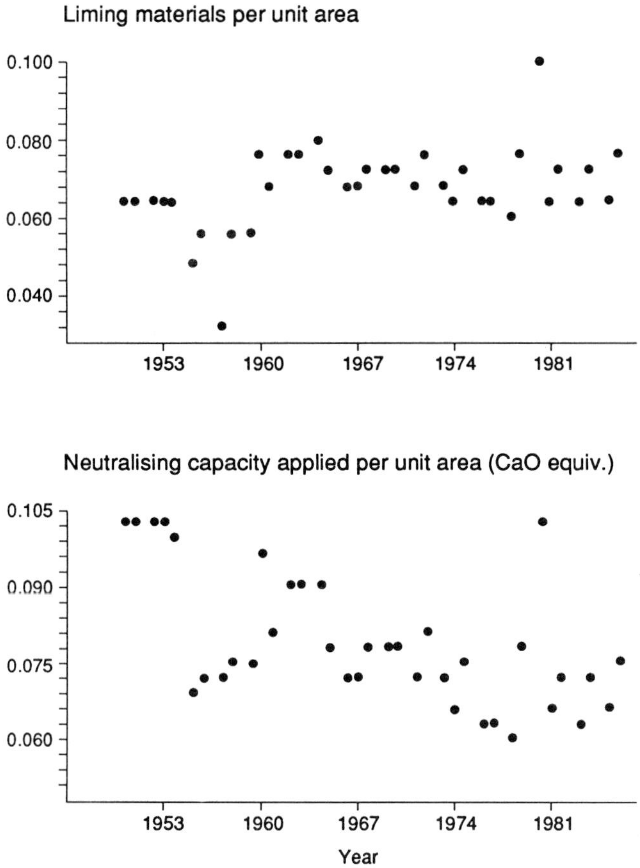

Fig. 6A.5. Application rates of liming materials in the Upper Twyi catchment in the period 1950 to 1986 (Source 3).

ammonium ions.[6, 11, 45] However, associated inputs of limestone, either mixed with the fertilizer compounds or applied as a separate dressing to reclaimed or improved pastures, would neutralise the acidity derived from nitrification of ammonium ions. This would reduce the buffering capacity otherwise available for neutralising acidic precipitation.

Conifer afforestation has been associated with elevated Al^{3+} concentrations, reduced pH and impoverished fisheries status (see Chaps. 17 and 19). In the acid susceptible areas of Wales defined by Welsh Water.[40] three main land use types are feasible; namely, (i) afforestation, (ii) moorland and/or (iii) improved pasture. The trend data presented above show the current directions of land use change between these options in several case study regions. The extent to which these land use types alter the acid susceptibility of a catchment is of considerable interest to the water industry which has statutory responsibilities for the maintenance of drinking and salmonid

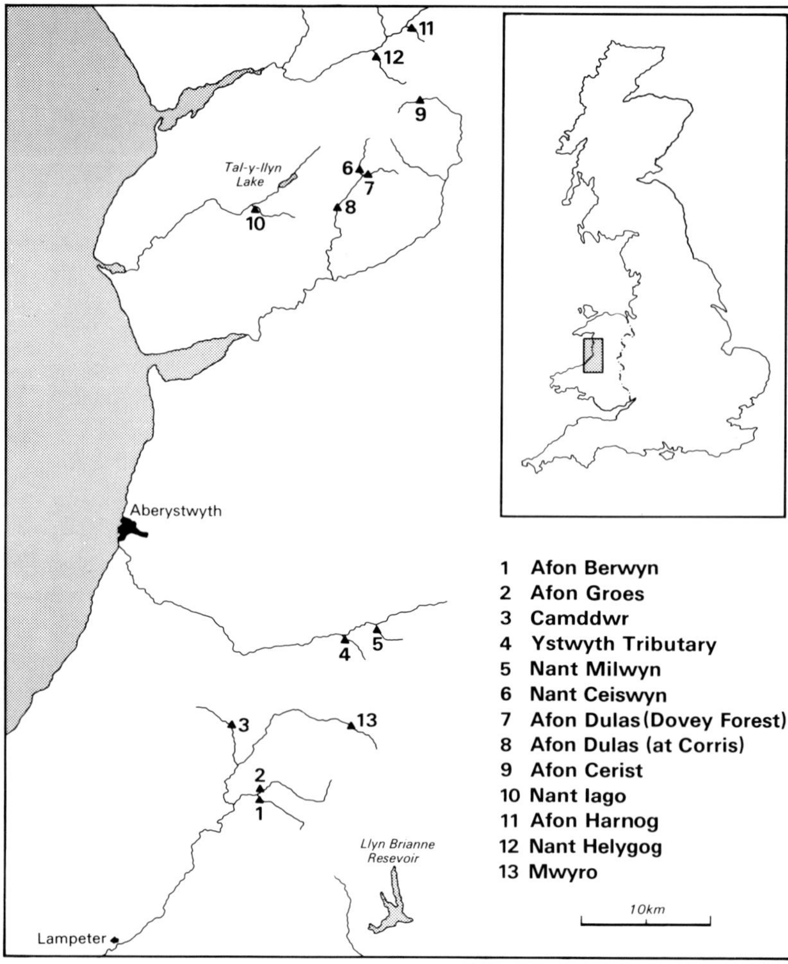

Fig. 6A.6. Thirteen study catchments in an acid susceptible region of West Wales (Source 24).

water quality.[14, 15, 16] The effects of land use on compliance rate with these E.C. directives have been studied by Kay and Stoner[24] for thirteen catchments in West Wales which all come within the acid susceptible region defined in Welsh Water[40] Fig. 6A.6 shows the locations of each catchment outlet and Table 6A.1 contains catchment characteristics and compliance rates for each of these sampling sites. A clear pattern is evident with high correlations between percentage catchment afforestation and non-compliance for pH and dissolved Al^{3+}. The percentage land in the best agricultural grade (IV) shows a set of negative correlations with the parameters characteristic of acidified surface waters.

These data suggest that contemporary land use change can alter the susceptibility of a catchment to acidic inputs. The extent to which both

Table 6A.1. Catchment characteristics, and percentage of samples exceeding water quality standards within an area of acid susceptibility in West Wales (Source 24).

Catchment	% Forest area	% Grade IV land	% Exceed 0.05mgl^{-1} Al	% Exceed 0.2mgl^{-1} Al	% pH<6.0	% FC>20 dl^{-1}
BERWYN	66	13	40	4	24	77
GROES	12	22	9	0	5	92
CAMDDWR	10	45	28	0	0	96
YSTWYTH TRIB	88	0	67	38	69	26
NANT MILWYN	9	1	36	0	40	83
NANT CEISWYN	1	0	21	0	26	7
AFON DULAS (N)	45	0	46	28	31	0
AFON DULAS	83	2	26	8	32	38
AFON CERIST	63	3	11	0	7	71
NANT IAGO	58	1	93	59	100	24
AFON HARNOG	27	3	44	4	16	59
NANT HELYGOG	0	6	36	0	15	62
NANT MWYRO	0	0	53	0	35	57

Percentage exceedence (i.e. non-compliance) values are based on weekly spot samples where n=37 to 47 for chemical parameters and n= 15 to 30 for *E.coli* (FC). The parameter values chosen are based on sources 14, 15, 16. For a full explanation see source 24.

historical land use change and anthropogenic pollution have each caused acidification is best examined using techniques of Quaternary environmental reconstruction. This *long-term* perspective is outlined in Chaps. 3 and 6B.

Water quality improvement strategies

The relationships reported above between land use and water quality in this acid susceptible area of West Wales are not static. Forestry Commission planting guidelines[2, 26] are one attempt to manage these interactions and reduce some of the worst effects of afforestation. Until recently, however, little hard information was available on which to base policy regarding acceptable land management in the uplands for either afforestation or agriculture. Process studies underway at Llyn Brianne, Plynlimon and Beddgelert described elsewhere in this book have started to identify the key areas in acidified catchments to which remedial measures should be targeted.[19]

Conclusions

The results presented in this chapter suggest;

(i) Upland land use change is responsive to government policy through subsidies and price support.

(ii) Upland land use has a significant effect on the susceptibility of an upland catchment to acidification.
(iii) Deliberate management of upland catchment land use can have significant effects on the acidification of surface waters.

No system exists for land use planning in upland Britain. Decisions on land use change are taken by individual farmers and commercial forestry interests. It is clear that these decisions affect society in many ways including aesthetic impacts and access restrictions which have caused past controversy. Impacts on surface water quality, important for potable and salmonid waters in Wales, should be added to these other considerations. Given this range of effects a strong case exists for a more formal system of upland land use planning in Britain.

References

1. A.D.A.S. (1984). Agricultural Development and Advisory Service. *National Lime Demonstration.* Welsh Office, Cardiff. 36p.
2. Binns, W.O. (1986). *Forestry and Freshwaters – Problems and Remedies* In Solbe J.F. de L. (Ed.) *Effects of Land Use on Freshwaters.* Ellis Horwood, Chichester, 568p.
3. Boon, R. and Kay, D. (1988). *Liming in the Upper Twyi Catchment.* Final Report to Welsh Water (in Press).
4. Bowers, J.K. (1984). *The economics of upland land use.* Mimeographed Paper available from Leeds University, School of Economics. 83p.
5. Bradley, R.T. (1983). *Forestry and Upland Land Use in Wales.* Welsh Conservation Conference, Cardiff. pp. 217–224.
6. Brady, N. (1974). *The Nature and Properties of Soils.* 8th Edition, Macmillan, New York. 639p.
7. Calder, I.R. and Newson, M.D. (1979). Land use and water resources in Britain: a strategic look. *Water Resources Bulletin 15(6)*, 1628–1638.
8. C.A.S. (1980). Centre for Agricultural Strategy *Strategy for the U.K. Forest Industry.* Reading. (CAS Report No. 6)
9. Catt, J.A. (1985). Natural Soil Acidity. *Soil use and Management*, 1(1): 8–10.
10. Clout, H.D. (1972). *Rural Geography.* George Allen and Unwin.
11. E.R.L. (1983). Environmental Resources Limited. *Acid Rain: a Review of the Phenomenon in the EEC and Europe.* Graham and Trotman, London. 159p.
12. European Community (1975a). Council Directive on Mountain and Hill Farming in Certain Less Favoured Areas (75/268/EEC). *Official Journal*, 128(19): 1–25.
13. European Community (1975b), Council Directive of 28 April 1975 Concerning the Community List of Less Favoured Farming Areas Within the Meaning of Directive No 75/268/EEC (United Kingdom) (75/276/EEC). *Official Journal*, 128(19): 231–266.
14. European Community (1975c). Council Directive 75/440/EEC of 6th June 1975 Concerning the Quality Required of Surface Waters Intended for the Abstraction of Drinking Waters. *Official Journal*, L194: 26–31.
15. European Community (1978). Council Directive 78/659/EEC of 18th July 1978 concerning the quality of fresh waters needing protection or improvement in order to support fish life. *Official Journal,* L222: 1–10.
16. European Community (1980). Council Directive 80/788/EEC of 15th July 1980 relating to the quality of water intended for human consumption. *Official Journal*, L194: 26–31.

17. F.C. (1977). Forestry Commission *The Wood Production Outlook for Britain: a review* The Forestry Commission, Edinburgh.
18. F.C. (1988). Forestry Commission *Annual Report and Accounts*. The Forestry Commission, Edinburgh.
19. Good, J.E.G. (Ed.) (1987). *Environmental aspects of plantation forestry in Wales*. I.T.E. Symposium No. 22. Proceedings. I.T.E., Bangor. 77p.
20. Green, D.A. and Dyfri Jones, W. (1986). *Forestry as an alternative to farming in the hills and uplands of Wales*. Department of Agricultural Economics, University College Aberystwyth. 121p.
21. H.M. Treasury (1972). *Forestry in Great Britain: an Interdepartmental Cost-Benefit Study*. H.M.S.O.. London.
22. Hornung, M., Le Grice, S., Brown, N. and Norris, D. (1988). Geology and soils. *Acid waters in wales* (Ed. by Edwards, R.W., Gee, A.S. and Stoner, J.H.) Dr. W. Junk Publishers, The Hague. Netherlands.
23. Jones, V.J., Stevenson, A.C. and Batterbee, R.W. (1986). Lake acidification and the Land-Use Hypothesis: a Mid-Post-Glacial analogue. *Nature*, 322(6075): 157–158.
24. Kay, D. and Stoner, J.H. (1988). The Effects of Catchment Land Use on Stream Water Quality in an Acid Susceptible Region in West Wales: the Implications for Compliance with EEC Drinking Water and Fisheries Directives. *Appl. Geog. 8(2)*: 191–205.
25. MacEwen, M. and Sinclair, G. (1983). *New Life for the Hills*. Council For National Parks, London. 45p.
26. Mills, D.H. (1980). The management of forest streams. *Forestry Commission Leaflet No. 78*. HMSO. London.
27. N.C.C. (1986). Nature Conservancy Council *Nature Conservation and Afforestation in Britain* Nature Conservancy Council, London. 108p.
28. Ogilvie, R.S.D. and Lamb, R. (1986). Whither Forestry? the Scene in 2025 a Forester's View. *Symposium on Trees and Wildlife in the Scottish Uplands*. (Ed. by D. Jenkins) Institute of Terestrial Ecology, Symposium No. 17.
29. Ormerod, S.J. and Edwards, R.W. (1985). Stream acidity in some areas of Wales in relation to historical trends in afforestation and the usage of agricultural limestone. *J. envir. Mgmt.* 20: 189–197.
30. Parry, M.L., Bruce, A. and Harkness, C.E. (1982). Changes in the Extent of Moorland and Roughland in the Brecon Beacons National Park. University of Birmingham. *Surveys of Moorland and Roughland Change No. 7.*
31. Parry, M.L. and Sinclair, G. (1985). *Mid-Wales Upland Study*. Countryside Commission, Cheltenham. 59p.
32. Patrick, S.T. (1987). The Significance of Land Use and Land Management Changes. *Palaeoecological evaluation of the recent acidification of Welsh Lakes*. (Ed. by A.C. Stevenson and S.T. Patrick). Palaeoecology Research Unit, U.C.L., London. 47p.
33. Rosenqvist, I.Th. (1978). Alternative sources for acidification of river water in Norway. *Science of the Total Environment* 10: 39–49.
34. Stoner, J.H. and Gee, A.S. (1985). Effects of Forestry on Water Quality and Fish in Welsh Rivers and Lakes. *J. Instn Wat. Engrs*, 39: 27–45.
35. Stoner, J.H., Gee, A.S. and Wade, K.R. (1984). The Effects of Acidification on the Ecology of Streams in the Upper Tywi Catchment in West Wales. *Envir. Poll. (Ser. A)*, 35, 124–157.
36. Taylor, J.A. (1983). The Peatlands of Britain and Ireland. *Mires: Swamp, Bog, Fen and Moor, B* (Ed. by A.J.P. Gore). pp 1–46. Elsevier Scientific Publishing Company, Amsterdam.
37. Tompkins, S.C. (1986). *The Theft of the Hills*. The Ramblers' Association and the World Wildlife Fund. 32p.
38. Wathern, P., Young, S.N., Brown, I.W. and Roberts, D.A. (1986). The E.E.C. Less Favoured Areas Directive: Implementation and Impact on Upland Land use in the U.K.. *Land Use Policy*, July, 205–212.
39. Wathern, P., Young, S.N., Brown, I.W. and Roberts, D.A. (1988). Recent Upland Land

Use Change and Agricultural Policy in Clwyd, North Wales. *Appl. Geog. 8(2)*, 147–164.
40. Welsh Water (1987a). *Acid Waters in Wales.* Welsh Water, Brecon. April 1987. 13p.
41. Welsh Water (1987b). *Afforestation in Areas Sensitive to Acidification.* Welsh Water, Brecon. 27.2.87. 6p.
42. Welsh Water (1987c). *Llyn Brianne Acid Waters Project.* Welsh Water, Brecon. 91p.
43. W.O.A.D. (1984). Welsh Office Agriculture Department *Less Favoured Areas Map, Scale 375,000:1* AS13(W) Welsh Office Agriculture Dept, Aberystwyth.
44. W.O.A.D. (1986a). Welsh Office Agriculture Department *Farming in Less Favoured Areas. The H.L.C.A. Scheme* AS13(W) Welsh Office, Cardiff.
45. W.O.A.D. (1986b). Welsh Office Agriculture Department *Grassland Improvement* AS13(W) Welsh Office, Cardiff.
46. Younger, G. (1980). Statement by the Secretary of State for Scotland to the House of Commons on 10.12.80. 61st Annual Report and Accounts of the Forestry Commission 1980–81.
47. Zehetmayr, J.W.L. (1987). Influences Shaping the Development of Plantation Forestry in Wales. *Environmental Aspects of Plantation Forestry in Wales* (Ed. by J.E.G. Good) pp. 5–8. Institute of Terrestrial Ecology, Bangor Research Station.

6B. Acidified Welsh lakes: the significance of land use and management

S.T. PATRICK and A.C. STEVENSON

Introduction

The acidification of surface waters is frequently ascribed to the impact of acidic precipitation in areas of acid susceptible lithology.[3, 6, 17, 20] However, other explanations have been forwarded that centre upon changes in land use and management within lake catchments.

The Palaeoecology Research Unit at University College London is investigating the history and causes of acidification in ten lakes in mid and north Wales (Fig. 3.1). The major thrust of this project involves the utilization of palaeoenvironmental techniques to reconstruct the historical water quality (particularly pH) of these lakes and the history of atmospheric deposition of metals and combustion products as recorded in the lake sediments (Chap. 3).

The work described in this chapter represents a separate theme within the project: the use of pollen analysis and documentary sources to examine the significance of land use and land management changes in lake catchments. The potential timescale of acidification is adequately covered by the combined use of pollen analysis of lake sediments to give a long term (ie. 10,000 year) to mid term (ie. 100–400 year) perspective and historical sources which in Wales cover the past 150–200 years. The latter timescale is particularly important since results from Galloway, south west Scotland[2, 16] and Wales[7, 8, 23, 24, 31, 32] (see also Chap. 3) suggest that the most acute phase of acidification has occurred within the past 50–150 years.

The potential acidification mechanisms discussed here are frequently described separately or together as 'land use hypotheses'. This chapter deals with 'land use' and 'land management' change and it is important that these terms be clearly defined. 'Land use' is the broad use to which the land is put and categories may include rough grazing, improved farmland or forest. 'Land management' includes practices which if of low intensity may remain independent of land use, but if intensively practised may represent a mechanism responsible for land use change (eg. enclosure, manipulation of stocking densities and regimes, drainage, burning).

R.W. Edwards et al. (eds) Acid Waters in Wales. 81–91.

Study sites

The location of the ten study sites is indicated in Fig. 3.1. They lie between 280–530m on acid susceptible lithologies.[13] Lake and catchment specifications are given in Table 6B.1. All the lakes lie in moorland areas. Eight of the catchments comprise open moorland utilised for rough grazing, while two (Llyn Berwyn and Llyn Cwm Mynach) have been afforested. The lake waters are acid, pH ranging from 4.8 to 5.6 in unlimed lakes (Table 6B.1). All have been acidified to a greater or lesser extent over the past 150 years.[7, 8, 14, 15, 23, 24, 31, 32] Fishery decline is well documented in Llyn Berwyn, Llyn Conwy, Llyn Gamallt and Llyn Hir. The fishery status of other lakes is less certain.

'Land use hypotheses'

Rosenqvist[26, 27, 28, 29, 30] considered that ion exchange reactions in raw humus layers and the uptake of cations by plant growth are the most important factors in determining the acidity of run-off waters. The acidification of lakes in southern Norway was attributed to the increased formation of acid humus resulting from the enhanced biomass of acidic heathland (particularly *Calluna*) and forest species, consequent upon a decline in the intensity of pastoral agriculture in the region. Documentary evidence was produced to illustrate the decline in agricultural activity that followed an era, in the 18th and 19th centuries, when increased farming activity had resulted in the replacement of heathland by grassy vegetation. Particular importance was attributed to the demise of burning which had oxidised much of the humus cover.[29] Streams flowing in areas of burnt forest were shown to possess higher pH and conductivity than streams flowing in unburnt areas.

The Rosenqvist hypothesis had found limited support as an explanation of contemporary acidification[16] although Pennington[25] invoked similar mechanisms to explain the long term acidification of upland soils in Cumbria.

Although the concept that lake acidification results from catchment change is most commonly associated with Rosenqvist, several other proposed mechanisms may be classified under the broad heading of 'land use hypotheses'. These include:

i. the effects of afforestation[5, 9, 33]
ii. drainage of organic soils and peats[1, 10, 11]
iii. the demise of agricultural liming[19]
iv. the impact of mining[4]

Against this background it is important to know whether such changes have occurred in the catchments of acidified lakes and whether, because of their intensity and timing, they can account for the acidification.

Table 6B.1. Lake and catchment characteristics.

Lake	Altitude m.	Lake area ha.	Catchment area excluding lake ha.	Lake volume m^3	pH 1984–85	pH 1986[1]	pH 1987[2]	Mean depth m.
Berwyn	438	13.04	83.70	417,655	4.1–4.5	5.9[3]		3.25
Conwy	450	40.10	95.90	3,073,846	4.85[4]	4.8	4.8	7.66
Cwm Mynach	285	5.89	131.54	50,141	5.0–6[4]	5.3	5.6	2.18
Dulyn	526	1.98	51.67	37,787	4.3–5.4	5.0	5.2	1.90
Eiddew Bach	380	1.38	10.94	37,990	4.3–5.4	4.9	4.9	2.76
Gamallt	465	13.00	55.47	41,694[5]	4.35[4]	4.4		1.50[5]
Gynon	430	25.26	286.16	533,835		5.3	5.2	2.18
Hir	435	4.89	17.93	136,367	4.5–5.1	6.1[3]		2.79
Llagi	375	5.86	162.16	331,734	4.5–5.2	5.2	5.3	5.80
Y Bi	455	2.71	45.01	42,638	4.3–5.3	4.9	5.3	1.58

[1] April 1986.
[2] May 1987.
[3] Post liming of lake.
[4] 1983–1984 mean.
[5] Llyn Gamallt Bach only.

Pollen analysis

Pollen analysis has been conducted on lake sediments from six lakes with unafforested catchments in the study.[7, 8, 23, 24, 31, 32] Long-term changes in catchment vegetation for three of these lakes (Llyn Gynon, Llyn Dulyn and Llyn Eiddew Bach) have been identified. The results demonstrate that until about 5000 B.P., the catchments of each lake were dominated by an oak/elder deciduous woodland. From 5000 B.P. the replacement of this woodland by open, acidic heathland and blanket mire communities is clearly seen.[23, 31, 32] Unfortunately, problems were encountered with hiatuses within the sediment cores. This prevented a detailed examination of the entire postglacial vegetational history of the catchments. However, in the case of Llyn Dulyn the diatom flora of the pre-heathland sediments shows no appreciable difference from that of the post-woodland community[32] indicating that the development of acid heathland in lake catchments is not an important cause of lake acidification. Similar results have been obtained from south west Scotland.[12]

An appreciation of the mid term changes in land use over the last 200 years is gained by examination of the pollen record of the sediments of two typical acidified, upland moorland lakes (Llyn Hir and Llyn Llagi) (Fig. 3.1).

In the case of Llyn Hir there appears to have been no appreciable vegetation change within the catchment over the last 200 years (Fig. 6B.1). The present dominance of the catchment by *Nardus stricta* and *Festuca* spp. is reflected by the high Gramineae values. In the case of Llyn Llagi (Fig. 6B.2) a clear change in vegetation is indicated, but instead of a change from Gramineae to *Calluna* as would be required under the land use hypothesis, the reverse trend occurs.

In general, of all the six unafforested lakes examined only one, Llyn y Bi, shows a trend to *Calluna* dominance over the last 200 years. Since all the lakes studied show a progressive acidification after c. 1850, the pollen data suggest that this acidification cannot be accounted for by vegetation change.

Documentary investigations

Whilst pollen analytical techniques can serve to illustrate broad continuous trends over long term periods, documentary sources can be employed to reconstruct a more detailed history for individual lake catchments. It is a fundamental criticism of Rosenqvist's original 'land use hypothesis' that it was formulated on the basis of regional rather than catchment-specific information.[21]

Full details of sources employed in the context of upland Wales may be found in Patrick.[21] They include estate papers, tithe survey documents, small scale Ordnance Survey maps, land utilisation and vegetation surveys, air photographs, annual parish agricultural returns and a series of corroborative and miscellaneous sources.

Land use

With the exception of the afforestation of the Llyn Berwyn and Llyn Cwm Mynach catchments, there has been no net land use change at the study sites. Manganese and slate were exploited in the Llyn Cwm Mynach catchment but this was a small scale and short lived feature of the 19th century.

At Llyn Berwyn it is suggested that acidification was enhanced by afforestation and in particular by the deep drainage of the maturing forest in 1974.[14] At Llyn Cwm Mynach the onset of acidification is dated to the 1930's[15] and therefore predates afforestation.

Fish decline in Llyn Conwy in the early 1960's was temporally coincident with the first exploitation of the lake as a reservoir and the onset of regular water level fluctuations. pH reconstruction, a more stringent measure of water quality change than fish decline, is not complete for this site so the significance of that coincidence cannot yet be determined.

Land management

All catchments have experienced changes in management practices, but of a varying nature, extent and historical occurrence. Enclosure in certain catchments may have resulted in a rationalisation and possible decrease in sheep grazing intensity. However the early date (late 18th–early 19th century) of such enclosure suggests that it was not an important process. The regional and local trends towards the replacement of wether flocks by ewes and lambs from the mid-19th century and a decline in winter stocking densities[21] were realised in all catchments. However, in the post-1945 period some catchments have experienced a significant increase in summer stocking densities. For example, at Llyn Hir[7] the unenclosed, undrained *Nardus/Festuca* grassland of the catchment has a long history of sheep grazing, which over the period of major acidification (c. 1940–1980), increased in intensity.

With the exception of the Llyn Eiddew Bach and Llyn Hir catchments which have never been regularly burnt, there has been a decline or cessation of burning in the past c. 20 years. However, only in the Llyn Conwy, Llyn Gamallt, Llyn y Bi and the unafforested section of the Llyn Cwm Mynach catchments is there any evidence of recent regeneration of heathland species, notably mature, leggy *Calluna* stands (*Pteridium* is not present in any catchment). With the exception of Llyn Gamallt, effects of the demise of burning have been accentuated by a decline in grazing intensity. In the eight lakes where pH reconstruction is complete the onset of acidification commenced well before the decline of burning. pH reconstruction is not complete for Llyn Conwy or Llyn Gammallt, but well documented fishery decline from the early 1960's[18, 22] suggests that acidification commenced before burning ceased in the mid/late-1960s.

Catchment drainage has only occurred at Llyn Conwy (1920's) and Llyn

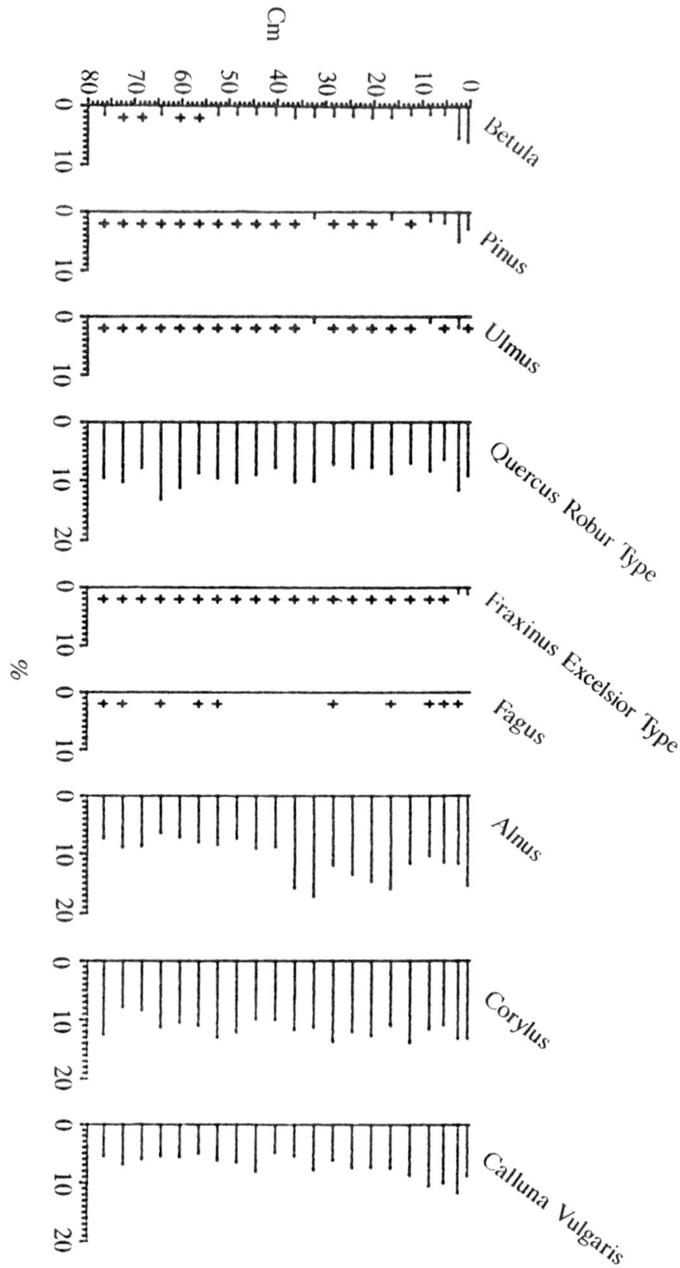
Cm
0
10
20
30
40
50
60
70
80
Betula
Pinus
Ulmus
Quercus Robur Type
Fraxinus Excelsior Type
Fagus
Alnus
Corylus
Calluna Vulgaris
%

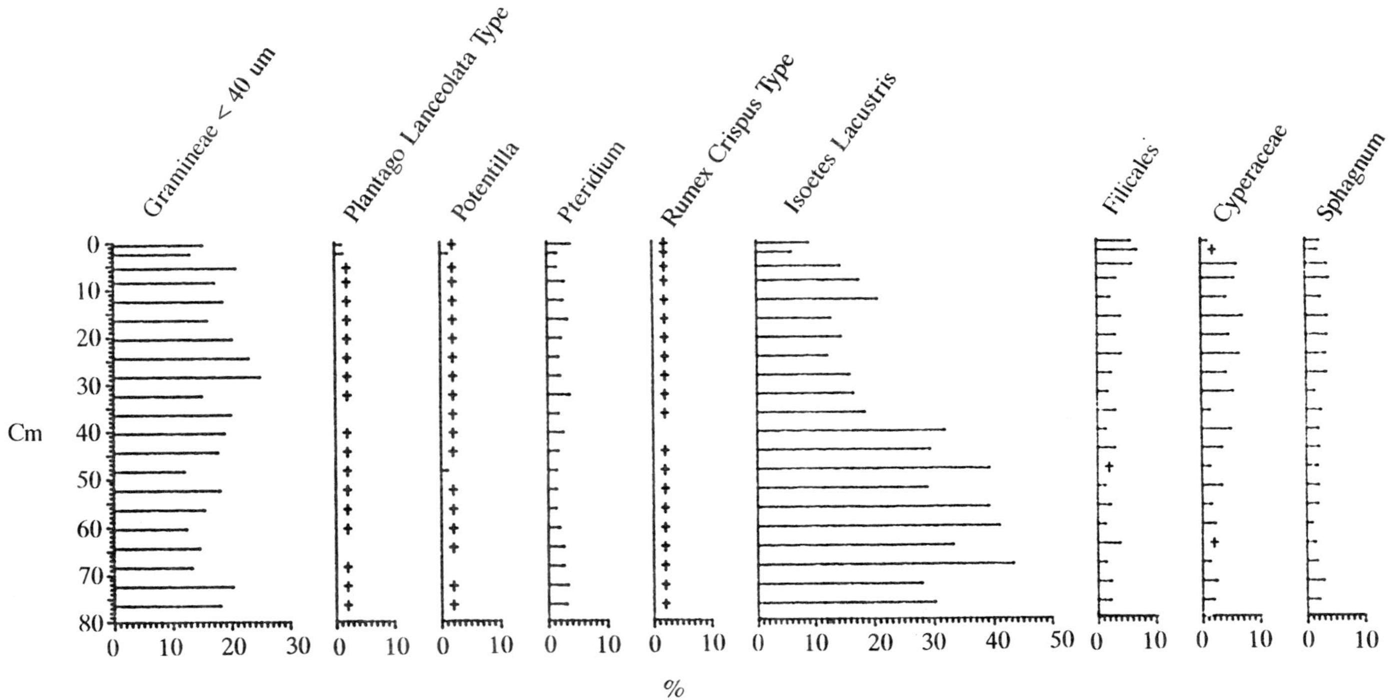

Fig. 6B.1. Summary pollen diagram for Llyn Hir (Core 1). Trees expressed as a percentage of the arboreal pollen. All other groupings as a percentage of the arboreal pollen plus peatland indicators.

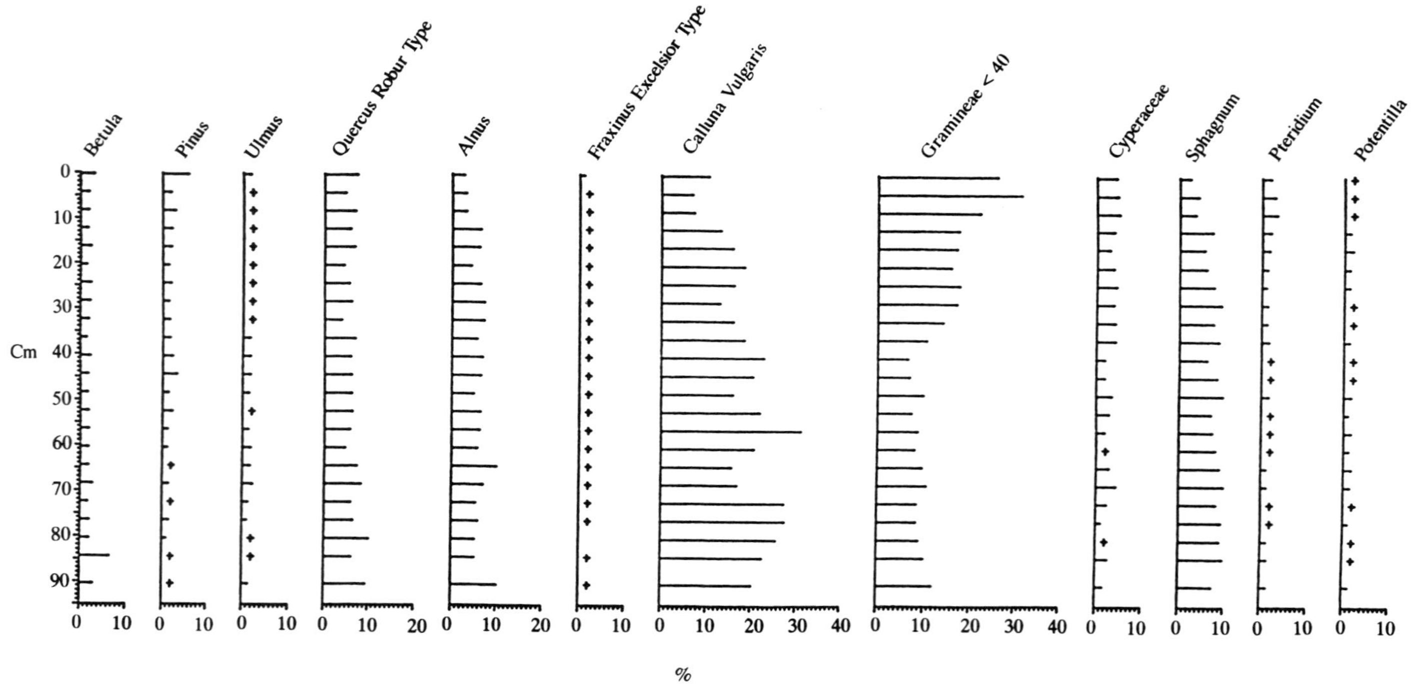

Fig. 6B.2. Summary pollen diagram for Llyn Llagi (see Fig. 6B.1 for details).

Gamallt (1940's), well before documented fishery decline in both cases (see above). At Llyn Gamallt drainage was only undertaken in the catchment of Llyn Gamallt Fawr, yet the lake has a very similar water chemistry to that of Llyn Gamallt Bach.[18, 22]

There is no evidence for, nor rational expectation of, the practice of agricultural liming in any catchment. All sites were remote from sources of affordable lime in the 19th century and have been classified as inherently unimprovable. The cessation of liming may thus be excluded as a potential acidification mechanism.

Overview

Diatom based pH reconstruction and atmospheric deposition histories[7, 8, 14, 15, 23, 24] indicate that the acidification of eight of the study lakes (pH reconstructions are not yet available for Llyn Conwy and Llyn Gamallt) is a recent phenomenon that has occurred within the past 120 (Llyn Hir, Llyn Dulyn, Llyn Llagi) to 40 (Llyn Gynon) years.

Apart from afforestation at two sites and water level manipulation at Llyn Conwy there has been no land use change in the Welsh study catchments. At the afforested sites, acidification was shown to have commenced before plantation. However, further work is in progress to assess whether afforestation enhanced acidification by scavenging processes and to what extent the exposure and subsequent oxidation of organic peats by forestry management practices have further contributed to the decline in pH of the lake waters.

Changes in land management have occurred in all catchments. However, this has involved different processes operating over different periods and at different magnitudes. There is no evidence that management change has resulted in lake acidification at any one site, much less that any general process of land use or land management has given rise to a general acidification mechanism. Together with evidence from carbonaceous particle and metals analyses (Chap. 4), the failure to attribute acidification of Welsh lakes to catchment processes strongly argues for an explanation in terms of acid precipitation.

References

1. Bache, B.W. (1985). Soil acidification and aluminum mobility. *Soil Use and Management*, 1: 10–13.
2. Battarbee, R.W. and Flower, R.J. (1985). Palaeoecological evidence for the timing and causes of lake acidification in Galloway, south west Scotland. Palaeoecology Research Unit, University College London, Research Paper 8.
3. Battarbee, R.W., Flower, R.J., Stevenson, A.C. and Rippey, B. (1985). Lake acidification in Galloway: a palaeoecological test of competing hypotheses. *Nature*. 314: 350–352.
4. Brugum, R.B. and Lusk, M. (1986). Diatom evidence for neutralisation in acid surface mine

lakes. In: Smol, J.P. and Battarbee, R.W. (Eds.), Diatoms and lake acidity. Dordrecht. pp. 115–129.

5. Burns, J.C., Coy, J.S., Tervet, D.J., Harriman, R., Morrison, B.R.S. and Quine, C.P. (1984). The Loch Dee project. I, Preliminary investigations. *Fisheries Management*, 15: 145–167.
6. Flower, R.J., Battarbee, R.W. and Appleby, P.G. (1987). The recent palaeolimnology of acid lakes in Galloway, south west Scotland: diatom analysis, pH trends and the role of afforestation. *J. Ecol.*, 75: 797–824.
7. Fritz, S.C., Stevenson, A.C., Patrick, S.T., Appleby, P.G., Oldfield, F., Rippey, B., Darley, J. and Batterbee, R.W. (1986). Palaeoecological evaluation of the recent acidification of Welsh Lakes. I. Llyn Hir, Dyfed. Palaeoecology Research Unit, University College London, Research Paper, 16.
8. Fritz, S.C., Stevenson, A.C., Patrick, S.T., Appleby, P.G., Oldfield, F., Rippey, B., Darley, J., Battarbee, R.W., Higgitt, S.R. and Raven, P.J. (1987). Palaeoecological evaluation of the recent acidification of Welsh Lakes. VII. Llyn y Bi, Gwynedd. Palaeoecology Research Unit, University College London, Research Paper, 23.
9. Harriman, R., and Morrison, B.R.S. (1982). Ecology of streams draining forested and non-forested catchments in an area of central Scotland subject to acid precipitation. *Hydrobiologia*, 88: 251–263.
10. Hornung, M. (1984). Precipitation canopy and soil water interactions. Institute of Water Pollution Control and Institution of Water Engineers and Scientists, joint meeting on acid rain, Birmingham.
11. Hornung, M. and Newson, M.D. (1986). Upland afforestation influences on stream hydrology and chemistry. *Soil Use and Management*, 2: 61–65.
12. Jones, V.J., Stevenson, A.C. and Battarbee, R.W. (1987). A palaeolimnological evaluation of peatland erosion. Report (First phase) to the NCC. Palaeoecology Research Unit, University College London, Research Paper, 28.
13. Kinniburgh, D.G. and Edmunds, W.M. (1986). The susceptibility of United Kingdom groundwaters to acid deposition. British Geological Surveys, NERC. Wallingford.
14. Kreiser, A., Stevenson, A.C., Patrick, S.T., Appleby, P.G., Rippey, B., Darley, J. and Battarbee, R.W. (1986). Palaeoecological evaluation of the recent acidification of Welsh lakes. II, Llyn Berwyn, Dyfed. Palaeoecology Research Unit, University College London, Research Paper, 18.
15. Kreiser, A., Patrick, S.T., Stevenson, A.C., Appleby, P.G., Rippey, B., Oldfield, F., Darley, J., Battarbee, R.W., Higgitt, S.R. and Raven, P.J. (1987). Palaeoecological evaluation of the recent acidification of Welsh Lakes. X, Llyn Cwm Mynach, Gwynedd. Palaeoecology Research Unit, University College London, Research Paper, 27.
16. Krug, E.C. and Frink, C.R. (1983). Acid rain on acid soil a new perspective. *Science*, 221: 520–525.
17. La Zerte, B.D. and Dillon, P.J. (1984). Relative importance of anthropogenic versus natural sources of acidity in lakes and streams of central Ontario. *Can. J. Fish. Aquat. Sci.*, 41: 1664–1677.
18. Milner, N.J. and Jones, B.E. (1985). Sport fisheries and biology of brown trout in some North Wales lakes. Welsh Water, Brecon.
19. Ormerod, S.J. and Edwards, R.W. (1985). Stream acidity in some areas of Wales in relation to historical trends in afforestation and the usage of agricultural limestone. *J. Environ. Management*, 20: 189–197.
20. Overrein, L.N., Seip, H.M. and Tollen, A. (eds.) (1981). Acid precipitation – effects on forest and fish. Final reports of the SNSF projects. Oslo.
21. Patrick, S.T. (1987). Palaeoecological evaluation of the recent acidification of the Welsh lakes. V, The significance of land use and land management. Palaeoecology Research Unit, University College London, Research Paper, 19.
22. Patrick, S.T. and Stevenson, A.C. (1986). Palaeoecological evaluation of the recent acidifi-

cation of Welsh lakes. III, Llyn Conwy and Llyn Gamallt, Gwynedd (site descriptions, fishing and land use/management histories). Palaeoecology Research Unit, University College London, Research paper, 21.

23. Patrick, S.T., Fritz, S.C., Stevenson, A.C., Appleby, P.G., Rippey, B., Oldfield, F., Darley, J., Battarbee, R.W., Higgitt, S.R. and Raven, P.J. (1987a). Palaeoecological investigations of the recent acidification of Welsh lakes: VIII, Llyn Eiddew Bach, Gwynedd. Palaeoecology Research Unit, University College London, Research paper 24.
24. Patrick, S.T., Stevenson, A.C., Fritz, S.C., Appleby, P.G., Rippey, B., Oldfield, F., Darley, J., Higgitt, S., Battarbee, R.W. and Raven, P.J. (1987b). Palaeoecological investigations of the recent acidification of Welsh lakes: IX, Llyn Llagi, Gwynedd. Palaeoecology Research Unit, University College London, Research Paper, 25.
25. Pennington, W. (1984). Long term natural acidification of upland sites in Cumbria. Freshwater Biological Association, Annual Report, pp. 28–46.
26. Rosenqvist, L. Th. (1977). Acid Soil Acid Water. Oslo.
27. Rosenqvist, L. Th. (1978). Alternative sources for acidification of river water in Norway. The Science of the Total Environment. 10: 39–49.
28. Rosenqvist, L. Th. (1980). Influence of forest vegetation and agriculture on the acidity of fresh water. In: Pfafflin, J.R. and Ziegler, N. (eds.), Advances in Environmental Science and Engineering. New York. pp. 56–79.
29. Rosenqvist, L. Th. (1981). Importance of acid precipitation and acid soil in freshwater lake chemistry. *Vann*. 4: 402–409.
30. Rosenqvist, L. Th., Jorgensen, P. and Rueslatten, H. (1980). The importance of natural H^+ production for acidity in soil and water. In: Drablos, D. and Tollen, A. (eds.). Ecological Impact of Acidic Precipitation. SNSF Project Report, Oslo. pp. 240–241.
31. Stevenson, A.C., Fritz, S.C., Patrick, S.T., Appleby, P.G., Rippey, B., Oldfield, F., Darley, J., Higgitt, S., Battarbee, R.W. and Raven, P.J. (1987a). Palaeoecological investigations of the recent acidification of Welsh lakes: IV, Llyn Gynon, Dyfed. Palaeoecology Research Unit, University College London, Research Paper, 20.
32. Stevenson, A.C., Patrick, S.T., Fritz, S.C., Appleby, P.G., Rippey, B., Oldfield, F., Darley, J., Higgitt, S., Battarbee, R.W. and Raven, P.J. (1987b). Palaeoecological investigations of the recent acidification of Welsh lakes: VI, Llyn Dulyn, Gwynedd. Palaeoecology Research Unit, University College London, Research Paper, 22.
33. Stoner, J.H. and Gee, A.S. (1985). Effects of forestry on water quality and fish in Welsh rivers and lakes. *J. Inst. Wat. Engrs. Sci.*, 39: 27–45.

7. The role of acidity in the ecology of Welsh lakes and streams

S.J. ORMEROD and K.R. WADE

Introduction

Acidity and its associated chemistry is of great importance in the ecology of temperate fresh waters, affecting most trophic levels. Demonstrable effects on the distributions of algae, macrophytes, invertebrates and fish have been recorded and associated with the direct toxic action of factors accompanying low pH. Influences on ecosystem processes such as nutrient cycling, primary production and decomposition have also been suggested.[1, 29] Thus, in addition to having direct physiological effects, acidity and elevated metal concentrations can influence some organisms indirectly through functional links such as trophic pathways. These indirect effects also affect semi-aquatic organisms, such as mammals and birds, which depend on freshwaters for food (Fig. 7.1).

This chapter reviews ecological studies so far undertaken into the role of acidity, and related factors such as aluminium concentration, in the ecology of Welsh lakes and streams. Data are drawn mostly from a regional survey, undertaken by W.W.A. in 1984, or from the catchment-based study around Llyn Brianne, and relate mostly to river systems.

Biologically important chemistry in acid waters

Two general categories of surface waters in Wales are those which are well-buffered, or circumneutral, and those which are acidic, either occasionally or permanently. Circumneutral waters typically have mean pH > 5.7–6, and alkalinities sufficiently high (>5 mg l^{-1}) to prevent pH falling below pH 5.7; concentrations of heavy metals or aluminium are usually low, whilst values of calcium and magnesium are high (e.g. Ca > 3–4 mg l^{-1}). By contrast, acid waters have low alkalinity (0–5 mg l^{-1}), and pH permanently or frequently below pH 5.7–6.0; aluminium concentrations increase markedly with decreasing pH below 5.5. Calcium concentrations are typically low in acidic waters (< 2 mg Ca l^{-1}), although a slight increase may occur due to leaching in areas subject to acid deposition.

Three ions, H^+, Al^{3+} and Ca^{2+}, are particularly important, biologically, in

R.W. Edwards et al. (eds) Acid Waters in Wales. 93–119.

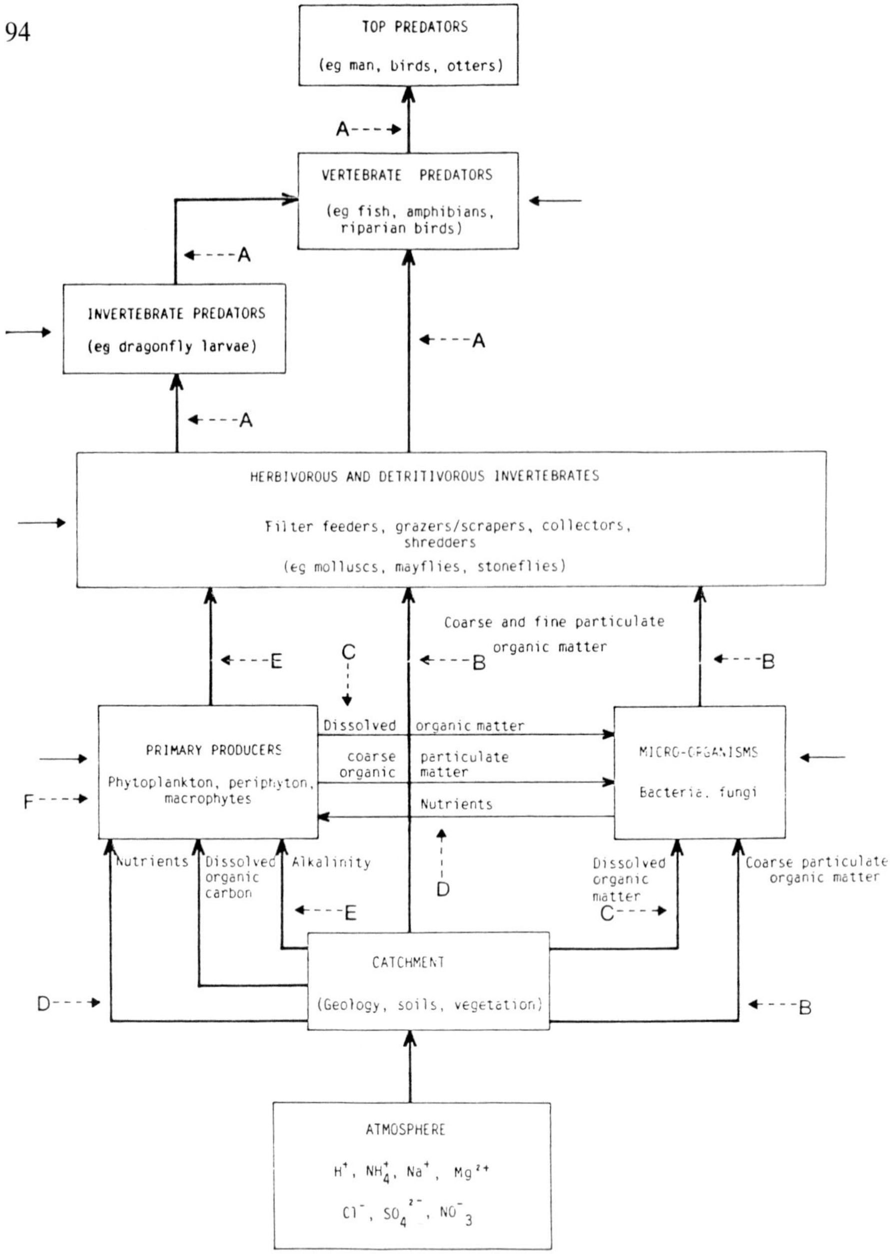

Fig. 7.1. Ecosystem processes likely to be affected by acidification. The arrowed boxes represent trophic levels where acidification has direct toxic actions, whilst broken arrows represent possible indirect effects. These indirect effects include: A. Loss of acid-sensitive fish and invertebrates, resulting in the proliferation of some other invertebrates, and hence alteration in the quality of food available for top predators. B. Reduced decomposition of coarse particulate organic matter, affecting the quality of food available for detritivores. C. Complexation of dissolved organic matter by metals, leading to loss of substrate for microbial action. D. Aluminium and phosphorus may complex, reducing nutrient availability for plants. E. Loss of alkalinity may reduce the availability of inorganic carbon, changing the quality of primary producers. F. Changes in the invertebrate fauna may alter the quality of grazing pressure. After Warren et al.[88]

acid waters. The first two may be toxic to many organisms, whilst calcium may ameliorate acid or aluminium stress (see Chap. 10). All three are often closely correlated, and hence their individual effects are difficult to determine from field studies alone. Other chemical characteristics, such as the concentrations of fulvic and humic acids, may be important in 'brown water' acid systems. These organic acids are able to sorb metals such as aluminium thereby mitigating their toxic effects.

The chemistry of surface waters is variable both between seasons and with changing flow. During heavy rainfall or snow-melt, the passage of water through catchments is rapid, reducing the time available for buffering reactions. As a result, 'episodes' of reduced pH, often with elevated aluminium concentration, occur in streams and in some lakes[13] (see Chap. 15). Such episodes expose organisms to extremes of water quality which may be important physiological stressors. However, the ecological effects of episodes are often difficult to separate, using field observations, from the long-term effects of chronic acid stress. This separation is not possible in many of the Welsh studies reviewed in this chapter, and no assumptions have been made over the respective importance of chronic and episodic influences, except where stated.

Primary procedures

In both lakes and streams, primary producers are planktonic, epiphytic or epilithic algae, bryophytes and higher plants. In fast-flowing rivers, however, higher plants are restricted, whilst most of the algae present are benthic. Besides fixing solar energy into a form which can be used by herbivores (i.e. primary production), plants have many important functions in aquatic ecosystems. They influence oxygen concentrations during photosynthesis and respiration, leach exudates rich in organic carbon, and act as a source of detritus after death and decomposition. In addition, macrophytes provide a surface area for epiphytic algae or bacteria, trap detritus from the water column, influence the characteristics of flow and sedimentation, and provide microhabitats in which animals avoid predators or turbulence. Some groups influence the flux or concentrations of metals or other ions through the ecosystem, either through fixation or by adsorption.[9]

The influence of acidity

Initial hypotheses concerning the effect of pH on primary productivity proposed that 'oligotrophication' could occur at reduced pH because of decreased nutrient release from sediments or because phosphorus co-precipitated with aluminum.[16, 25] In the few instances where measurements have been made, however, effects of low pH on primary productivity have not been pronounced.[74] Indeed, artificial acidification in both streams[30, 48, 61] and lakes has sometimes led to an increase in algae standing crop.[20] Reasons proposed for the increase at low pH include reduced grazing pressure because of changes to the invertebrate fauna, reduced decomposition, and a succession to acid tolerant species not readily usable by the grazers present. The biomass of macrophytes in acidic lakes can also be large,[74] though variable.

There is more information to indicate an influence of low pH on floral composition. Phytoplankton, periphyton and macrophytes usually show a reduced diversity with declining pH.[20, 34, 73] Acid tolerant taxa include the dinoflagellates *Peridinium* and *Gymnodinium* amongst the phytoplankton, filamentous chlorophytes such as *Mougeotia* or blue-green algae amongst the periphyton, and *Sphagnum* and *Juncus bulbosus* amongst macrophytes.[74] Some diatoms are particularly sensitive algae, often being restricted to *Cyclotella* spp. in the plankton or *Eunotia* and *Tabellaria* in the benthos of acid streams and lakes. This group has provided an effective indicator system for assessing trends in lake pH (see Chap. 3). Various physiological hypotheses have been proposed to explain why qualitative changes in floral composition accompany reduction in pH.[34, 74]

Clearly, even in the absence of quantitative differences in the flora of water-bodies of different chemistry, qualitative differences could have consequences for animals which preferentially graze certain plants, or are dependent on them for other functions such as habitat provision.

Studies in Wales

Welsh studies on primary producers have been restricted mostly to qualitative assessments, although several floral groups have been included in work on both lakes and streams. A study involving the reconstruction of past changes in pH using diatom remains in lake sediments is reported elsewhere (see Chap. 3).

Algae

Round (pers. comm.) compared the diatom floras in the benthos of streams of contrasting pH, and in acid lakes before and after liming. Only 'live' cells were counted. Prior to liming, the acidic lake Llyn Hir was dominated by *Eunotia rhomboidea* in all benthic habitats, with *Tabellaria quadriseptata* the second most common species. Liming increased pH from 4.8 (annual mean) to 6.4, whilst the calcium concentration increased from 0.92 mg l^{-1} to 2.63 mg l^{-1}. The most pronounced change in the diatom flora over the following twelve months was an increased abundance of *Acnanthes minutissima*, a reduction in *E. rhomboidea* and disappearance of *T. quadriseptata*. *A. minutissima* also occurred following the liming of Llyn Pendam and Llyn Berwyn, whilst acidobiontic species disappeared. These changes were generally the reverse of trends occurring during the acidification of these and other Welsh lakes in the twentieth century (see Chap. 3).

At Llyn Brianne, circumneutral streams draining moorland were dom-

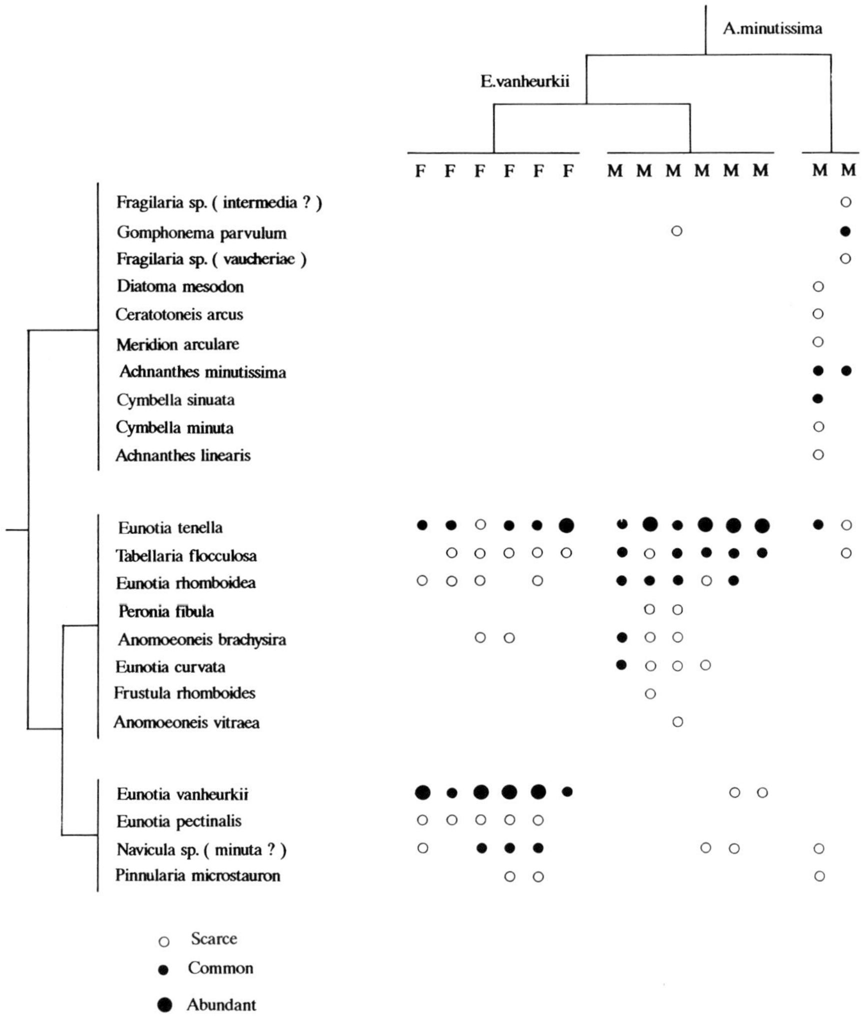

Fig. 7.2. TWINSPAN classification of sites around Llyn Brianne according to their diatom floras. Abundances were determined from the percentage composition of samples at each site. M = moorland streams, F = forest. See Table 7.1 for chemical characteristics.

inated by *A. minutissima*, *Cymbella sinuata* and *Gomphonema* spp, whilst acidic streams were dominated by *Eunotia tenella* and *Tabularia flocculosa* if they drained moorland, and by *Eunotia vanheurkii* if they drained forest (Fig. 7.2; Table 7.1). Interestingly, diatoms were sometimes abundant at low pH, particularly in streams draining moorland, but fewer were present in acidic streams draining forest.

In a study of a broader range of benthic algae, undertaken during the summer period, Proctor[64] assessed chlorophyll 'a' concentrations in the

Table 7.1. Some biological characteristics of streams around Llyn Brianne.

Stream type	pH	Ca	Al	Characteristic biota: Diatom	Characteristic biota: Macroinvertebrate	Fish m^{-2}
Circumneutral moorland (Streams 5, 6)	6.9	2.5–5.0	0.04–0.07	*Acnanthes minutissima*	*Baetis rhodani*	0.12–2.79
Circumneutral, deciduous woodland (Stream 14)	6.3	1.75	0.05	Not sampled	Stonefly 'shredders'	N/A
Acidic moorland (Streams 8–11)	5.0–5.9	0.8–1.2	0.07–0.16	*Eunotia tenella*	Stonefly 'shredders', *Leptophlebia marginata, Isoperla grammatica*	0–0.91
Acidic, conifer forest (Streams 1–4, 7)	4.6–5.3	1.05–1.80	0.17–0.48	*Eunotia vanheurkii*	Stonefly 'shredders'	Absent

Note: Only one deciduous woodland stream was included, which is impassable to fish.
All concentrations (except pH) are in mg. $litre.^{-1}$.

benthos of five streams at Llyn Brianne, but found no systematic differences in relation to pH or land use. Interestingly, chlorophyll 'a' concentrations were lowest in an acidic stream whose banksides had been recently cleared of conifer forest, indicating that this management action had not had the required effect of increasing primary production. Taxonomic differences between sites were more pronounced, with diatoms dominant in circumneutral moorland streams. Diatoms were still present in acidic streams, but filamentous chlorophytes or cyanophytes were most abundant (Fig. 7.3). Seasonal variations in these patterns have yet to be assessed.

These studies support those undertaken elsewhere in indicating altered algal floras with reduced pH, but little consistent difference between stream of different chemistry in overall numbers or biomass. The occurrence of diatoms under some acid conditions (e.g. Llyn Hir pre-liming; acid streams draining moorland at Llyn Brianne), and proliferation of filamentous algae in acidic streams, have interesting corollaries for the invertebrate fauna (see below).

Macro-flora

Macro-floral studies have been undertaken in Welsh lakes and streams across a range of chemistry. Lowther[43] surveyed macrophytes at fifty lakes in mid and south Wales, although her study included riparian as well as aquatic species; bryophytes with the exception of *Sphagnum* sp. were not identified. Using the TWINSPAN classification procedure (two-way indicator species analysis[37]), three groups of sites were identified which differed mostly in chemistry. An ordination derived by DECORANA[38] also correlated with pH more strongly than with any other variable. Several species were confined to two circumneutral site-groups (mean pH 6.3–7.6), resulting in reduced taxon richness at the most acidic sites (mean pH 4.9, total hardness 7.0 mg $CaCO_3^{-1}$; 0.2 mg Al 1^{-1}) (Fig 7.5). However, none of 27 sites common to surveys by Lowther[43] and Seddon[71] had shown any gross change in flora in the intervening period of more than two decades. Indeed, species expected[26, 65, 70] to disappear from acidifying lakes such as *Lobelia dortmanna*, *Isoetes* spp, *Littorella uniflora*, *Pilularia globulifera* were still present at some sites despite evidence of recent acidification (e.g. Llyn Hir, Llyn Gynon; see Chap. 3). Possibly, the proliferation of *Sphagnum* and *Juncus bulbosus*, known to occur in some acidifying lakes,[27, 70] has not yet progressed sufficiently in Welsh lakes to affect other soft-water macrophytes detrimentally. Alternatively, lentic systems in which the most pronounced macro-floral changes have occurred are in the Netherlands and are probably affected by ammonium deposition, leading to nitrification.[70] This situation is unlikely in Wales. Nevertheless, a more widespread study than the initial pilot survey by Lowther[43] would prove interesting, particularly in forest lakes where metal concentrations may be elevated.[75]

WWA surveyed macrophytes at 88 sites on Welsh streams (hardness < 25 mg $CaC0_3 1^{-1}$) as part of a regional survey of acid waters in 1984.[59] Bryophytes, macroscopic algae and aquatic angiosperms were all recorded,

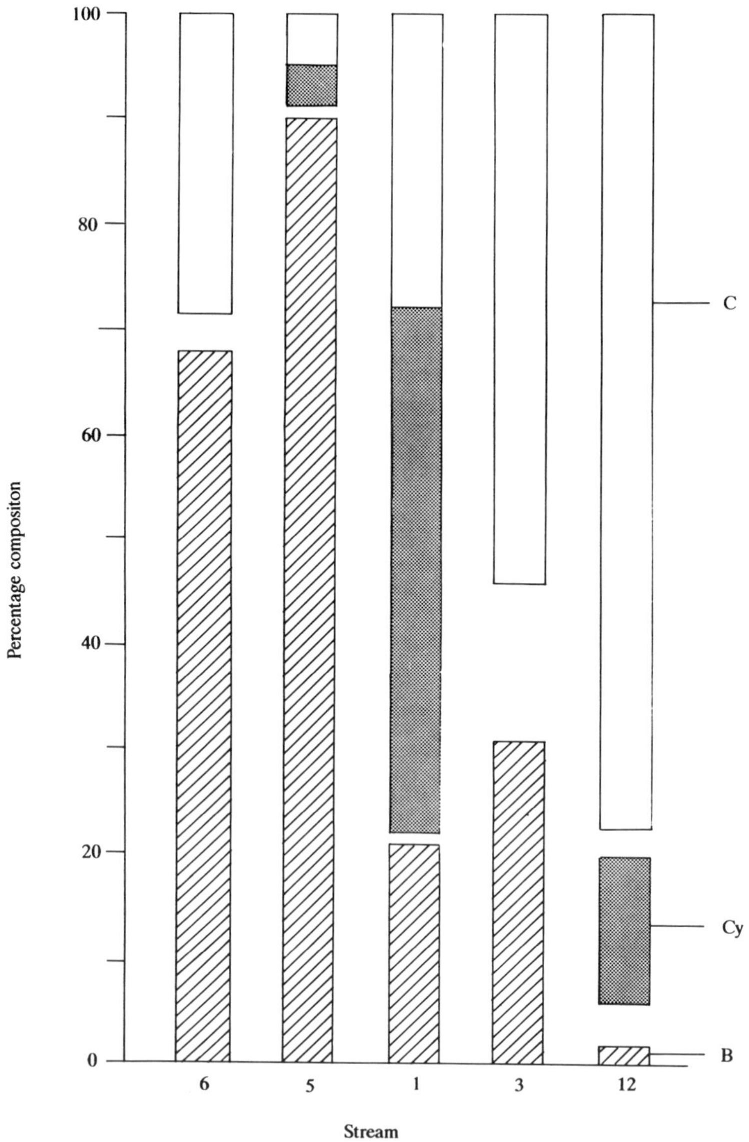

Fig. 7.3. The percentage composition of samples of major epilithic algae in five streams around Llyn Brianne during summer. C, Chlorophytes; Cy, Blue-greens; B, Diatoms. Other taxa are represented by the gaps. Streams 5 and 6 are circumneutral and drain moorland, 1 and 3 are acidic and drain forest, and 12 is acidic and drains moorland. See Table 7.1 for chemical characteristics.

although the latter were only present at 8% of the sites. Macro-floral composition differed markedly according to pH, with the alga *Lemanea*, the moss *Fontinalis squamosa* and the liverworts *Nardia compressa* and *Scapania undulata* characterizing progressive increase in acidity and aluminium con-

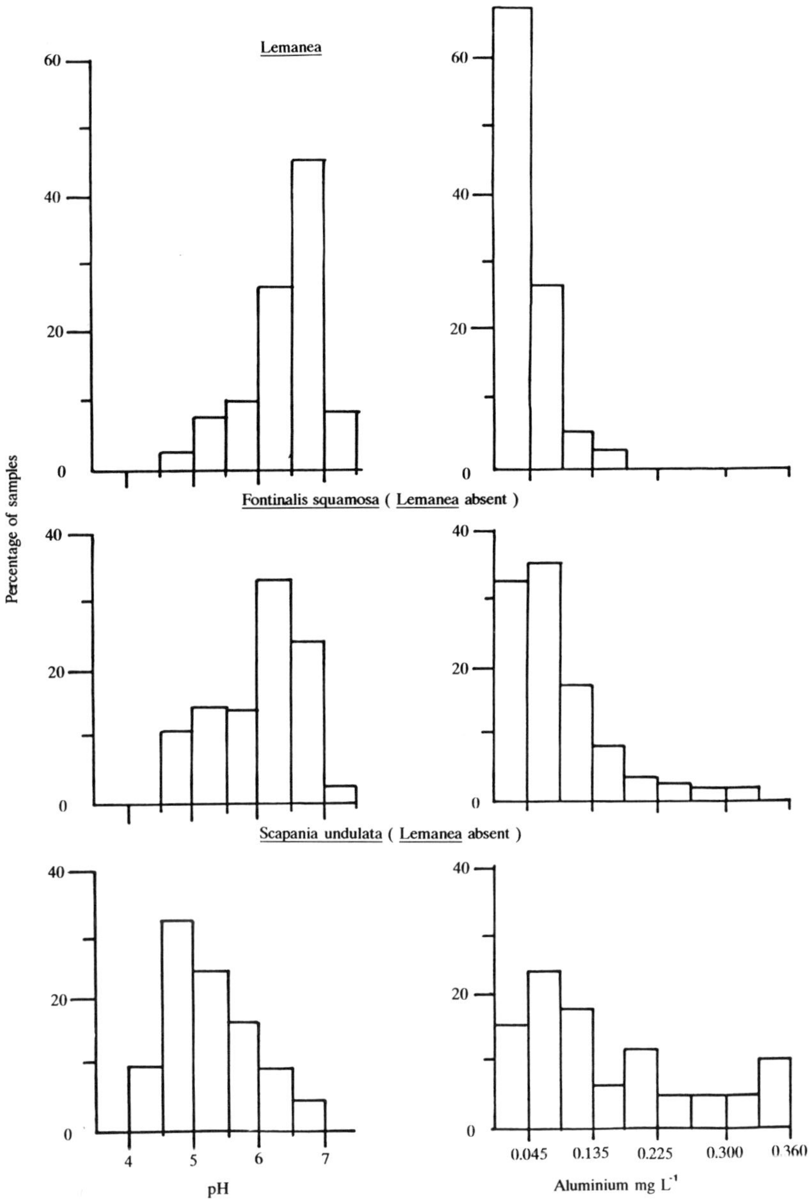

Fig. 7.4. Stream chemistry at sites occupied by three different species of macro-flora.

centration (Fig. 7.4). Chlorophytic algae such as *Mougeotia*, *Ulothrix* and *Stigeoclonium* were also most frequent at low pH. TWINSPAN classification of the flora revealed four site-groups (Fig. 7.5), which differed mostly in pH, aluminium and calcium concentration. Similar patterns were found in streams around Llyn Brianne[80], where the liming of catchment 9 has since led to

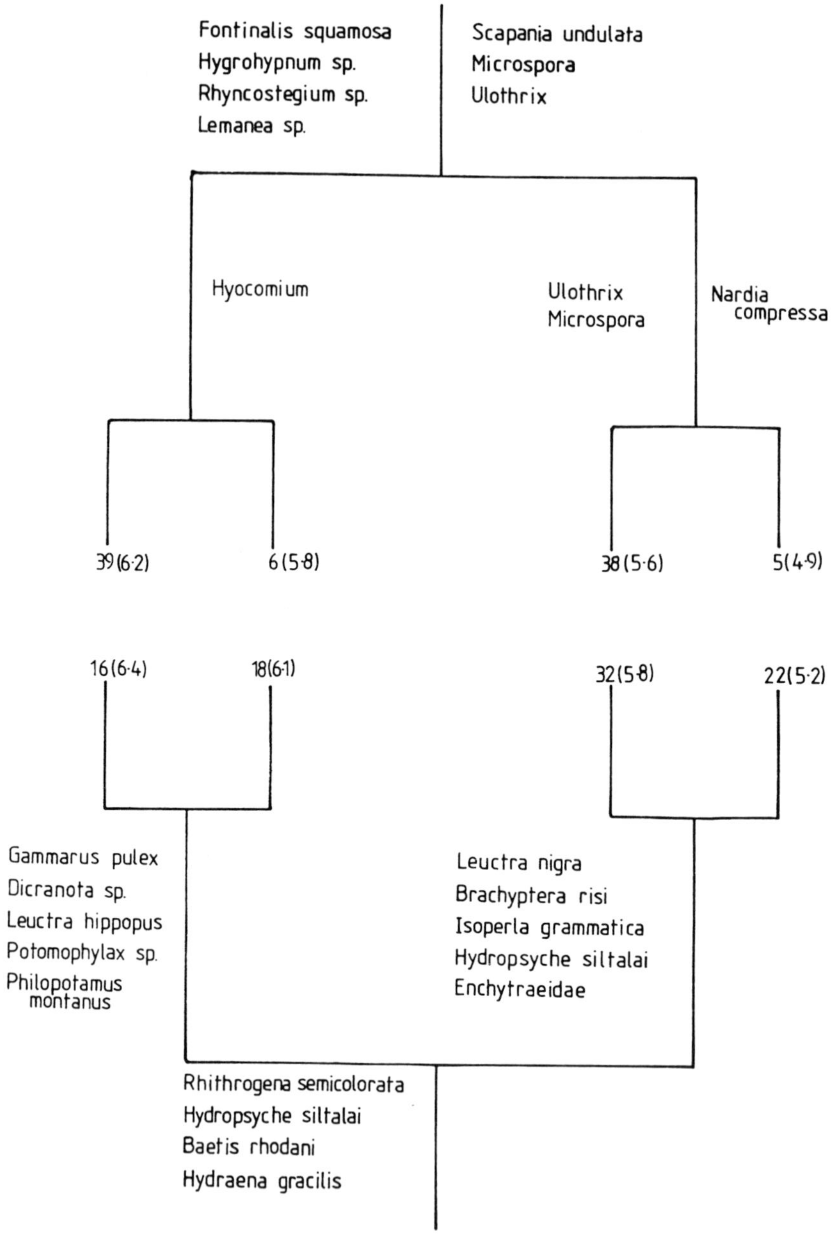

Fig. 7.5. TWINSPAN classification of macro-flora and macroinvertebrates at 88 sites on streams in upland Wales. The indicator species are given at each division. The number of sites in each category is shown, with the mean pH (October–March) for each site-group in parentheses.

increased pH and a large reduction in the abundance of *N. compressa*. The addition of lime to Llyn Hir and Llyn Berwyn also led to reductions in the abundance of *Sphagnum* and *Nardia compressa*, whilst *Fontinalis squamosa* and *Lobelia dortmanna* increased. These experimental manipulations support an effect by water chemistry on the distribution and abundance of such plants.

Bacteria and fungi

In addition to primary production, there is a large input of organic material from terrestrial to freshwater ecosystems. Indeed, leaves, woody debris and other organic matter from these 'allochthonous' sources can form the major energetic base of upland streams, particularly those which rise in areas of woodland.[21] Once in the aquatic system, an important part of the conversion of such plant litter to a form which can be used by invertebrates is degradation by fungi and bacteria. The fate of some aquatic primary production is also microbial decay. Leaves falling into fresh-water leach their soluble constituents rapidly, and are quickly colonized by freshwater fungi, particularly hyphomycetes.[3] These fungi produce all the enzymes necessary to degrade leaf cell walls and some also degrade woody material.[10] Bacteria become important as degradation progresses.[11, 40]

Bacteria and fungi, in combination with algae, are also important constituents of the epilithic layer on the stream bed. Such layers are complex mixtures of these organisms in a slime matrix, and may be important in removing dissolved organic material or nutrients from the water column, in producing or holding enzymes for heterotrophic activity, in acting as a site for adsorption of metals and other ions, and providing a food source for grazing invertebrates.[43] Once sloughed from the epilithon and suspended in the water column of lakes or streams, such bacteria sometimes utilize and concentrate dissolved organic matter, or act as a food source to filter-feeding invertebrates.[18]

The influence of acidity

Several studies have now revealed that the decomposition of plant litter is retarded at low pH in streams[11, 45] and lakes,[83] although not consistently.[69] Reduced shredding activity because of invertebrate scarcity has sometimes been suggested as a reason for impaired decomposition.[22] However, material such as wood, not normally shredded by invertebrates, also decays more slowly in acid waters.[83] Moreover, the abundance of shredding invertebrates sometimes increases when leaf decay is retarded.[45]

A more likely reason for reduced decomposition is through the effect of low pH on microbial activity. In streams in the English Lake District, slow decay of oak *Quercus robur* and alder *Alnus glutinosa* at pH 4.9 was associated with reduced counts of bacteria and fungi on leaves, and reduced species diversity of hyphomycetes.[11] Experimental acidification of a stream to pH 4 also reduced hyphomycete abundance.[30] Whilst such effects could be detrimental to invertebrates which favoured conditioned litter as food source, it might benefit those which feed on coarse leaf particles.[36, 44]

In lakes, there appears to be no consistent relationship between pH and counts of bacteria in the water column or sediments.[6, 83] Nevertheless, bacteria in the sediments of acidic lakes are sometimes important in generating alkalinity through the processes of denitrification and sulphate reduction.[68]

Few data are available from streams, but the epilithon at an acidic site in southern England was lower in organic carbon content, and contained fewer living cells, than at adjacent sites which were less acidic.[92] Elsewhere, counts of epilithic bacteria were also lower in acidic headwaters than in adjacent circumneutral streams,[24] although the streams in this study were unusually acidic ($pH < 4$) and also had reduced chlorophyll ‘a’. Suspended bacteria in circumneutral streams in north east England were only slightly more numerous than in acidic streams in Galloway, but those in the former were larger and more active.[67]

Studies in Wales

Only one study has been undertaken in Wales on the influence of pH on decomposition. Following similar work elsewhere[36] Stanyer[72] measured weight loss of cellulolytic material (rice grains) in streams of contrasting pH in the Wye catchment. Confirming previous work, decay rate was markedly reduced at low pH.

It is probable that the characteristics of litter supply and degradation in upland Welsh streams are similar to those in other upland areas of Britain, such as the Lake District. Few cases now remain where Welsh soft-water streams rise in deciduous woodland, instead draining either planted conifer forest (mostly sitka spruce *Picea sitchensis*) or *Nardus*/*Molinia*/*Festuca* moorland. These grasses, together with plants such as *Nardia compressa* or *Scapania* in the stream, are likely to provide the major sources of detritus in moorland. Some invertebrates such as *Gammarus pulex* are unable to feed adequately on such a food base.[91] The scarcity of deciduous trees along many Welsh stream-sides might also result in a low level of fungal inoculum, adding to the effect of low pH in retarding leaf decay.[11]

The value of litter inputs from the exotic conifers now present in many Welsh catchments has yet to be determined. Leaves from *Pinus* spp. and other conifers are usually colonized and conditioned only slowly by hyphomycete fungi, partly because of their thick cuticle, but also because they contain fungal inhibitors.[4] Nevertheless, some invertebrates are able to feed on conifer needles when they are conditioned.[77] It is not known whether the low pH and elevated concentrations of aluminium, typically found in Welsh forest streams, are suitable to support such conditioning.

The bacterial epilithon of five streams around Llyn Brianne was investigated by Kemmy[41] during one summer. Contrasting with patterns described above, total counts of bacteria per unit area of substratum were higher in three acidic streams than in two circumneutral moorland streams (Fig. 7.6). Counts on acid media differed particularly between the sites, indicating that bacteria in acid streams were acid tolerant. However, bacteria in the two circumneutral streams were largest, and were more active than in two of the acidic streams (see also Reference 66). Interestingly, bacterial count, activity (as glucose mineralization) and percentage viability were all greatest in an acidic forest stream (mean pH 4.8, 0.36 mg Al l^{-1}), which also had a large algal biomass (as chlorophyll 'a').

Overall, these data indicate that the large and active bacteria in circumneutral streams might be able to provide a greater energy source to grazing invertebrates than in acidic streams. However, the presence of acid tolerant algae, sometimes prolific even at $pH < 5$, could clearly be important in supporting populations of epilithic bacteria. Differences between streams in algal populations could explain at least some of the contrasts between bacterial counts in acid streams in Wales and those in southern or northern England.[24, 92] The role of macrofloral differences between Welsh streams[59] in influencing populations of epiphytic or suspended bacteria, in turn affecting some invertebrates, has yet to be determined. Even in acidic streams, epiphytic bacteria can be abundant on plants such as *Nardia compressa*, although their contributions to the diets of grazing or filter-feeding invertebrates will be dependent on bacterial activity and hence doubling time.[67]

Secondary and tertiary production: macroinvertebrates

Many macroinvertebrate groups are represented in lakes and streams, although flatworms, true worms, molluscs, crustaceans, and immature insects (dragonflies, may flies, stoneflies, caddisflies and true flies) are most numerous. Their roles in aquatic ecosystems are many and varied. Some convert primary production directly into animal tissue by eating plants. Others are part of the detrital processing system, 'shredding' coarse particulate organic matter, or 'collecting' fine organic matter by grazing bacteria or filtering suspended material from the water column. Others are predators, whilst all at some stage may provide food for vertebrates such as fish, amphibians and birds. Besides these energetic roles, some groups have other important functions, such as assisting the interchange of oxygen and nutrients between sediments and the water column.

The influence of acidity

Relationships between acid-base status and the macroinvertebrate fauna of streams has been recognized for many years.[39, 44] Typically, taxon richness is

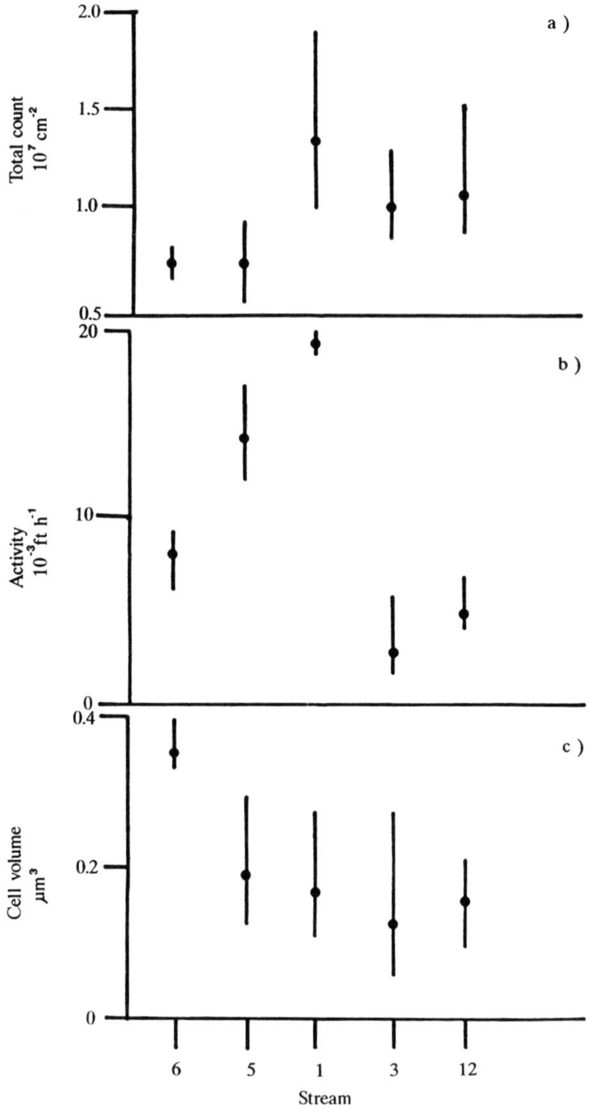

Fig. 7.6. Total counts (a), activity (b) and cell volume (c) of bacteria in five streams around Llyn Brianne during summer. Notation as in Fig. 7.3. The values are the means of 24 samples taken over a four-week period. Stream 1 had abundant chlorophytic algae. See Table 7.1 for chemical characteristics.

reduced at low pH. Mayflies, philopotamid or hydropsychid caddis, molluscs and crustaceans are scarce under acidic conditions (< pH 5.7), despite widespread occurrence elsewhere. By contrast, stonefly nymphs are often abundant.[79] Patterns of total abundance have been less clearly related to pH in soft-waters, and invertebrate biomass is sometimes similar in acid and circumneutral streams.[31]

In lakes, molluscs and zooplankton show reduced diversity at low pH[23, 49] whilst acid-tolerant taxa include dragonfly nymphs, alder-fly larvae (*Sialis*), and corixid bugs.[78] In some cases, acid lakes have increased diversity and abundance of nektonic organisms (e.g. Corixidae), a feature ascribed to decreased predation by fish.[5, 19]

Much of the debate concerning the impact of low pH on aquatic invertebrates has centred on whether effects are direct (i.e. physiological) or indirect (e.g. through trophic pathways). Animals scarce at low pH are often grazing herbivores or filter-feeders,[79] whilst shredders are most abundant.[82] This might indicate either that acid tolerance varies systematically between feeding groups, or that the nature of the food base changes with pH reduction. However, toxicological investigations in the laboratory have confirmed that invertebrates which are scarce in acid waters are often pH sensitive.[1] Aluminium can sometimes exacerbate acid stress,[7] and sometimes ameliorate it.[33, 35]

Studies in Wales

Streams

Studies in Wales have generally confirmed patterns of macroinvertebrate abundances indicated elsewhere. Within groups such as mayflies, caddis and crustacea, abundances in Welsh streams correlated with acidity, but only over a wide pH range which also incorporated marked differences in calcium concentration.[50] Work confined only to soft-waters has not shown any simple relationship between overall density and pH. Some acid streams have markedly reduced density, whereas others have densities at least as great as in adjacent circumneutral streams. One possibility is that invertebrate abundance in upland Welsh streams is controlled by factors other than acidity, such as their spatey character, temperature regime or poor detrital food base. Streams which flow through deciduous woodland in lowland Britain have overall densities greater by an order of magnitude than upland Welsh sites, even at similar pH (Hildrew pers. comm.). Moreover, the experimental enhancement of leaf litter in two streams at Llyn Brianne led to an increase in the abundance of shredders, indicating that some organisms in these streams are limited by their food supply (Dobson unpublished data.).

Qualitative patterns with changing pH in Welsh streams are more pronounced. In particular, the taxon richness of molluscs, crustacea, mayflies, and all groups combined increased markedly with increasing pH and decreasing aluminium (Fig. 7.7; see also References[56, 76, 86, 89]). Patterns revealed by multivariate techniques such as ordination and classification have repeatedly shown strong correlations with pH or aluminium concentration, and these relationships have been of importance in modelling (see Chap. 17). Species scarcest at low pH included the crustacean *Gammarus pulex*, mayflies such as *Baetis rhodani* and *Rhithrogena semicolorata*, caddis such as *Philopotamus montanus* and *Hydropsyche* spp., and beetles such as *Hydraena gracilis* (Fig. 7.5). By contrast, species tolerant of low pH (< 5.5–5.8) included the

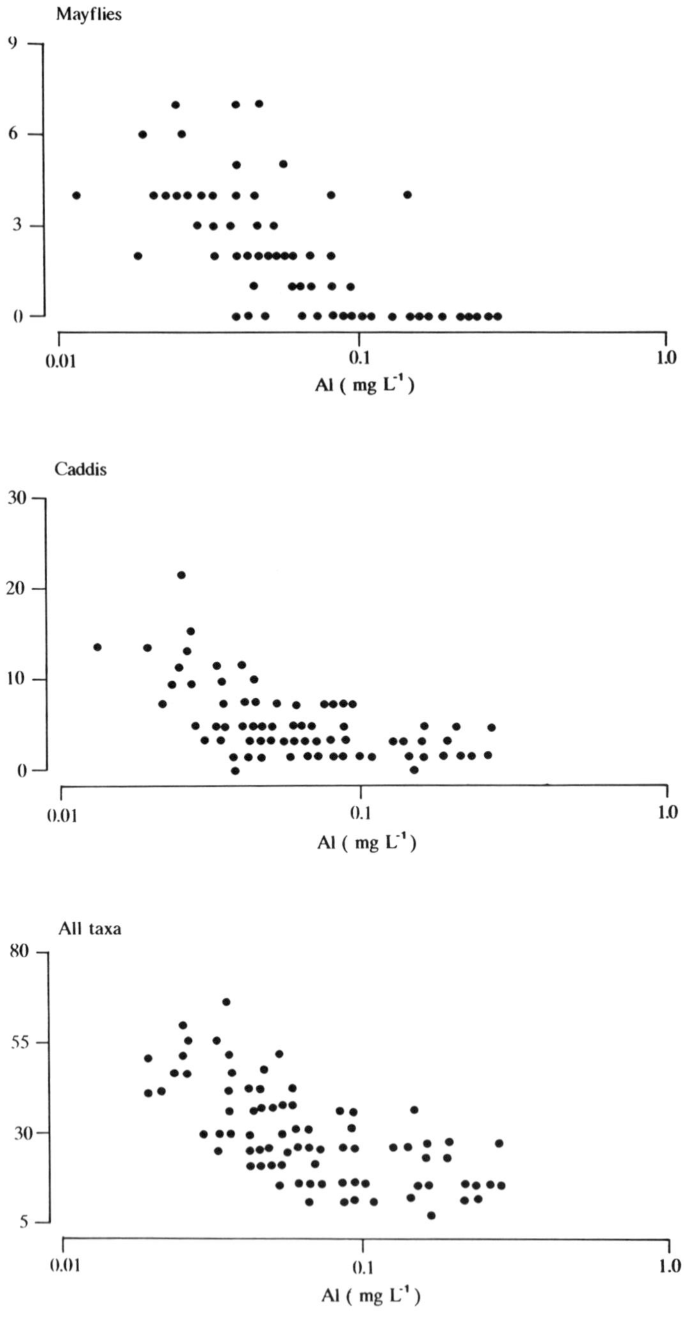

Fig. 7.7. The richness of invertebrate taxa in relation to the mean aluminium concentration at sites in upland Wales.

stoneflies *Amphinemura sulcicollis*, *Chloroperla torrentium*, *Leuctra hippopus*, and *Nemoura cinerea*, the dragonfly *Cordulegaster boltonii* and the mayfly *Leptophlebia marginata*. However, even some acid tolerant species were absent from acidic streams draining afforested catchments, which often had low taxon richness.[86, 89] These forest streams had elevated concentrations of aluminium (see Table 7.1), although forest presence might also have affected their habitat quality and discharge regime.[54, 60]

Differences between streams in the quality of epilithic algae and bacteria (e.g. Figs 7.2, 7.3, 7.6) are sometimes suggested as a reason for the scarcity of grazers such as mayflies at low pH.[79] Chlorophytes or cyanophytes are often assumed to be toxic, or refractory to invertebrate digestion, and their filaments may be difficult for mayflies to ingest.[27] By contrast, diatoms are a favoured and easily digested food.[27] However, food availability alone is unlikely to explain the scarcity of grazer/scrapers in acidic Welsh streams. Bacterial growth on filamentous algae at low pH could provide a food source (see Fig. 7.6). Additionally, grazing mayflies were scarce in acidic Welsh streams draining moorland, where diatoms were sometimes abundant. There is evidence that trophic links could exist between herbivores and macroscopic plants, such as *Scapania*, which were present down to pH 4.5 (Fig 7.4; see also Reference 59). At Llyn Brianne, grazing organisms such as *Baetis rhodani* and *Rhithrogena semicolorata* occur only in streams dominated by the diatom *Acnanthes minutissima* and it would be tempting to infer a trophic dependence between these organisms. However, this small diatom often proliferates in the presence of heavy grazing,[62] and it is clearly difficult to determine whether the grazing fauna is a consequence of the floral composition, or the reverse.

As an alternative explanation, Ormerod, Wade and Gee[59] hypothesized that the significant concordance between invertebrate and floral assemblages (Fig. 7.5) in Welsh streams might indicate dependence by some invertebrate on habitat provision by plants. However, Thomas[80] subsequently showed that, whereas bryophytes did support higher densities of *Baetis* and *Ephemerella ignita* than the surrounding substratum at a circumneutral site, the same plant species (e.g. *Scapania*) at an acid site did not hold these species. Moreover, *Baetis* is often abundant in circumneutral streams with impoverished macrophyte floras.

Some of these features might indicate direct physiological limitation of acid-sensitive invertebrates. Ecotoxicological results from the experimental manipulation of Welsh streams show that *Baetis rhodani*, *Ecdyonurus* and *Gammarus pulex* respond to episodes of low pH and elevated aluminium by drifting or dying (Chap. 10). Such short-term effects could not be mediated through the food supply, indicating the importance of direct physiological effects.[30] Nevertheless, such a direct pathway offers only a proximate explanation for the qualitative differences between streams of different pH, and does not illustrate why some insect groups should have evolved acid tolerance (e.g. stonefly nymphs), whilst others have not (e.g. mayfly nymphs).

Lakes

Only one study has been undertaken on invertebrates and acidity in Welsh lakes. Thomas et al.[81] sampled several habitats in sixteen lakes in April 1984, and classified and ordinated the combined fauna using TWINSPAN and DECORANA. The study included the identification of chironomids to species in view of their important contribution to the benthos of oligotrophic lakes.[17] pH provided the strongest correlate with taxon richness, and differences in pH, calcium and aluminium concentrations were also apparent between lake groups. Species confined to circumneutral lakes included molluscs, the leech *Erpobdella octoculata*, and the mayflies *Cloeon simile*, and *Caenis* spp. By contrast, the stonefly *Nemoura cinerea* and the chironomid *Zalutschia (humphriesiae?)* were found only in the more acidic lakes. Other *Zalutschia* spp. also occur at low pH in Canada and Scandinavia.[87] *Z. (humphriesiae?)* disappeared from the Welsh lake Llyn Berwyn following liming, further indicating the importance of pH as an influence on its occurrence.

Interestingly, the taxon richness of nektonic insects (corixid bugs and some beetles) was highest in an acidic lake group, a feature which has been noted elsewhere and related to decreased fish predation (see above). However, one of the lakes with high nekton diversity had been stocked with acid tolerant brook char *Salvelinus fontinalis*, whilst another had a large population of eel *Anguilla anguilla*. Some corixids are highly tolerant of low pH and high aluminium concentrations, mobilising salts from internal sources to counter the effects of osmotic stress.[93] The adaptive value of this physiological mechanism has yet to be determined, but it may allow corixids to avoid predation whilst at the same time permitting the utilization of resources such as slowly decaying organic matter.

Tertiary production: fish and amphibia

Fish

Fish are represented at several trophic levels in aquatic systems, from detritivore and herbivore to piscivorous carnivore. In upland Welsh waters vulnerable to acidification, most fish are salmonids, eels, and cottids, with perch *Perca fluviatilis*, minnow *Phoxinus phoxinus* and coregonids less frequent. All these species are likely to be predominantly carnivorous, with the possible exception of the minnow, which sometimes feeds on algae.

The influence of acidity

Amongst fish, salmonids have received most attention in relation to acidification and there is now evidence to show that they are directly affected by acid related factors. Much of this evidence has been reviewed elsewhere in the volume (see Chap. 8). There is no indication that brown trout, at least, are limited by food scarcity in acid streams, partly because differences in invertebrate abundance between streams of different pH are not pronounced, but also because they can feed on insects entering the stream from the surrounding terrestrial habitats.[85] The quality and quantity of prey arriving through this pathway is likely to reflect the nature of the surrounding vegetation.[46] Prey inputs from exotic conifers, which often occupy the banks of acidic Welsh streams[76] have not been examined in detail, though one study has shown that the overall biomass of inputs from conifers is reduced by comparison with moorland or deciduous woodland, and the annual timing of inputs was considerably different (Ormerod and Tyler, unpublished data). Whilst salmonids are probably excluded from many afforested streams because of low pH and elevated aluminium, prey inputs from conifers may require consideration in those cases where activities such as liming restore water quality.

Bullheads (= sculpins, *Cottus gobio*) are often absent from acidic Welsh streams and, unlike salmonids, most of their prey are benthic insects.[90] One possibility might be that they are limited in acidic streams because of quantity or quality of invertebrate prey, and this requires examination. Ecotoxicological studies show that bullheads are sometimes able to tolerate acid episodes, although gravid females are sensitive (see Chap. 10).

There is experimental evidence from river systems outside Wales that fish sometimes exert a 'top-down' predatory effect on the benthic fauna, restricting the proliferation of large bodied benthic organisms such as *Plectrocnemia conspersa* (Schofield, Hildrew and Townsend, pers. comm.). These authors suggested that such benthic organisms can be abundant in acid streams either because of reduced predatory load, or because they replace fish as top predators in acid systems.

Amphibians

Amphibians occupy a unique niche in temperate regions, being dependent on fresh waters at the spawning and larval stages, but generally occurring at other times in terrestrial systems. Their ecological importance is often overlooked, but they can be an abundant component of aquatic systems, influencing algal or zooplankton composition and affecting energy or nutrient transfer to surrounding habitats.[63] In the Hubbard Brook forest of the northeastern USA, amphibian biomass was twice that of birds and equal to that of small mammals.[8]

The influence of acidity

Most work on amphibians in relation to pH has centred on physiological effects. Many species show increasing mortality only below pH 4, but lethal and sub-lethal responses, including suppressed growth and developmental abnormality, vary widely between species.[63] The European common frog (*Rana temporaria*) shows poor development at metamorphosis when previously exposed to pH < 4.8 and 0.8 mg Al l^{-1} of aluminium.[15] The ecological significance of such physiological effects have yet to be determined.

In mid-Wales, Slater (pers. comm.) could find no statistically significant relationship between the amphibian fauna of ponds and pH across the range 4.8–7. Smooth *Triturus vulgaris* and palmate newts *T. helveticus* occur throughout this range, and common frogs breed often in peaty pools at pH < 4.5. The crested newt *Triturus cristatus*, however, confined to the area east of the River Wye and south of the River Irfon where ponds are mostly calcareous and circumneutral. Leeches sometimes form the bulk of its diet,[28] and are also scarce in acid systems.

Tertiary production: birds and mammals

Throughout the world, many species of birds are wholly dependent on production in freshwaters for food. Most occur on and around lakes or wetlands, but several species of wading and perching birds are particularly associated with rivers. These include some sandpipers (Tringinae), American water-thrushes and flycatchers (Parulidae and Tyrannidae), pipits and wagtails (Motacillidae), and one especially adapted family, the dippers (Cinclidae). Fewer mammals feed exclusively from fresh-waters, but those which often do, such as the otter *Lutra lutra* and water shrew *Neomys fodiens*, are of importance to conservation.

The influence of acidity

Only in isolated cases has acidity been demonstrated as an influence on the ecology of semi-aquatic vertebrates such as birds or mammals. Impaired fish recruitment in acidic lakes probably affects some lake dwelling piscivorous birds, but insectivorous birds sometimes benefit from reduced competition in lakes where fish are scarce.[84] None of the affected birds breed around Welsh lakes (e.g. osprey *Pandion haliaetus*; divers *Gavia* spp.). Nevertheless, two Welsh studies have considered relationships between vertebrates and stream acidity, and are important in providing evidence for effects through trophic pathways.

River birds in Wales

Dippers

Dippers *Cinclus cinclus* belong to a family unique amongst perching birds in being anatomically and physiologically adapted to swim and dive into rivers, feeding almost exclusively on benthic invertebrates and small fish. Throughout the year, they occupy linear territories along fast-flowing reaches and nest at densities of 2–16 pairs per 10 km.[55]

Features such as stream slope exert a major effect on dipper distribution, but pairs are scarce along Welsh rivers which have low mean pH (< pH 5.7–6.0) and elevated concentrations of aluminium (> 0.1 mg Al l^{-1}), despite there being suitable habitat.[55] In one documented case from the River Irfon, a sharp fall in the dipper population accompanied a pH reduction between the 1960's and 1980's when the catchment was heavily afforested.[52, 57] Moreover, by comparison with pairs along circumneutral streams (> pH 6), dippers at low pH begin breeding later, lay fewer eggs, and rear smaller broods with slower growing young.[55]

It now seems probable that the distribution and breeding performance of dippers reflects food availability. Some of the invertebrates which are scarce at low pH, such as caddis larvae and mayfly nymphs, are particularly important to dippers in rearing young.[55] Birds on acidic streams are able to feed on stonefly nymphs, but the overall size and abundance of key prey is reduced by comparison with circumneutral streams. Adults on acidic streams are, therefore, able to provision their young less effectively (Fig. 7.8). Whilst reduced food abundance *per se* could be responsible for these patterns, it is possible that the scarcity of minerals could also impose limits to egg formation and brood provisioning at low pH. Prey rich in calcium such as fish, molluscs and *Gammarus* are all scarce below pH 5.7–pH 6. The shell-thickness and weight of dipper eggs are both significantly reduced at low pH, which would be consistent with low availability of calcium.[51]

Other riverine birds

Other characteristic birds along Welsh rivers include grey wagtails *Motacilla cinerea*, common sandpipers *Actitis hypoleucos*, kingfishers *Alcedo atthis* and goosanders *Mergus merganser*. Amongst these, only the grey wagtail has been investigated in detail in Wales, and shows no influence of pH on either distribution or breeding performance.[55] These and common sandpipers are less dependent than dippers on prey of riverine origin.[56, 94] Both goosanders and kingfishers are piscivores, and likely to be restricted from areas of reduced fish abundance. Goosanders only recently began breeding in Wales and their distribution or ecology has not been examined in relation to pH. Kingfishers in Wales are confined mostly to rivers of low gradient which are not vulnerable to acidification. They are, however, scarce along acidic rivers even where gradients are suitable, a pattern consistent with reduced food abundance.[58]

Otters in Wales

After declining in most areas during the 1950's and 1960's, probably because of toxic effects by organochlorine pesticides, especially dieldrin,[13] some otter

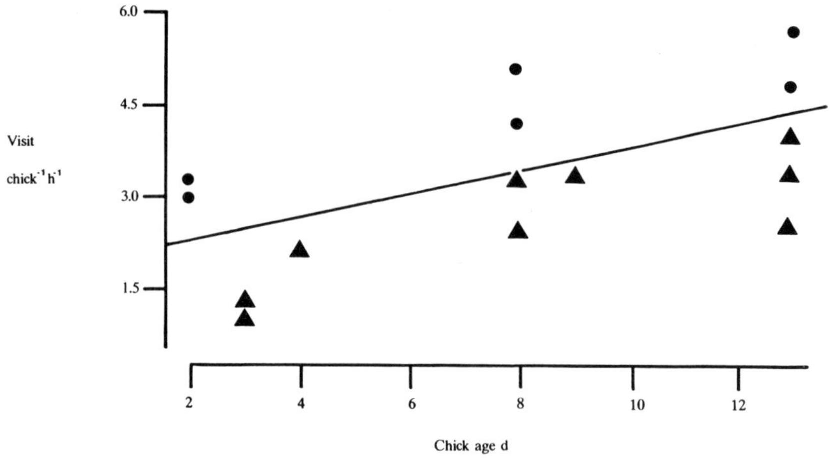

Fig. 7.8. The number of feeding visits made by five pairs of Dippers to nestlings of different age. (●, circumneutral sites; ▲ acidic sites). The line was fitted by least squares and relates the number of feeding visits to chick age.

populations are now recovering. In Wales, the number of sites showing positive signs of otter presence increased in most areas between 1977–78 and 1985, with some increase in known range.[2] Thus, after Ireland and Scotland, Wales is the major stronghold of otters in the British Isles.[12, 14]

Because fish usually provide the major component of otter diet,[12] reduced fish density might be expected to exert some influence. Indeed, Mason and Macdonald[46] have shown that they were absent from acidic headwaters of the Severn, but present along adjacent streams which were circumneutral. The acidic tributaries drained forest, and had reduced fish abundance. Elsewhere in Wales, otter signs were scarce along other acidic headwaters (e.g. Conwy, Mawddach, upper Tywi and Camddwr, upper Ystwyth, upper Irfon, Elan), possibly reflecting reduced food abundance.[2] However, these gaps in the distribution of otters occur against a background of generally increasing populations, and may be outlaying areas which are yet to be colonized. Additionally, the types of headwater vulnerable to acidification may be infrequently used by otters, except in crossing watersheds, because of generally impoverished food supplies.

Concluding comments

Overall, Welsh studies have revealed strong relationships between pH, or its related chemistry, and the ecology of Welsh lakes and streams. Most evidence is of qualitative differences between water bodies of different pH, apparent in several trophic levels from primary producers to aquatic and semi-aquatic carnivores. In some cases, such as fish, low pH or elevated aluminium concentration probably act directly on the organism. Other groups, such as some riverine birds, are probably affected indirectly through their food supply. In invertebrates, both direct and indirect effects are possible, and may well be inter-related.

Qualitative biological differences between acidic and circumneutral freshwaters provide several options for using 'biological indicators' for assessing spatial and temporal patterns in surface water chemistry. Some such indicators are already in practical use (see Chap. 3), whilst others are in advanced stages of development.[55, 59, 86] However, careful appraisal, and further research, is required so that biological indicator systems can be developed for detecting both episodic and chronic patterns in stream chemistry.

The major implication of ecological differences between acidic and circumneutral systems is that they can illustrate some of the changes which might have occurred with surface water acidification. Such change clearly has profound economic, conservational and aesthetic consequences.

References

1. Altshuller, A.P. and Linthurst, R.A. (1984). *The acidic deposition phenomenon and its effects: critical assessment and review papers. Volume 2, Effects Sciences*. North Carolina State University Acid Deposition Programme, EPA, Washington.
2. Andrews, E. and Crawford, A.K. (1986). Otter survey of Wales, 1984–85. Vincent Wildlife Trust, London.
3. Barlocher, F. and Kendrick, B. (1974). Dynamics of the fungal population on leaves in a stream. *Journal of Ecology*, 62: 761–791.
4. Barlocher, F., Kendrick, B. and Michaelides, J. (1978). Colonization and conditioning of *Pinus resinosa* needles by aquatic hyphomycetes. *Archiv für Hydrobiologie*, 81: 462–274.
5. Bendell, B.E. and McNicol, D.K. (1987). Fish predation, lake acidity and the composition of aquatic insect assemblages. *Hydrobiologia*, 150: 193–202.
6. Boylen, C.W., Schick, M.O., Roberts, D.A. and Singer, R. (1983). Microbiological survey of Adirondack lakes with various pH values. *Applied and Environmental Microbiology*, 45: 1538–1544.
7. Burton, T.M. and Allan, J.W. (1986). Influence of pH, aluminium, and organic matter on stream invertebrates. *Canadian Journal of Fisheries and Aquatic Sciences*, 43: 1285–1289.
8. Burton, T.M. and Likens, G.E. (1975). Salamander populations and biomass in the Hubbard Brook experimental forest, New Hampshire. *Copeia*, 1975: 541–546.
9. Caines, L.A., Watt, A.W. and Wells, D.E. (1985). The uptake and release of some trace metals by aquatic bryophytes in acidified waters in Scotland. *Environmental Pollution* (series B), 10: 1–18.
10. Chamier, A-C. (1985). Cell-wall degrading enzymes of aquatic hyphomycetes: a review. *Botanical Journal of the Linnaean Society*, 91: 67–81.
11. Chamier, A-C. (1987). Effect of pH on microbial degradation of leaf litter in seven streams of the English Lake District. *Oecologia*, 71: 491–500.
12. Channin, P.R.F. (1985). *The natural history of otters*. Croon Helm, London
13. Channin, P.R.F. and Jefferies, D.J. (1978). The decline of the otter *Lutra lutra* in Britain : an analysis of hunting records and discussion of causes. *Biological Journal of the Linnaean Society*, 10: 305–328.
14. Chapman, P.J. and Chapman, L.L. (1982). Otter survey of Ireland, 1980–1981. The Vincent Wildlife Trust, London.

15. Cummins, C.P. (1986). Effects of aluminium and low pH on growth and development in *Rana temporaria* tadpoles. *Oecologia*, 69: 248–252.
16. Dickson, W. (1978). Some effects of the acidification of Swedish Lakes. *Internationale Vereinigung fuer Theoretische und Angewandte Limnologie Verhandlungen*, 20: 851–856.
17. Dillon, P.J., Yan, N.D. and Harvey, H.H. (1984). Acidic deposition: effects on aquatic ecosystems. 13: 167–194.
18. Edwards, R.T. (1987). Sestonic bacteria as a food source for filtering invertebrates in two southeastern blackwater rivers. *Limnology & Oceanography*, 32: 221–234.
19. Eriksson, M.O.G., Henrikson, L., Nilsson, B.I., Nyman, G., Oscarson, H.C., Stenson, A.E, and Larsson, K. (1980). Predator-prey relations important for the biotic changes in acidified lakes. *Ambio*, 9: 248–249.
20. Findlay, D.L. and Kaisan and S.E.M. (1986). Phytoplankton community responses to acidification of Lake 223 experimental lakes area, northwestern Ontario. *Water, Air & Soil Pollution*, 30: 719–726.
21. Fisher, S.G. and Likens, G.E. (1973). Energy flow in Bear Brook, New Hampshire: an integrated approach to stream ecosystem metabolism. *Ecological Monographs*, 43: 421–439.
22. Friberg, F., Otto, C., and Svensson, B. (1980). Effects of acidification on the dynamics of allochthonous leaf material and benthic invertebrate communities in running waters. *Ecological Impact of Acid Precipitation*. (Ed by D. Drablos and A. Tollan) pp. 304–305. SNSF, Oslo, Norway.
23. Fryer, G. (1980). Acidity and species diversity in freshwater crustacean faunas, *Freshwater Biology*, 10: 41–45.
24. Goulder, R. (1987). Evaluation of the saturation approach to measurement of V_{max} for glucose mineralization by epilithic freshwater bacteria. *Letters in Applied Microbiology*, 4: 29–32.
25. Grahn, O., Hultberg, H., and Landner, L. (1974). Oligotrophication: a self accelerating process in lakes subject to excessive supply of acid substances. *Ambio*, 3: 93–94.
26. Grahn, O. (1977). Macrophyte succession in Swedish Lakes caused by deposition of airborne acid substances. *Water, Air & Soil Pollution*, 7: 295–305.
27. Gregory, S.V. (1983). Plant-herbivore interactions in stream systems. *Stream Ecology* (Ed. J.R. Barnes and G.W. Minshall) pp. 157–189. Plenum, New York.
28. Griffiths, R.A. and Myloote, V.J. (1987). Microhabitat selection and feeding relations of smooth and warty newts, *Triturus vulgaris* and *T. cristatus*, at an upland pond in mid Wales. *Holarctic Ecology*, 10: 1–7.
29. Haines, T.A. (1981). Acidic precipitation and its consequences for aquatic ecosystems: a review. *Transactions of the American Fisheries Society*, 110: 669–707.
30. Hall, R.J., Likens, G.E., Fiance, S.B. and Hendrey, G.R. (1980). Experimental acidification of a stream in the Hubbard Brook experimental forest, New Hampshire. *Ecology*, 61: 976–989.
31. Harriman R. and Morrison, B.R.S. (1982). Ecology of streams draining forested and non-forested catchments in an area of central Scotland subject to acid precipitation. *Hydrobiologia*, 88: 251–263.
32. Hasselrot, B., Anderson, I.B., Alenas, I. and Hultberg, H. (1987). Response of limed lakes to episodic events in southwestern Sweden. *Water, Air & Soil Pollution*, 32: 341–362.
33. Havas, M. (1985). Aluminium bioaccumulation and toxicity to *Daphnia magna* in soft water at low pH. *Canadian Journal of Fisheries and Aquatic Sciences*, 42: 1741–1748.
34. Havas, M. (1986). *Effects of aluminium on aquatic biota. Aluminium in the Canadian Environment* (Eds M. Havas and J. F. Jaworski) pp. 79–127. National Research Council of Canada, Ottawa.
35. Herrmann, J. (1987). Aluminium impact on freshwater invertebrates at low pH: a review. *Speciation of metals in water, sediment and soil systems*. (Ed L. Lander) pp. 157–175. Springer Verlag, Berlin.
36. Hildrew, A.G., Townsend, C.R., Francis, J. and Finch, K. (1984). Cellulolytic decomposi-

tion in streams of contrasting pH and its relationship with the invertebrate community. *Freshwater Biology*, 14: 323–328.
37. Hill, M.O. (1979a). *TWINSPAN–A FORTRAN programme for arranging multivariate data in an ordered two-way table by classification of the individuals and attributes*. Department of Ecology and Systematics, Cornell University, Ithaca, New York.
38. Hill, M.O. (1979b). *DECORANA–a FORTRAN programme for detrended correspondence analysis and reciprocal averaging*. Ecology and Systematics, Cornell University, Ithaca, New York.
39. Jones, J.R.E. (1948). The fauna of four streams in the Black Mountain district of South Wales. *Journal of Animal Ecology*, 17: 51–65.
40. Kaushik, N.K. and Hynes, H.B.N. (1971). The fate of the dead leaves that fall into streams. *Archiv fur Hydrobiologie*, 68: 465–515.
41. Kemmy, F. (1986). An investigation into the nature of the bacteria in the epilithon of acid and non-acid upland streams in mid-Wales. *Unpublished MSc thesis*. University of Wales.
42. Lock, M.A., Wallace, R.R., Costerton, J.W., Ventullo, R.M. and Charlton, S.E. (1984). River epilithon: towards a structural-functional model. *Oikos*, 42: 10–22.
43. Lowther, R. (1988). Macrophyte assemblages in Welsh Lakes. Unpublished Report, University of Wales Institute of Science and Technology.
44. Macan, T.T. (1970). *Biological studies of the English Lakes*. Longmans, London.
45. Mackay, R.J. and Kersey, K.E. (1985). A preliminary study of the aquatic insect communities and leaf decomposition in acid streams near Dorset, Ontario. *Hydrobiologia*, 122: 3–11.
46. Mason, C.F. and Macdonald, S.M. (1982). The input of terrestrial invertebrates from tree canopies to a stream. *Freshwater Biology*, 12: 305–311.
47. Mason, C.F. and Macdonald, S.M. (1987). Acidification and otter *Lutra lutra* distribution on a British river. *Mammalia*, 51: 82–87.
48. Mullholland, P.J., Elwood, J.W., Palumbo, A.V. and Stevenson, R.J. (1986). Effect of stream acidification on periphyton composition, chlorophyll and productivity. *Canadian Journal of Fisheries and Aquatic Sciences*, 43: 1846–1858.
49. Okland, J. (1983). Factors regulating the distribution of freshwater snails (Gastropoda) in Norway. *Malacologia*, 24: 277–288.
50. Ormerod, S.J. (1985). The distribution of macroinvertebrates in the catchment of the River Wye in relation to ionic composition. *Unpublished PhD thesis*. University of Wales.
51. Ormerod, S.J., Bull, K., Cummins, C., Tyler, S.J. and Vickery, J.A. (1988a). Egg mass and shell thickness in dippers *Cinclus cinclus* in relation to stream acidity in Wales and Scotland. *Environmental Pollution*, 55: 107–121.
52. Ormerod, S.J. and Edwards, R.W. (1985). Stream acidity in some areas of Wales in relation to historical trends in afforestation and the usage of agricultural limestone. *Journal of Environmental Management*, 20: 189–197.
53. Ormerod, S.J. and Edwards, R.W. (1987). The ordination and classification of macroinvertebrate assemblages in the catchment of the River Wye in relation to environmental factors. *Freshwater Biology*, 17: 533–546.
54. Ormerod, S.J., Mawle, G.W. and Edwards, R.W. (1987). The influence of forest on aquatic fauna. *Environmental aspects of forestry production in Wales* (Ed by J.E. Good). ITE symposium number 22, Institute of Terrestrial Ecology, Grange over Sands.
55. Ormerod, S.J. and Tyler, S.J. (1987a). Dippers *Cinclus cinclus* and Grey Wagtails *Motacilla cinerea* as indicators of stream acidity in upland Wales. *The Value of Birds* (Ed. A.W. Diamond and F.L. Filion) pp. 191–208. International Council for Bird Preservation, Cambridge.
56. Ormerod, S.J. and Tyler, S.J. (1987b). Aspects of the breeding ecology of Welsh grey wagtails *Motacilla cinerea*. *Bird Study*, 34: 43–51.
57. Ormerod, S.J., Tyler, S.J. and Lewis, J.M.S. (1985). Is the breeding distribution of dippers influenced by stream acidity? *Bird Study*, 32: 33–39.
58. Ormerod, S.J., Tyler, S.J., Pester, S.J. and Cross, A.V. (1988b). Censussing distribution

8. Effects of acidification on fish and fisheries in Wales

N.J. MILNER and P.V. VARALLO

Introduction

Welsh freshwater fisheries are dominated by the salmonids: Atlantic salmon (*Salmo salar*), sea trout and non-migratory brown trout (*S. trutta*). They constitute an important national resource generating, on first estimates,[33] a gross annual revenue expenditure in excess of £15m. In terms of numbers, rod catches of Welsh salmon and sea trout contributed on average 45% and 56% respectively of the total catch for England and Wales between 1979 and 1985. Brown trout fisheries are also important, trout anglers being the single largest group of licence buyers (52% of total between 1983 and 1985).

These resources are vulnerable to acidification because salmonids are particularly susceptible to low pH, and because their spawning and rearing habitats are typically found in the upper reaches of gravel bedded flashy streams. Such waters, and a multitude of lakes supporting trout fisheries, are located in the uplands of Mid and North Wales; precisely those areas where, through characteristics of high annual deposition and poor buffering capacity, acidification can be anticipated.

Natural reproduction supports the bulk of Welsh salmonid fisheries, but supplemental stocking of streams and lakes is also an important management tool. The failure of stocking, in some cases marked by fishkills, has sometimes been the first indication of a problem. The responses of stocked and wild fish are therefore of equal concern to the manager.

Acid water can also be detrimental to conservation of fish species in general. Survival of arctic charr (*Salvelinus alpinus*) and certain non-salmonid stocks, such as upland perch populations, is also potentially compromised by deteriorating water quality. In addition, the extinction of some isolated stocks and consequent loss of genetic diversity may be harmful to the long term survival of some species particularly brown trout.[6]

Fisheries response to acidification is the cumulative result of many separate impacts on individuals. These may kill sensitive stages directly or act sublethally by modifying physiological and behavioural mechanisms that regulate natural population structure and abundance. The use of experimentally derived concentration-response relationships to link observed population

R.W. Edwards et al. (eds) Acid Waters in Wales. 121–143.

parameters with field chemistry is a traditional procedure in water quality impact assessment. However, its effectiveness depends on the ability to establish either uniquely diagnostic symptoms or to separate the influence of other factors which cause similar response. Acidification typically proceeds slowly, at least in its early stages, with gradual progression from chronic to increasing frequency of acute exposure. This presents real difficulties in identifying and measuring (or predicting) population response in the field. In this chapter we examine evidence of diagnostic symptoms in individual fish and describe some responses of salmonid populations, in order to define the regional extent of impact.

Effects on individuals

Ionoregulation

Trout stocked into acid lakes exhibit typical ionoregulatory dysfunction that leads to changes in concentration of some plasma ions, principally loss of chloride (Cl_p) and sodium (Na_p), and consequent reduction in osmolarity (Fig. 8.1). Measurement of these changes can be useful in monitoring the degree and timescale of response. Greatest effects typically occur between 36 and 48 hours after stocking. Studies in Llyn Conwy, North West Wales, have shown that yearling farm-reared brown trout can withstand step transfer from pH 7.0 to pH 4.6, even at low calcium ($1.3 mgl^{-1}$). However, use of more sensitive stages, marginally more acidic conditions or the presence of toxic metals, particularly aluminium, can cause high mortalities.[22] Ionoregulatory effects are stongly H^+-related, but are profoundly modified by presence of other cations notably calcium and aluminium.[38] Thus water quality changes following liming (increased Ca and pH, and reduced Al) dramatically reduced ionoregulatory dysfunction in stocked brown trout in Mid Wales lakes.[34]

Great care is required in interpreting variation in blood composition in experimental studies. Hypochloremia, for example, is a secondary stress response initiated by several stressors.[19] Even confinement in cages at densities below those required by conventional protocol can greatly increase Cl_p loss (Fig. 8.1) and exaggerate the apparent acid water effect. Handling and transport are well-established stress factors that necessarily accompany most stocking operations, but can cause rapid substantial changes, even in the absence of subsequent acid exposure (Fig. 8.2). It is likely that some historical post-stocking fish kills, attributed at the time to acute acid stress, were the result of cumulative stress factors potentiating mild acid exposure.

Although significant differences in Cl_p can be detected between wild trout living in waters of different pH (unpublished data), the variation is minor and seasonally inconsistent. Therefore such measurements are of limited value in diagnosis of sublethal acid exposure, although they have proved useful in confirming the occurence of acute exposure leading to a fish kill (see below).

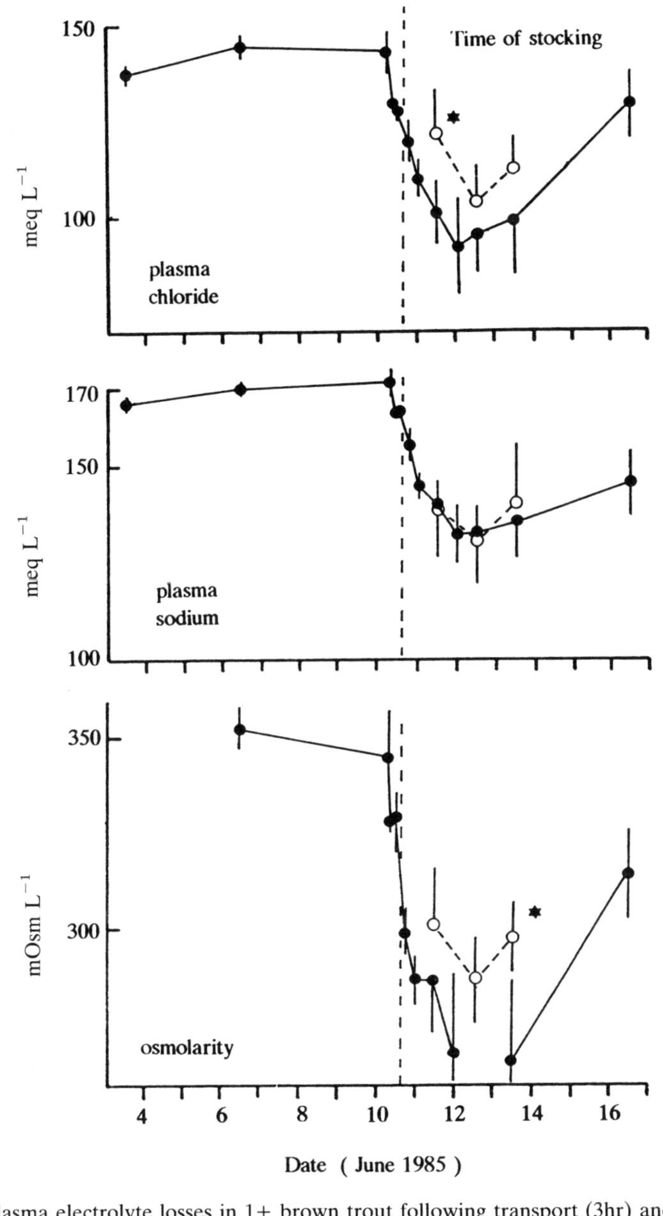

Fig. 8.1. Plasma electrolyte losses in 1+ brown trout following transport (3hr) and stocking into Llyn Conwy (water quality changes: pH 7.3 to 4.6, Ca 14 to 1.3 mgl^{-1}). Note changes occurring during transport immediately prior to stocking. Solid lines and symbols – fish held in cages, dashed lines and open symbols – fish freeswimming in the lake, sampled periodically by fyke net, asterisks indicate significant difference ($P < 0.05$) between groups. Means ± 95% confidence limits.

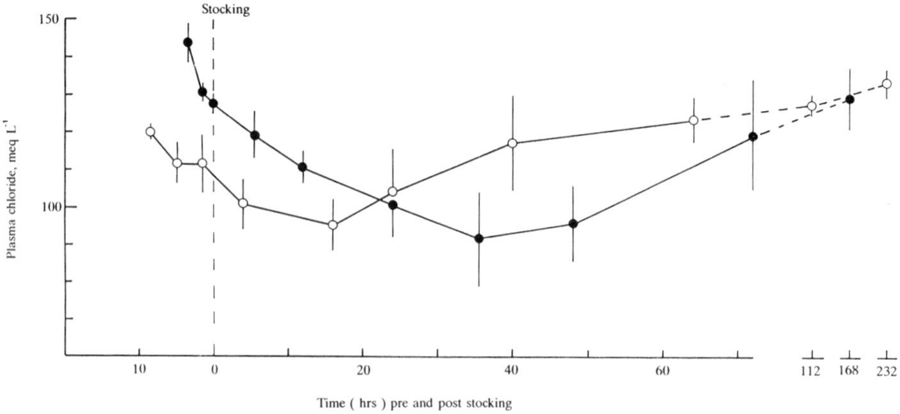

Fig. 8.2. Plasma chloride loss in brown trout subject to both transport and stocking (solid symbols) and to transport alone (open symbols). Means ± 95% confidence limits.

Reproduction, growth and condition

Several North American studies have reported impaired oogenesis, failure to ovulate, or reduced fecundity in non-salmonids.[25] However, studies on brown trout lake populations in Gwynedd (North Wales) have revealed that whilst there were significant differences in fecundity, egg size and incidence of atresia these did not correlate with water quality.[22]

Acid exposure has been shown to reduce growth of salmonids in experimental dosing studies,[17] with aluminium exerting a stronger influence than pH[28] and having effects at concentrations as low as $0.027 mgl^{-1}$.

However, field studies give conflicting results. Trout in acid Welsh lakes grew as well as fish in less acid conditions (Fig. 8.3) and it has not been possible to explain growth variation in terms of water quality. It is likely that any physiological effects are masked by environmental and biological variables (e.g. temperature, morphoedaphic factors and density) that are more important determinants of growth rate. A further factor affecting apparent growth responses in adults may be the early selective mortality of sensitive individuals through growth and feeding impairment at the early fry stage, effects which have been reported in lake trout (*Salvelinus namaycush*) alevins.[9] Where recruitment failure has reduced population density, there may be reduced competition for food which could account for the significant increase in condition factors reported in Norwegian,[27] Scottish[18] and Welsh[22] acidified lakes. In contrast, Turnpenny et al.[32] found no significant association between water quality and trout growth or condition in Welsh streams.

Metal accumulation and demineralisation

Acidification generally leads to mobilisation of metals and an increase in their concentration in surface waters[3]. Biological availability of metals may also

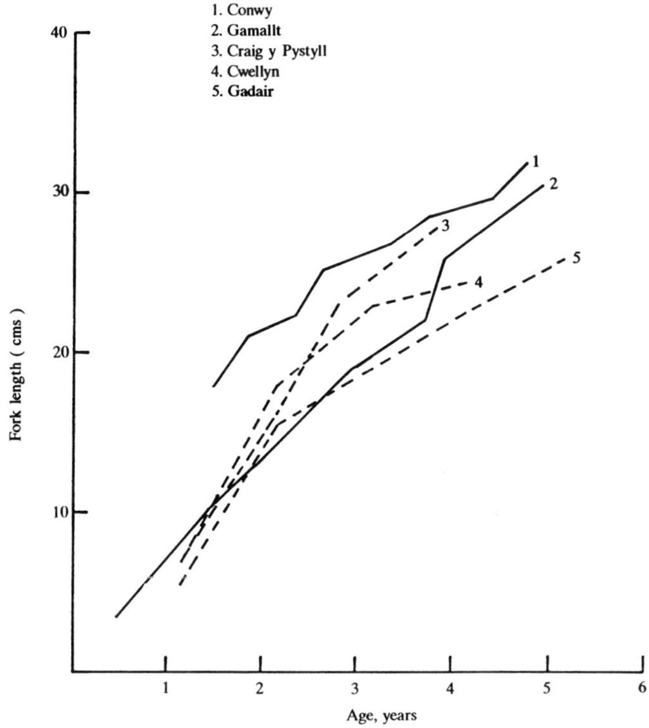

Fig. 8.3. Growth of trout in acid-impacted lakes (solid lines, based on length at age measurement) and non-impacted lakes (dashed lines, based on back calculations).

increase and accumulation in fish tissues has been used as an acidification index.[23] Calcium metabolism may be impaired in acid waters and Beamish et al.[2] suggested decalcification could be a cause of skeletal deformities in white suckers.

An investigation of relationships between metals (Zn, Mn and Ca) in trout hard tissues (scales, opercula, vertebrae) and water chemistry in 14 Gwynedd lakes (Table 8.1) demonstrated no decalcification. pH did not significantly influence Mn or Zn concentration in any tissue, but Mn in scales and vertebrae was positively correlated with water Mn. Opercula Zn was inversely correlated with lake conductivity, Ca and Mg. There were no significant correlations amongst the tissue metal concentrations.

There are no comparative published data for brown trout, but these results follow the pattern reported by Moreau[23] for Mn and Zn in brook trout opercula and scales from Canadian lakes (pH 5.2–7.0, Ca 1.3–3.5 mgl^{-1}). Neither lake concentrations of Mn or Zn were significantly correlated with pH in Gwynedd lakes, but were strongly correlated in a larger survey of stream chemistry (Table 8.2) throughout Wales.[34] These results show that tissue Mn and Zn concentrations vary with availability of these elements in water, which in turn is likely to be influenced by acidification. Tail deformities

Table 8.1. Correlations between water chemistry and metal concentrations in trout tissues from 14 lakes.

Water quality		Tissue concentration								
		Manganese			Zinc			Calcium		
	(Range)	Opercula	Scales	Vertebrae	Opercula	Scales	Vertebrae	Opercula	Scales	Vertebrae
pH	4.1–6.4	0.353	−0.153	−0.214	−0.310	−0.023	−0.069	−0.013	0.278	0.114
Cond $\mu S s^{-1}$	42–71	−0.116	−0.132	−0.081	−0.669*	−0.462	0.365	−0.099	0.380	0.059
Ca mgl^{-1}	0.8–5.2	0.235	−0.180	−0.203	−0.594*	−0.313	0.156	−0.071	0.314	0.103
Mg mgl^{-1}	0.5–1.4	0.312	−0.151	−0.270	−0.804**	−0.485	−0.280	0.119	0.128	−0.386
Al mgl^{-1}	0.03–0.14	−0.257	0.231	0.187	0.322	0.104	0.390	−0.454	−0.395	−0.083
Mn mgl^{-1}	0.06–0.33	−0.173	0.648*	0.663**	0.339	0.238	0.451	−0.387	0.196	−0.096
Zn mgl^{-1}	0.01–0.05	−0.411	0.006	0.093	−0.229	−0.193	0.336	0.005	0.288	0.052
Geometric mean range ($mgkg^{-1}$)		13–351	67–1546	46–760	217–543	532–2529	159–1046	159–241	49–245	78–220

* $P < 0.05$
** P 0.01 − 0.05

Table 8.2. Product moment correlation coefficients relating fish density to habitat and water quality variables (logs in Tables 8.2 & 8.3 are to $base_{10}$).

	1	2	3	4	5	6	7	8	9	10	11	12	13	14
1. log trout $100m^{-2}$														
2. log salmon $100m^{-2}$	—													
3. log altitude (m)	—	*												
4. log gradient	—	—	—											
5. log average daily flow (m^3s^{-1})	***	*	***	—										
6. log gravel (%)	***	—	—	***	**									
7. percentage catchment afforested	—	—	—	—	—	—								
8. pH	***	—	—	—	—	**	**							
9. log Al	***	—	—	—	**	**	***	***						
10. log Zn	***	—	—	**	—	—	—	***	***					
11. log Pb	—	—	*	—	—	—	—	—	—	—				
12. log Mn	***	—	*	—	***	**	**	***	***	***	—			
13. log Fe	**	—	***	—	*	—	—	*	*	**	—	***		
14. log hardness as $CaCO_3$	***	—	—	—	*	***	—	***	*	***	—	**	—	
15. log UV absorbance at 225nm (humic acids)	—	—	***	—	—	—	*	—	—	—	—	***	**	—

Key: $P_{(0.001)}$ = ***; $P_{(0.01)}$ = **; $P_{(0.05)}$ = *; — = Not significant reported in trout from acidified Scottish lochs,[18] and attributed to decalcification, have not so far been recorded in Wales.

Effects on populations and fisheries

Fishkills

There are few documented examples of fishkills in Britain that can be directly attributed to acid water. The best example in Wales occurred in the Afon Glaslyn, September 1984, when 117 freshrun sea trout and salmon were killed during floods following a drought. In the main river, pH of 5.4 to 5.9 and aluminium concentrations of 0.18 to 0.25 mgl^{-1} were recorded after the fishkill was reported. Mortalities occured over 2–3 days and concentrations of Cl_p and Na_p in survivors were low (means of 99 and 115 meq^{-1} respectively), commensurate with acid stress.[20] Kills following the same pattern (floods after drought) were also reported on the Afon Glaslyn in 1966 and 1976 although no data were collected to demonstrate the nature of toxicity. Freshrun fish are likely to be particularly sensitive due to the major physiological adjustments they have to make on entry to freshwater, and fishkills of similar type have been reported in Cumbrian[10] and Norwegian rivers.[13]

Extensive acid water fishkills of resident juvenile salmonids have not been reported in Wales, although they have occured in Cumbria.[10] Difficulties of observation may in part account for this, but it is also likely that the long-term pattern of acidification is such that the progressive increase in intensity and frequency of toxic episodes would gradually deplete populations rather than cause catastrophic mortality. However, 1,500 salmon parr were stocked into the upper Tywi in 1981 during low summer flows, yet when this stream was electrofished four weeks later no fish could be found. Acid conditions were subsequently shown to prevail in this stream.[29] Deaths of fish stocked into acid lakes have been reported in Pendam, Blaenmelindwr and Syfydrin, mid-wales,[30,31] and in Upper Lliw reservoir.[35]

Lake populations

The typical response of acid impacted lake trout populations is a reduction in stock accompanied by an increase in average size as recruitment failure gradually eliminates the younger age classes.[13, 27] This effect is exhibited in several Welsh lakes where significant recruitment failure is reliably confirmed (Fig. 8.4), but in some waters the situation is complicated by stocking. Thus in Llyn Conwy the population is maintained almost entirely by stocking, only 12% being derived from natural recruitment during 1984 and 1985.

Progressive recruitment failure is reflected in angling catches. Unfortunately, good catch records are rarely kept for upland trout fisheries, but data are available for Llyn Conwy in Gwynedd, North Wales, (Milner and Jones,

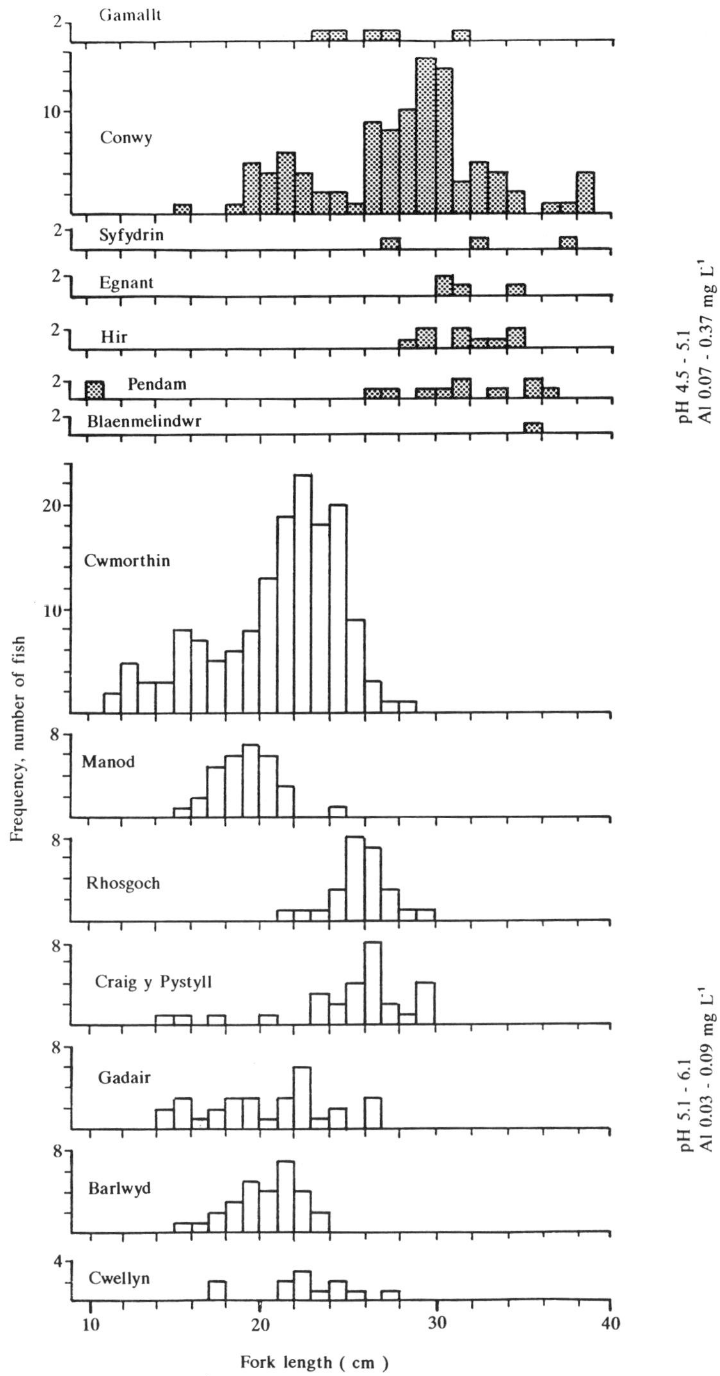

Fig. 8.4. Length-frequency distributions for Welsh lake brown trout populations (fyke net samples, October 1983). Shaded and unshaded histograms represent acid-impacted and non-impacted lakes respectively.

in prep.) and a group of Mid Wales lakes. In Conwy (Fig. 8.5) and the nearby Gamallt lakes (based on limited catch data and reliable anecdotal evidence) high catch rates evident in the 19th Century declined rapidly in the early 1950's and average size increased steadily. These trends continued until stocking began in Conwy in 1978, but by 1984 no fish were caught by angling in Gamallt, although a small residual population was left. Other factors (e.g. metal pollution, habitat change) can lead to similar symptoms and need to be eliminated before causal links between acidification and fishery decline can be proposed. In the case of Gamallt and Conwy, Milner and Jones[22] reviewed the likely alternatives and concluded that acidification was the only impact that could not be discounted.

Considering pH and calcium alone the distribution of affected fisheries (as manifested in declining recruitment) matches well with fry survival thresholds (<4.5 pH and <1 mgl^{-1} Ca) reported from experimental studies[15] (Fig. 8.6). Only two out of ten lakes with critical water quality possessed apparently unaffected fisheries. A further four lakes with Ca > 1mgl^{-1} had also suffered decline, but all had Al > 0.25 mgl^{-1} and three were afforested (Pendam, Blaenmelindwr, and Upper Lliw). Affected lakes occurred over the Al range 0.07 to 0.37 mgl^{-1} (Fig. 8.6b); but of those having Al <0.16mgl^{-1} (seven lakes) all also experienced pH <4.5 and Ca <1 mgl^{-1} and could have been fishless on that account. Within the Al range 0.07 to 0.16 mgl^{-1}, twenty-two (76%) lakes supported fisheries. The role of forestry in mediating acid water toxicity by increasing aluminium levels is well established.[30] These data demonstrate that aluminium toxicity can cause fishery damage in areas where effects due to pH toxicity alone would not be anticipated, e.g. the Penrhyncoch lakes of Mid Wales.

There have been regional differences in timing of fishery decline that seem to be related to catchment acid vulnerability (see Chap.5) and land use, principally forestry. Thus in unafforested catchments of North Wales, exemplified by Gamallt and Conwy, decline was evident in the 1950's and coincided with declines in unafforested Scottish lochs[11] that have catchments of similar high acid vulnerability (soils dominated by rankers and deep peats). In contrast, declines in afforested Mid Wales lakes situated in less acid vulnerable areas[31] were not evident until the late 1960's, a time postdating canopy closure of forests planted in the late 1940's. Aluminium mobilisation is thought to increase at canopy closure,[11, 14] so effects on fish would have been expected at this time.

Several populations were apparently unaffected at aluminium levels nominally higher than those regarded as toxic (Fig. 8.6b), and two main reasons may account for this. Firstly, these data take no account of aluminium speciation. Organic complexes may reduce aluminium toxicity from that indicated by its nominal 'dissolved' concentration.[4] The toxic labile monomeric fraction (aluminium (L)) in Gamallt, for example, varied between 56 and 100% (mean 75%) of dissolved aluminium over a year. Secondly, observable recruitment decline probably represents a rather advanced stage

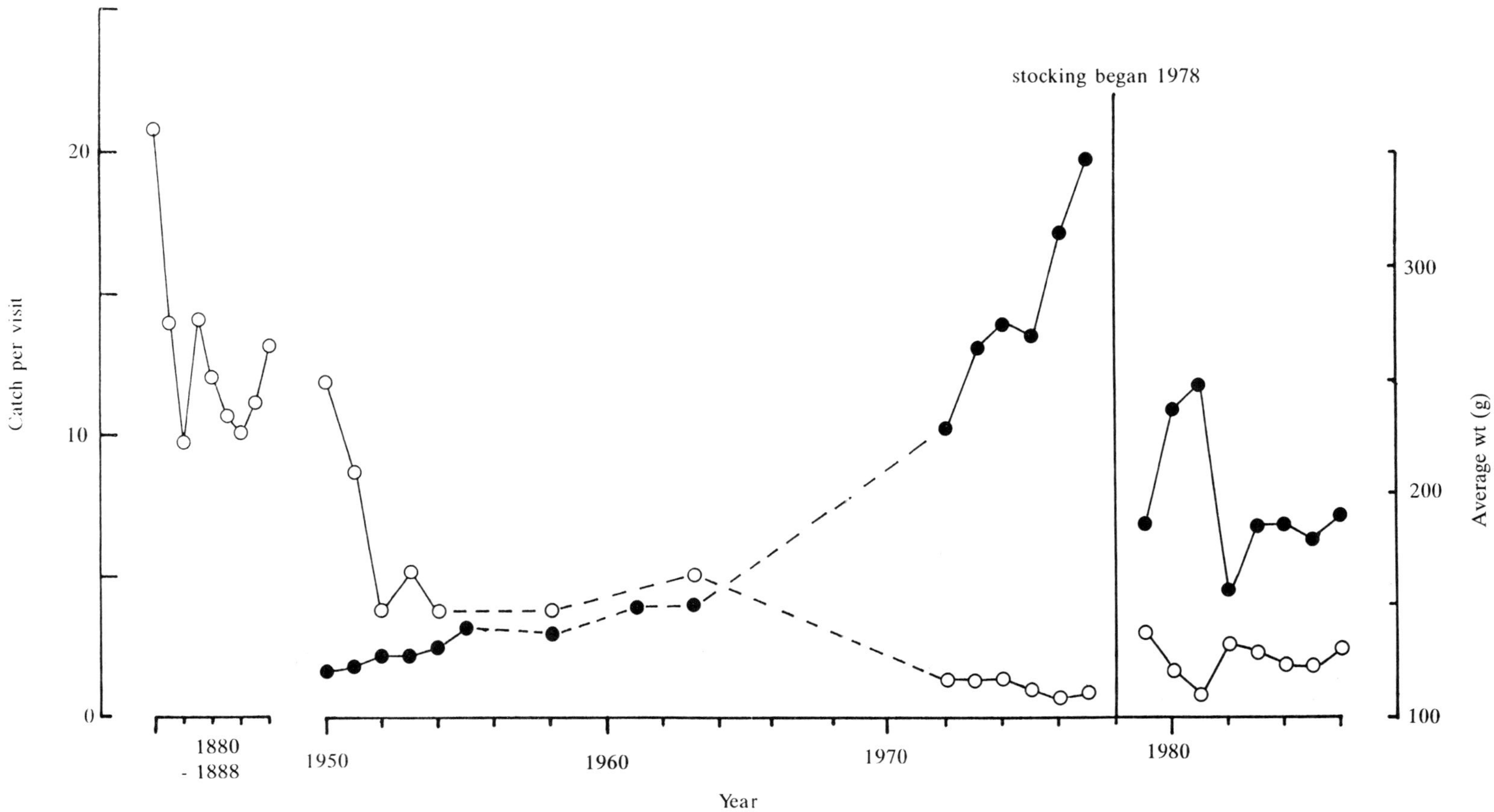

Fig. 8.5. Historical changes in brown trout catch per visit (open symbols) and average weight (closed symbols) between 1880 and 1986 in Llyn Conwy, Gwynedd, North Wales.

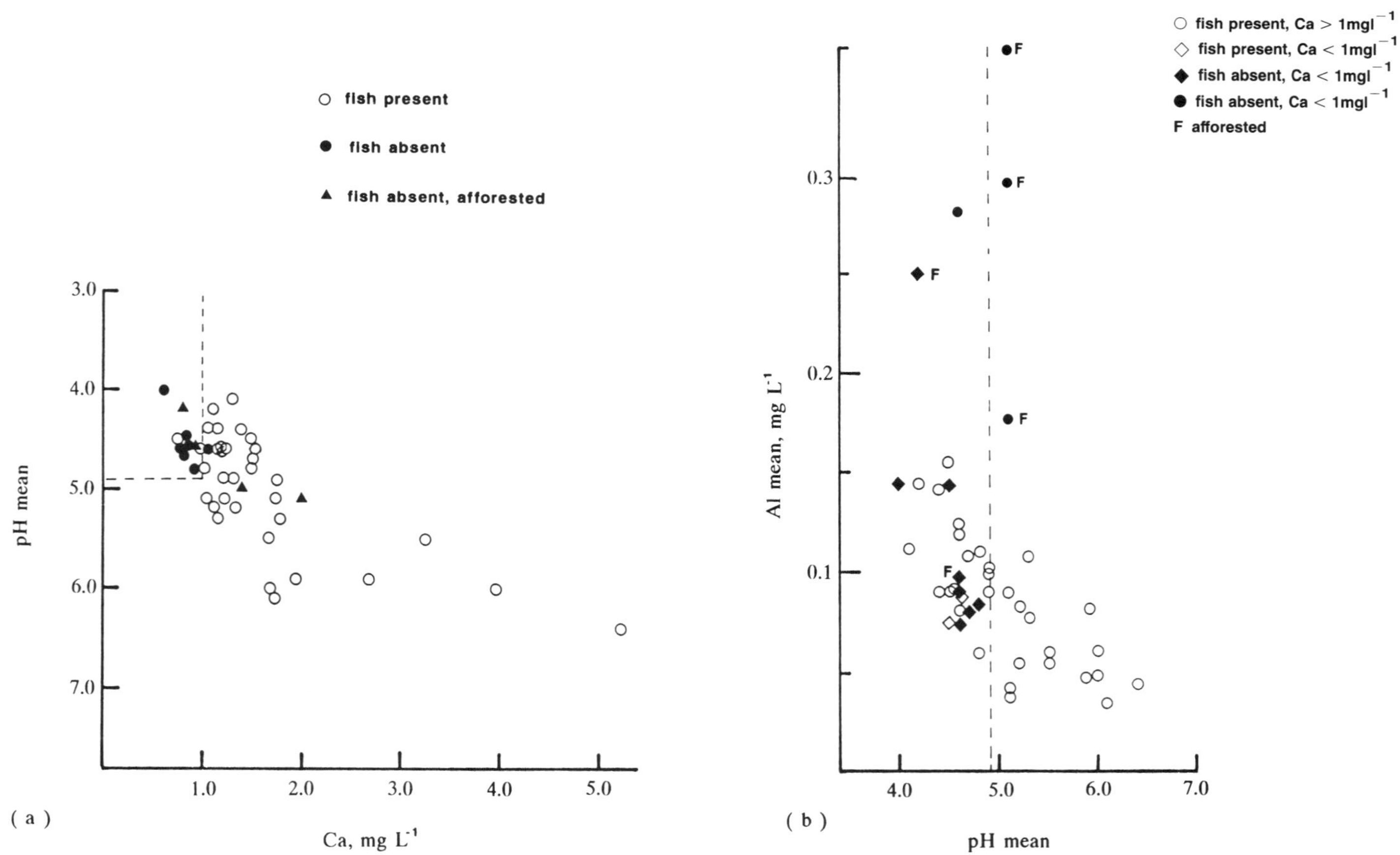

Fig. 8.6. pH/Ca (a) and Al/pH (b) plots for 46 Welsh lakes, showing values for fishless (or virtually fishless) lakes. Based on 6 bimonthly samples. Dashed lines indicate notional threshold values for trout survival, *viz* Ca/mgl^{-1} and pH 4.5. (pH$_{mean}$ 4.9 corresponds to pH$_{min}$ 4.5, by regression).

of response that occurs when a population's capacity to counteract diminishing recruitment (through feedback mechanisms) is exhausted. As acidification is a progressive deterioration in water quality, present day acidities in evidently impacted lakes may be substantially worse than those initiating the earliest effects, possibly years previously.

Stream populations

Relating population structure and abundance of stream salmonids to water quality poses similar problems to lakes, with the additional difficulty that habitat[21] and random climatic factors[5] are major determinants of salmonid abundance. Also, because episodes are far more pronounced in streams, mean chemistry may grossly underestimate the worst toxic conditions.

In a WWA regional survey of eighty three sites carried out in 1984, only 44% of the forty six sites accessible to migratory fish contained juvenile salmon.[36] Although salmon are especially sensitive to aluminium and pH,[8] their abundance did not correlate with water quality. Higher densities were associated with larger, lower altitude streams and may have been mainly determined by availability and distribution of spawners. In contrast, trout were found at 89% of sites, and densities were significantly correlated with several environmental variables including water quality (Table 8.2, Fig. 8.7).

Stepwise multiple regression produced models containing terms for aluminium and pH which explained 50% and 47% respectively of trout density variance (Table 8.3). Part of the unexplained variance is attributable to sampling and environmental factors, but the effects of potentially toxic metals in addition to aluminium may also be important. Significant negative correlations were obtained between trout density, zinc, iron and manganese (Table 8.2). Zinc concentrations were thought not to exceed toxic levels, due to the ameliorating influence of magnesium.[39] However, mixtures of metals, each at nominally sublethal concentrations, may be limiting. Turnpenny et al.[32] found that good populations (biomass $>5g\ m^2$) only occurred at Al(L) $<0.04\ mgl^{-1}$ and when the concentrations of zinc, lead and copper were <0.4 of their combined median lethal threshold concentrations (t-LC50). Populations were poor or absent at Al(L) 0.04 or Cu-Pb-Zn toxicity 0.7 t-LC50. Manganese and iron have been regarded as the proximate toxic factors in some acidic Swedish streams.[1] The toxic roles of these metals are not adequately understood, and may be more important than is presently recognised in the metalliferous catchments common in parts of upland Wales.

Trout populations were poor or absent when dissolved Al $>0.1mgl^{-1}$ occured with pH <5.4 (Fig 8.7), a concentration which corresponds to $0.09mgl^{-1}$ of labile inorganic aluminium in Llyn Brianne streams.[37] This is twice the limiting level proposed by Turnpenny et al.,[32] who excluded afforested sites. In contrast, 34% of the WWA study sites were heavily afforested, which could have resulted in higher mean aluminium values.

Although the deleterious effect of coniferous forestry on stream salmonids

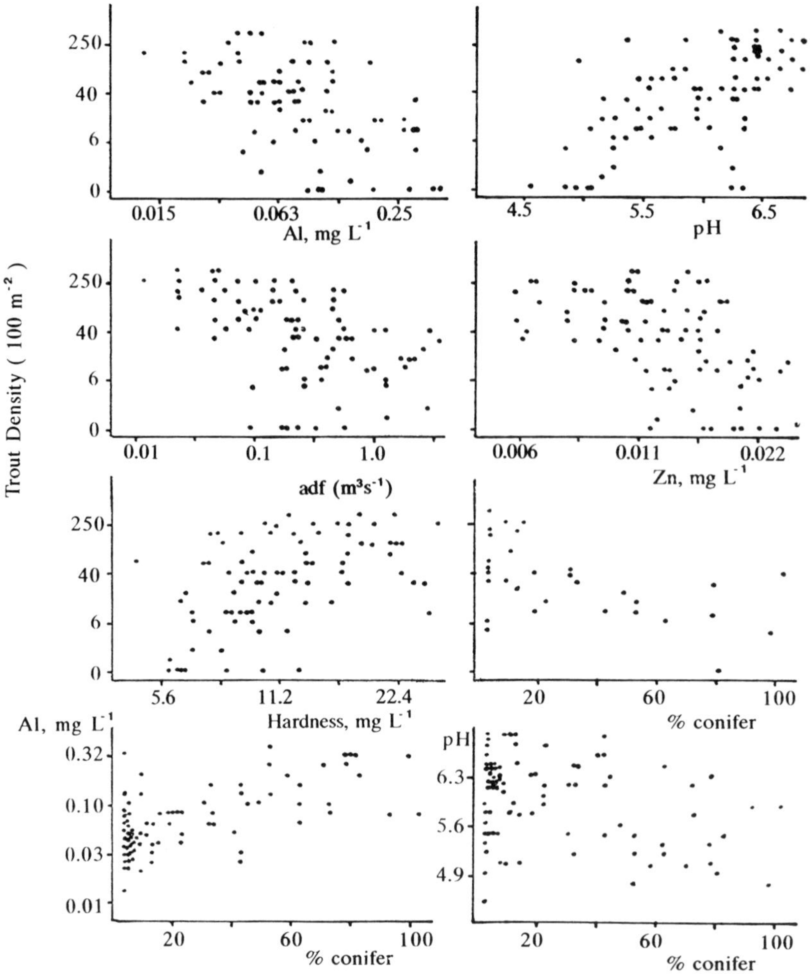

Fig. 8.7. Plots of trout density against pH, aluminium, adf, hardness and zinc concentrations, and influence of coniferous forestry on aluminium and pH. Note log scales (except conifer %).

has been unequivocably demonstrated in localised areas,[30] effects on a regional basis were only detectable in data classified by catchment hardness. A significant negative effect of forestry on trout abundance was found for the most sensitive range (hardness 8–12 mgl^{-1} $CaCO_3$), but not for ranges 12–15 mgl^{-1} or 15–25 mgl^{-1}. Allowance for crop age, species and catchment soil type will need to be made to improve these predictions.[14]

Regional distribution of acid water effects

Fisheries response is characteristically site specific because, in addition to biological and environmental variables, catchments themselves may differ in

Recording Chinese mitten crabs

Report your sightings by email or phone

Please tell us **where** and **when** you saw a Chinese mitten crab, your **contact details** and, if possible:

- was it male or female?
- take a photograph:
  - take photo from above
  - use a 50p coin to show scale
  - try to show claws clearly

MALE

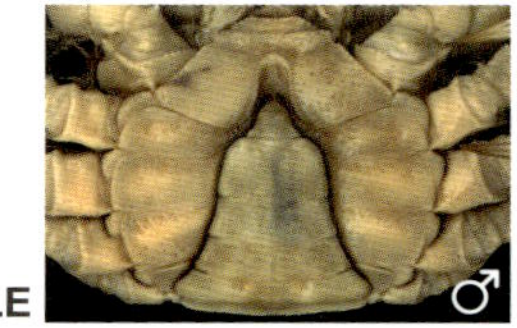

FEMALE

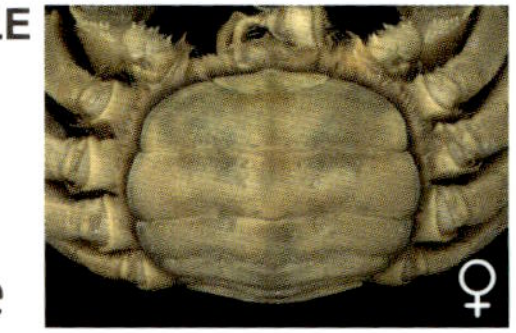

You can also report sightings by text or picture message to 07806 938789 or online at

www.mittencrabs.org.uk

Have you seen a Chinese mitten crab?

Chinese mitten crab
(*Eriocheir sinensis*)

- Grey-green to dark brown crab
- Long walking legs
- Squarish body up to 86 mm across
- Dense brown 'fur' on the white-tipped claws
- Habitat: rivers, brackish water estuaries, rarely along the marine inshore coast
- 4+4+4 pattern of teeth around front of shell
- Juveniles may lack 'fur' on claws

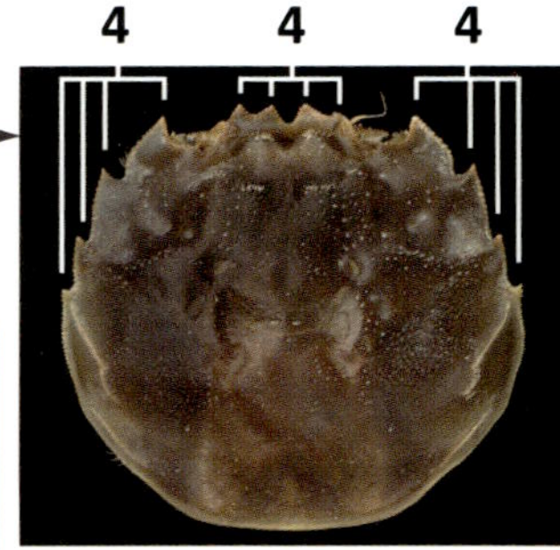

Powerful claws

Handle with care

Not to be confused with:

Common shore crab
(*Carcinus maenas*)

- Lacks 'fur' on claws
- Shell shape more triangular
- Found on the sea shore and in estuaries but never in freshwater rivers

The broad-clawed porcelain crab
(*Porcellana platycheles*)

- 2 long antennae
- Only 3 pairs of walking legs apparent
- Found on the sea shore but never in freshwater rivers

Table 8.3. Multiple regression relationships between trout density, water quality and catchment features.

1. log trout density =	$-3.59 - 0.307$ log av. daily flow $+ 0.492$ pH -1.01 log Zn ($R^2 = 47.1\%$ n = 82)
2. log trout density =	$-1.24 - 1.08$ log Al $+ 1.33$ log hardness -0.22 log average daily flow ($R^2 = 50\%$ n = 82)
3. log Al =	$0.465 - 0.28$ pH $+ 0.00552\%$ catchment afforested $+ 0.128$ log average daily flow ($R^2 = 70.6\%$, n = 82)
4. pH =	$2.43 - 0.0107\%$ catchment afforested $+ 2.71$ log hardness $- 0.653$ log UV absorbance at 225nm ($R^2 = 71.9\%$, n = 82)

Trout densities $100m^{-2}$
Average daily flow as m^3s^{-1}
Zn, Al and Hardness as mgl^{-1}

their vulnerability to acid deposition. However, it is possible to establish regional patterns of catchment vulnerability and a five point classification system (1, most acid, to 5 least acid vulnerable) has been developed for Wales (see Chap. 5). Stratifying fishery data by these classes allows broad regional patterns of impact to be assessed. Two sources of data are available: a questionnaire survey of fishery status in Welsh stillwaters carried out in 1986, and combined electrofishing results from four surveys between 1984 and 1986 covering 200 stream sites.

Acid classes 1 and 2 represent 20% of Wales, and are mountainous areas containing 28% of all stream lengths (>2nd order) and 54% of the 810 lakes (>1Ha) considered to be significant usable water for salmonids. Within this pattern there is a further concentration of lakes, with 93% of class 1 lakes lying in Gwynedd.

The variation in existing lake fishery status amongst different classes is not great (Table 8.4), although as might be anticipated most 'good' fisheries are found in the least acid areas. The only slightly higher proportion of 'poor or none' fisheries in class 1 lakes (35%) compared with class 5 (33%) suggests that the number of fisheries where fishing quality has been severely affected is still small. However, trends in fishery status (Table 8.4) indicate that fisheries in classes 1 and 2 have suffered substantial reductions in fishing quality compared with 5, and exhibit a much lower potential for increase. Lakes showing strong evidence of acid impacted fisheries are all found in the two most acid classes. Differences are also seen in stream salmon and trout populations, with relatively more impoverished and fewer flourishing sites in classes 1 and 2 compared with class 5. These patterns do not demonstrate causal links between acid vulnerability and fishery status (see Chap. 2), and in the absence of reliable historical data on chemistry or fish populations, the direct influence of water quality cannot be assumed. However, severe reduction in abundance is confirmed in some afforested waters, which can be compared with adjacent, unafforested, but otherwise similar catchments e.g. tributaries of the upper Wye, Tywi, Cothi, Ystwyth, Rheidol, Dyfi, Mawddach,

Table 8.4. Description of lake trout fishery and stream fish population status within acid vulnerability zones (1–5).

Acid vulnerability zones	No. lakes (%)* surveyed	Lakes (based on questionnaire replies) Existing fishery status (%)			Trends in fishery status (%)				Stream populations (electrofishing data)** Salmon (% sites)				Trout (% sites)			
		good	fair/ moderate	poor/ none	increase	no change	some decline	severe decline	No. sites	fishless	low	medium/ high	No. sites	fishless	low	medium/ high
1	119 (48.8)	4	60	35	5	38	37	20	32	59	3	38	57	12	47	40
2	62 (25.4)	2	56	42	3	18	44	36	64	30	22	48	87	5	21	75
3	12 (4.9)	0	58	42	38	12	25	25	–	–	–	–	–	–	–	–
5	51 (20.9)	15	52	33	38	38	12	12	66	11	5	85	76	0	14	86

* % of all surveyed lakes

** density (N/100m^2) data classed as: low, <10 0+ and < 5 1++
med/high, >10 0+ or > 5 1++

Dysynni, Glaslyn and Conwy. Evidence for acid-related declines in unafforested catchments is more equivocal, but the recent demonstration of historical pH reduction in lakes (see Chap. 3) and the absence of other major environmental impacts in the uplands, points to the likelihood of acidification as a significant factor. These uncertainties will be reduced with improved understanding of relationships between habitat variables and salmonid abundance, which are the subject of present research.

The potential scale of the problem is large, however (Fig. 8.8). In a pilot survey of 37 Gwynedd lakes located in class 1 and 2 catchments between 8 and 21% were thought to have suffered fishery decline that could be acid water related. Extrapolating to Wales as a whole, which has 437 lakes in class 1 and 2 (54% of all stillwaters >1Ha), the regional total becomes 35 to 92 waters >1Ha, and considerably more smaller waters. These are provisional estimates that require confirmation by more extensive field data. Turning to rivers, impacted afforested catchments have been noted above, but on water quality criteria (mean pH < 5.4, mean Al >0.09mgl^{-1}) and observed poor fish abundance, many of the unafforested headwaters of the rivers listed above could also be experiencing acidification.

Economic losses

Estimates of economic loss require measures of resource loss (i.e. juvenile production or fishery yields) and means to derive their value. However, estimates of resource loss are constrained by the ability to distinguish acid water effects from variation due to other factors whilst the accuracy of the value of that loss, is restricted by current economic evaluation methods. Accepting these limitations, loss to Welsh fisheries was placed in one of the ranges <£1m, £1–5m, £5–25m or greater, as described below.

Resource loss

The current fisheries status of upland Welsh lakes and streams has been derived from extensive electrofishing, netting and questionnaire surveys (see above). On the basis of published toxicity and observed relationships between aluminium, pH and fish population status, it has been assumed that salmonid populations may be reduced if threshold values of 0.09mgl^{-1} aluminium and pH 5.4 are exceeded. Using a linear regression derived from the regional survey, the aluminium threshold corresponds to a stream trout density of 23 fish 100m^{-2}. It has therefore been assumed that all stream sites with densities of <30 salmonids 100m^{-2} (to allow for greater sensitivity of salmon), and all lakes which are virtually fishless lying in acid classes 1 and 2, are affected by acidification. The extent of present-day actual resource loss was then estimated by two methods. In method 1 resource loss was estimated in terms of (i) acid-affected lake area (lake brown trout) (ii) acid-affected stream area

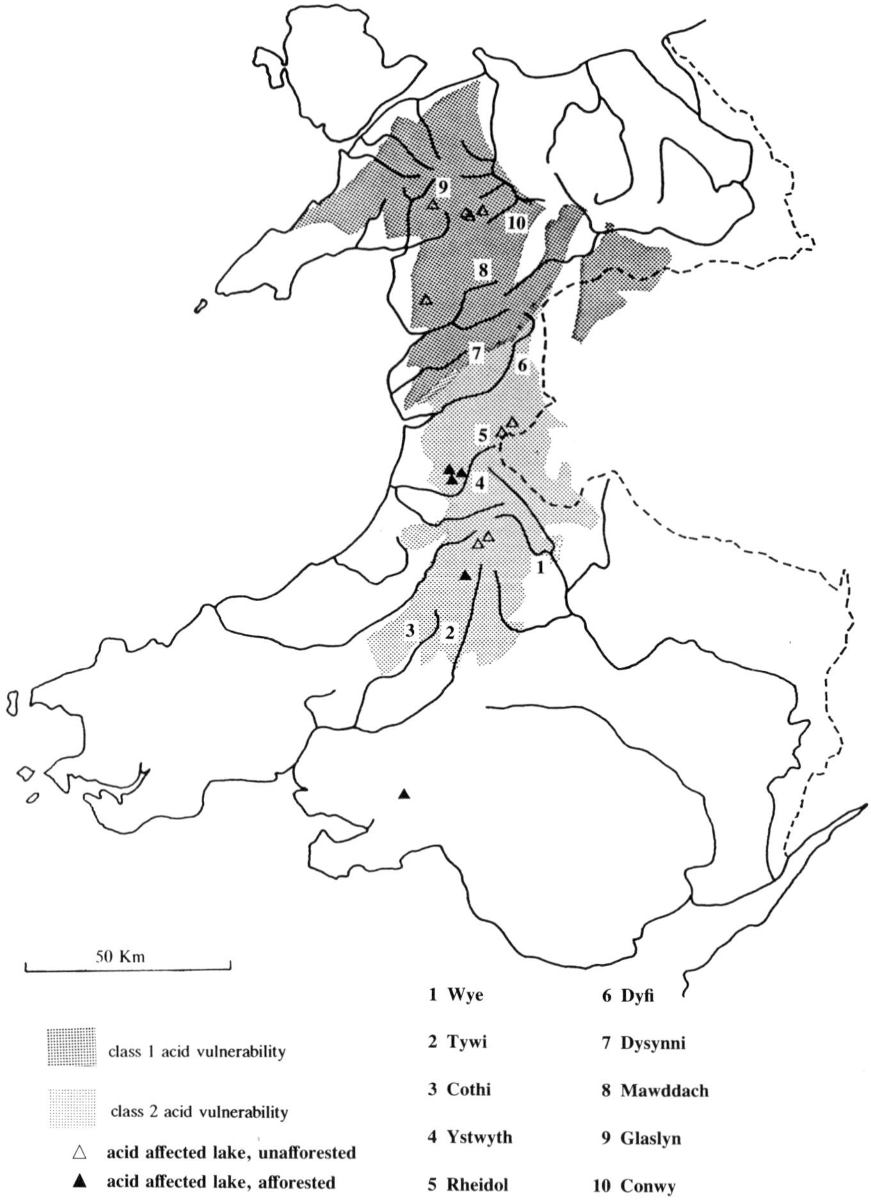

Fig. 8.8. Location of known acid affected lake fisheries and major Welsh rivers in relation to land acid vulnerability class.

weighted by observed proportional reduction in populations (stream brown trout) and (iii) rod and net catch reductions estimated to result from lost smolt production (salmon and sea trout). In method 2, lake trout loss remained unchanged, but loss of stream salmonids was expressed as the proportion of total available stream length considered to be acid-affected. In addition, as acidification is believed to be a continuing process, it was assumed that all lakes and streams in acid classes 1 and 2 could eventually become fishless. These additional losses were combined with estimated present day loss to give potential resource losses.

Economic value

The net economic value of a fishery is taken to represent the loss in benefit to society which would result if the fishery ceased to exist. Consequently there are many components of net economic value that have to be assessed, including the value of a fishery to the owner, to the anglers and to the local economy, as well as more intangible benefits to the well-being of the community e.g. conservation value.

Estimates 1 and 2 (Table 8.5) are alternative ways of calculating loss to the fishery owners (economic rent). In estimate 1 economic loss has been calculated by applying, to the appropriate components of resource loss, capital values of (i) migratory rod catch losses (based on Radford's data for the Mawddach and Wye,[26] and assuming one salmon = four sea trout), (ii) migratory fish commercial losses and (iii) upland lake and stream brown trout fisheries (based on recent purchases). For estimate 2 economic loss was calculated by applying the proportion of affected stream length to the estimated total economic rent of the migratory salmonid and stream brown trout fisheries (as above), whilst lake brown trout loss remained as (iii). These calculations indicate that actual loss to Welsh fishery owners is probably £1–5 million and potential loss is £5–25 million.

An additional component of net economic loss is the value of salmon and sea trout to anglers (consumer surplus), which may be estimated by capitalising annual expenditure data.[26] This yields unit values of £2.5k and £0.63k for salmon and sea trout respectively (Mawle, pers comm) which, when applied to resource losses in estimate 1, indicates an actual consumer surplus loss of £0.5 million and potential loss of £11.7 millions (Table 8.5). Whilst not all economic values have been calculated, if estimates of economic rent and consumer surplus are combined it can be seen that the actual capital loss in net economic value of Welsh fisheries is probably £1–5m and the potential loss is £5–25million (Table 8.5). Despite their imprecision these estimates provide an economic justification for protecting Welsh upland fisheries.

Discussion

The present day pattern of fishery impacts in Wales is a mosaic of many populations in various stages of response. Although severe effects are still localised, and mainly confined to Mid and North Wales, preliminary estimates indicate that the potential regional scale of damage in biological and economic terms may be considerable. The management consequences are

Table 8.5. Estimates of fisheries economic loss (£million) resulting from acidification.

	Economic rent		Consumer surplus*		Total	
	actual	potential	actual	potential	actual	potential
Estimate 1	0.5	6.3	0.5	11.7	1.0	18.0
Estimate 2	1.0	11.6	1.9	12.0	2.9	23.6

* Salmon and sea trout only.

already significant. In addition to *in situ* losses, deterioration of water quality in presently inaccessible uplands reduces the scope for future management that might utilise these areas for migratory fish stock enhancement. Indeed, this has already constrained fisheries mitigation options in relation to some large scale engineering schemes such as Llyn Brianne Reservoir and the Afon Conwy Tunnel Crossing Scheme (a highway development). Acidification effects may thus have consequences far removed from the site of their immediate impact.

Forestry exerts a major influence on the distribution of acidification, in sensitive areas invariably operating through increasing aluminium concentrations. Although aluminium toxicity is a major component of the acid-water phenomenon, in some catchments, more common in North Wales, pH seems to be the limiting factor; and certainly most of the distribution of affected fisheries in unafforested lakes can be explained in terms of pH/Ca interaction alone. Matching observed fish populations and water quality with experimental concentration-response criteria is more difficult in the case of aluminium, particularly in streams. Issues of speciation and the character of natural exposure regimes undoubtedly contribute to this, but can be resolved by improved measurement. In addition, habitat variables and other pollutants, especially heavy metals which are a feature of Welsh upland surface water quality, must be better understood before direct effects of pH/Al toxicity can be clearly identified.

A significant biological problem lies in understanding the mechanisms by which individual responses are translated into population change. Catastrophic losses of parr and adults are rare and progressive recruitment failure appears to be the principal cause of decline, a conclusion reached for most other acid impacted areas.[13, 24] However, it is not clear if this is initiated by reproductive deficiencies in the parents or, perhaps more likely, by direct effects on sensitive egg and fry stages. Recent work at Llyn Gamallt, North Wales, indicates that naturally deposited trout eggs suffered high mortality during early incubation (B.E. Jones, pers. comm.), but this is the only known example of such an effect in Wales.

Sublethal effects in salmonids have been difficult to detect against natural background variation. This has confounded attempts at early diagnosis of stressed populations, on the basis of symptoms in individual fish. It is worth noting however that some aspects, such as trophic interactions, behavioural responses and developmental biology have scarcely been studied and warrant investigation. Fisheries impacts can, at present, only be recognised when recruitment failure is sufficient to be reflected in population characteristics; by which time acidification will be well advanced. It would be very useful to be able to identify biological changes that mark the early stages of response and to test the effects of minor but persistent recruitment failure. Mathematical models exist[7, 16] which could incorporate survivorship terms based on concentration-reponse relationships. The accuracy of these relationships will be central to the success of models, which will have to be directed at the appropriate critical life stages.

References

1. Andersson, P. and Nyberg, P. (1984). Experiments with brown trout (*Salmo trutta* L.) during spring in mountain streams at low pH and elevated levels of iron, manganese and aluminium. *Report of the Institute of Freshwater Research, Drottningholm*, 61: 34–47.
2. Beamish, R.J., Lockart, W.L., Van Loon, J.C. and Harvey, H.H. (1975). Long-term acidification of a lake and resulting effects on fishes. *Ambio*, 4: 98–102.
3. Dickson, W. (1980). Properties of acidified water. *Proceedings of an International Conference on the Ecological Impact of Acid Precipitation*, pp. 75–83. SNSF Project, Norway.
4. Driscoll, C.T., Baker, J.P., Bisogni, J.J. and Schofield, C.L., (1980). Effect of aluminium speciation on fish in dilute acidified waters. *Nature, London*, 284: 161–164.
5. Elliot, J.M. (1987). Population regulation in contrasting populations of trout *Salmo trutta* in two Lake District streams. *Journal of Animal Ecology*, 56: 83–98.
6. Ferguson, A. (1986). Genetics and the management of natural salmonid stocks. Proceedings of the Institute of Fisheries Management Annual Study Course. Coleraine, 1986.
7. Gee, A.S. and Radford, P. (1982). The regulation of stock characteristics in a simulated Atlantic Salmon population. *Fisheries Research*, 1: 105–116.
8. Grande, M., Muniz, I.P. and Anderson, S. (1979). The relative tolerance of some salmonids to acid waters. *Vereinigung fuer Theoretische und Angewandte Limnologie Verhandlungen*, 20: 2076–2084.
9. Gunn, J.M. and Noakes, D.L.G. (1987). Latent effects of pulse exposure to aluminium and low pH on size, ionic composition and feeding efficiency of lake trout (*Salvelinus namaycush*) alevins. *Canadian Journal of Fisheries and Aquatic Sciences*, 44: 1418–1424.
10. Harper, E., Crawshaw, D.H., Prigg, R.F., Diamond, M., Robinson, J.F. and Cragg-Hine, D. (1986). Episodic variations in stream water chemistry associated with acid rainfall and runoff and the effect on aquatic ecosystems, with particular reference to fish populations in N.W. England. *Final Report to European Commission, North West Water Authority*, Contract No. Env. 867 UK (H).
11. Harriman, R., Morrison, B.R.S., Caines, L.A., Collen, P. and Wyatt, A.W. (1987). Long term changes in fish populations of acid streams and lochs in Galloway South West Scotland. *Water, Air and Soil Pollution*, 32: 89–112.
12. Harvey, H.H. (1982). Population responses of fisheries in acidified waters. *Acid rain/ Fisheries*, (Ed. by R.E. Johnson), pp. 227–242. Proceedings of an international symposium on acidic rain and fishery impacts on North Eastern North America. American Fisheries Society, Bethesda, Maryland.
13. Henricksen, A., Skogheim, O.K. and Rosseland, B.O. (1984). Episodic changes in pH and aluminium speciation kill fish in a Norwegian salmon river. *Vatten 40*, 225–260.
14. Hornung, M. (1984). Precipitation–canopy and soil–water interactions. *Symposium on "Acid Rain"*, 28p, Institute of Water Pollution Control.

15. Howells, G.D. (1983). Fishery status and water quality in areas affected by acid deposition. *Water Science and Technology*, 15: 67–80.
16. Lacroix, G.L. (1987). Model for loss of Atlantic Salmon stocks from acidic brown waters of Canada. *Acid Rain: Scientific and Technical Advances*, Proceedings of Lisbon Conference. P and H Selper Ltd., London, 508–515.
17. Leivestad, H., Hendrey, G., Muniz, I.P. and Snekvik, E. (1976). Effects of acid precipitation on freshwater organisms. *Effects of acid precipitation on forest and freshwater ecosystems in Norway*, pp. 87–111. SNSF Project Research Report, Vol.6.
18. Maitland, P.S., Lyle, A.A. and Campbell, R.N.B. (1987). Acidification and fish in Scottish Lochs. Institute of Terrestrial Ecology Report, 71pp.
19. Mazeaud, M.M., Mazeaud, F. and Donaldson, E.M. (1977). Primary and secondary effects of stress in fish: some new data with a general review. *Transactions of the American Fisheries Society*, 106: 201–212.
20. Milner, N.J. and Hemsworth, R.J. (1984). Fisheries investigation on Afon Glaslyn fish kill, September 1984. Welsh Water internal report 84/2.
21. Milner, N.J., Hemsworth, R.J. and Jones, B.E. (1985). Habitat evaluation as a fisheries management tool. *Journal of Fish Biology, Supplement A*, 27: 85–108.
22. Milner, N.J. and Jones, B.E. (1985). Sport fisheries and biology of brown trout in some North Wales lakes: with special reference to the effects of acid water. Welsh Water internal report, 85/1.
23. Moreau, G., Barbeau, C., Frenette, J.J., Saint-Onge, J. and Simoneau, M. (1983). Zinc manganese and strontium in opercula and scales of brook trout (*Salvelinus fontinalis*) as indicators of lake acidification. *Canadian Journal of Fisheries and Aquatic Sciences*, 40: 1685–1691.
24. Muniz, I.P. and Leivestad, H. (1980). Toxic effects of aluminium on the brown trout, *Salmo trutta* L. *Proceedings of an International Conference on the Ecological Impact of Acid Precipitation*, pp. 320–321. SNSF Project, Norway.
25. Peterson, R.H., Daye, P.G., Lacroix, G.L. and Garside, E.T. (1982). Reproduction in fish experiencing acid and metal stress. *Acid rain/Fisheries*, (Ed. by R.E. Johnson), pp. 177–196. Proceedings of an international symposium on acidic rain and fishery impacts on North Eastern North America. American Fisheries Society, Bethesda, Maryland.
26. Radford, A. (1984). The economics and value of recreational salmon fisheries in England and Wales: An analysis of the Rivers Wye, Mawddach, Tamar and Lune. Internal Report, Marine Resources Research Unit, Portsmouth Polytechnic. 250pp.
27. Rosseland, B.O., Sevaldrud, I., Svalastog, D. and Muniz, P. (1980). Studies on freshwater fish populations – effects of acidification on reproduction, population structure, growth and food selection. *Proceedings of an International Conference on the Ecological Impact of Acid Precipitation*, SNSF project, Norway.
28. Sadler, K. and Lynam, S. (1987). Some effects on the growth of brown trout from an exposure to aluminium at different pH levels. *Journal of Fish Biology,* 31: 209–219.
29. Stoner, J.H., Gee A.S. and Wade, K.R. (1984). The effects of acidification on the ecology of streams in the Upper Tywi catchment in West Wales. *Environmental Pollution (Series A)*, 35: 125–157.
30. Stoner, J.H. and Gee, A.S. (1985). Effects of forestry on water quality and fish in Welsh rivers and lakes. *Journal of the Institute of Water Engineers and Scientists*, 39: 27–45.
31. Swales, S. and Fish, J.D. (1986). Angling catch returns as indicators of the status of upland trout lakes. *Aquaculture and Fisheries Management (1986)*, 17: 75–93.
32. Turnpenny, A.W.H., Sadler, K., Ashton, R.J., Milner, A.G.P. and Lynam, S. (1987). The fish populations of some streams in Wales and Northern England in relation to acidity and associated factors. *Journal of Fish Biology*, 31: 415–434.
33. Welsh Water (1985). Working group report on Welsh salmon and sea trout fisheries. *Brecon*, 152pp.
34. Welsh Water (1986a). The survival and physiological responses of brown trout stocked in

two acidic upland lakes, before and after liming and an assessment of the performance of trout fisheries re-established in them. Fisheries Technical Unit, South West. Report No. SW/86/30. 46pp.

35. Welsh Water (1986b). Water quality and fishery status of the upper Lliw Reservoir. Fisheries Technical Unit, South West. Report No. SW/86/23.

36. Welsh Water (1986c). Regional acid waters electrofishing survey interim report. Fisheries Technical Unit, South West. 56pp.

37. Welsh Water (1987). Llyn Brianne Acid Waters Project. *First Technical Summary Report*, April 1987.

38. Wood, C.M. and McDonald, D.G. (1987). The physiology of acid/aluminium stress in trout. *Annales de la Société Royale Zoologique De Belgique*, 117, suppl.1, 399–410.

39. Zitko, V. and Carson, W.G. (1976). A mechanism of the effects of water hardness on the lethality of heavy metals to fish. *Chemosphere, 5*: 299–303.

9. Acid deposition – drinking water quality and health

C.B. BUCKLEY and L. KEIL

Introduction

In the area administered by WWA, 271 sources are treated to produce 1000 Mld^{-1} of potable water for public supply. These comprise 75 impounding reservoirs, 90 direct river abstractions and 106 groundwaters, the latter providing only 6% of total output. The particular physical, chemical and biological processes used at a treatment site for water from these sources depend principally on the quality and quantity of the source water (and its variability), together with the geology and topography of the area. Treatment works range from unpowered sites serving a few households in remote areas to major treatment facilities producing up to 120 Mld^{-1} of treated water supplying 0.6 million people and industrial needs.

The problems associated with the acidification of water supplies are related almost exclusively to sources in upland areas which are vulnerable to episodic stream acidification. Such sources are either direct river abstractions or small shallow impoundments – both types of source being particularly affected by water quality variations during periods of run-off. These sources are also used most frequently to serve small populations in comparatively remote areas. In consequence the areas most vulnerable to acidification of water supplies are frequently those with poorest treatment facilities. Many rural dwellings use private supplies which are at particular risk as they receive little or no treatment.

This chapter deals with the standards set by EC legislation for those parameters affected by acidification, in both 'source' and 'treated' waters, along with the public health implications of water quality changes. It also considers the adverse consequences of acidification on treatment processes and costs both for WWA and owners of private supplies.

Quality Standards

Of the EC Directives concerned with water quality standards, the first, relating to the quality of surface water sources (75/440/EEC), specifies the

R.W. Edwards et al. (eds) Acid Waters in Wales. 145–158.

physical, chemical and biological qualities of surface water before being abstracted for drinking water according to the type of treatment required. Sources are divided into three quality categories, and the treatment appropriate to each quality category is specified as follows: A1 – simple physical treatment and disinfection; A2 – normal physical and chemical treatment and disinfection; A3 – intensive physical and chemical treatment and disinfection. The directive lays down mandatory quality standards (I) and guide levels (G) for many determinands, by which the waters are categorised. In effect, if the treatment works is not sufficient to treat the water, either the water quality or the level of treatment, or indeed both must be improved.

In Wales, source waters require either A1 or A2 treatment. Whilst larger works may be capable of providing A3 treatment, no sources actually fall into this water quality category. The determinands which are most influenced by acidification are pH and alkalinity, colour, iron, manganese and aluminium. Of these, standards and guide levels are quoted for all but alkalinity and aluminium (Table 9.1). In Wales there are many A1 treatment works, or those nominally A2 but without an efficient pH correction stage, which are unable to treat acidic source waters that are in some cases far below the pH guide values of 6.5 and 5.5 for these two categories of treatment.

The second pertinent Directive relates to drinking water for human consumption (80/778/EEC) and is applied at the point of delivery, essentially the customer's tap. Table 9.2 shows both the maximum admissible concentrations (MAC) and guide values (G) for those determinands affected by acidification.

A recent change in the way the United Kingdom interprets the Directive has resulted in individual samples, rather than average values, being tested for compliance against the MAC. This will have major implications for existing and future water treatment strategies, especially for those works which cannot cope with wide fluctuations in source water quality and/or where the supply infrastructure requires refurbishment.

Where standards cannot be met, either because of natural geological or climatic factors, or because of temporary inadequacies in treatment or distribution networks, derogations may be granted provided these are not related to microbiological or toxic aspects of quality. Welsh Water has been granted derogations for 32 sources mostly associated with those determinands affected by acidification (pH, aluminium, iron and manganese) (Table 9.3).

Effects of acidification on specific water quality parameters

1. pH and alkalinity

Acidification is defined as the loss of alkalinity rather than the decrease in pH.[11] In those upland waters which are susceptible to acidification, alkalinity levels (bicarbonate) may be insufficient to achieve effective aluminium sulphate coagulation. Waters with conductivities of less than 30 umhos are likely

Table 9.1. Mandatory standards (I) and Guide Values (G) in the 'Surface Water' Directive for selected quality parameters influenced by acidification. A1 and A2 treatment facilities are referred to in the text.

Parameter	A1		A2	
	I	G	I	G
Mn (mgl^{-1})	–	0.05	–	0.1
Fe (mgl^{-1})	0.3	0.1	2	1
pH	–	6.5–8.5	–	5.5–9.0
Colour (Hazen)	20	10	100	50

Table 9.2. Maximum Admissible Concentrations (MAC) Guide Values (G) in the 'Drinking Water' Directive for selected quality parameters influenced by acidification.

Parameter	MAC	G
Al (mgl^{-1})	0.2	0.05
Mn (mgl^{-1})	0.05	0.02
Fe (mgl^{-1})	0.2	0.05
pH	6.5–9.5	6.5–8.5
Colour (Hazen)	20	1

Table 9.3. Number of derogations, some temporary, granted to Welsh Water Authority.

	Colour	pH	Aluminium	Iron	Manganese
Number of derogations	1	4	5	10	19

to be sensitive to acidification.[16] In addition, the close pH control required for effective coagulation is difficult to achieve because of pH measurement problems particularly in cold weather.

Standards aim to minimise corrosion and the release of toxic metals within the distribution system – problems associated with unbuffered and hence unstable waters of low alkalinity, below pH 6.5.[6]

The direct health effect of pH of drinking water is impossible to ascertain because of effects of pH on other aspects of water quality.[18] Failure to stabilize aggressive water will result in the dissolution of plumbing materials, lead being of particular significance.

2. *Colour*

Humic and fulvic acids are the main contributors to the type of source water colour which is becoming an increasing problem in acid sensitive areas of upland Wales, as indeed it is in much of the rest of upland Britain. An important factor is the drying out of peat beds in the drought years of 1975, 1976 and 1984. The exact role of acidification and the relationship between high colour and contemporaneous increases in iron, aluminium and manganese have not been fully explored. Colour levels are influenced by climatic and seasonal variations with peaks occurring in autumn and early winter.

The polyanionic electrolytic properties of humic substances play an essential role in the dissolution, transport and deposition of positively charged inorganic ions with complexation rendering iron up to one thousand million times more soluble.[18] They may also complex with organic compounds, adsorbing them in amounts which can exceed their aqueous solubilities. Fulvic acids are the lower molecular weight fraction of humic material and are highly substituted with carboxyl, phenolic and aliphatic hydroxyl groups, leading to an increased biological activity.

The toxicity of many organic compounds can be related to their solubility in octanol since it is thought to mimic the solubility characteristics of biological membranes. For many ionizable compounds, membrane penetration is only possible for the uncharged form. For naturally occurring humic phenolic compounds with pK's between 9 and 11 (ie weakly dissociated; $pK = \log_{10}\frac{1}{K}$) a reduction in pH increases the amount in the undissociated form and hence increases octanol solubility.[15]

The primary importance of colour in drinking water is aesthetic with standards based on water becoming objectionable at a colour level above 15 Hazen units. There is however a danger that customers who receive highly coloured water may seek an alternative, possibly unsafe supply.[18] Whilst not in itself a health risk, colour which is not removed by coagulation and filtration may react with chlorine to give trihalomethanes.[20] Drinking waters from a wide range of source water types have been found to contain complex mixtures of organic compounds. Humic material acts as a precursor for organochlorine formation and the mutagenic activity of drinking water has been largely associated with chlorination. However, the weight of epidemiological evidence indicates that drinking water is at most a very minor contributor to human cancer, if indeed it has any influence at all.[10]

Acidification may mobilise metal species which complex with humic substances to form compounds which are difficult to remove at the treatment works. Humic acids have been reported to reduce the toxicity of aluminium and copper by complexation but to increase the toxicity of cadmium.[15] It has been suggested that the ligand formation between cadmium and humus leads to increased solubility of the complex. The ligand, once inside the cell, is broken and the metal released.

3. Manganese

Manganese is present in natural surface waters in both the dissolved and suspended form. Manganese levels above 0.15 mgl^{-1} produce unsightly staining of laundered goods and plumbing fittings and can impart an unpleasant taste to drinking water and beverages.[18] Manganese compounds which are in solution precipitate out upon oxidation and at levels as low as 0.02 mgl^{-1}

Mn, may form pipe coatings which can slough off as black particles, rendering the water supply aesthetically unacceptable. Acidification can cause the release of manganese and iron from soil and lake sediments with a resultant increase in dissolved metals in the source water.

Standards are based on consideration of the staining properties of manganese, the aim being to keep the concentrations of the soluble form to a minimum. Manganese presents no known health risk at the levels currently observed in public water supplies throughout Wales.

4. *Iron*

Iron occurs in water mainly as soluble ferrous or insoluble ferric salts. Iron is present in surface waters mainly in the ferric state with the ferrous form being present only in those groundwaters, lakes or reservoirs where reducing conditions prevail. As well as naturally occurring iron which has not been removed by treatment processes, iron may occur in drinking water due to the use of iron salts in the coagulation process and also as a result corrosion of cast iron mains.

The presence of elevated levels of iron in drinking water is objectionable due to the staining of laundry/plumbing fittings and unacceptable taste. Iron deposits in distribution systems reduce the flow and promote both microbial growth and biological activity. Changes in flow patterns are likely to cause severe local water quality problems.

As with manganese, acidification can increase concentrations in source water by releasing insoluble iron from soils and lake sediments. Further problems may occur if fluctuations in water quality interfere with the coagulation process or result in a corrosive water entering supply.

Standards are based upon aesthetic problems occurring in water supplies when the iron concentration approaches 0.3 mgl^{-1}.[18] Iron presents no known health risks at levels currently observed in public water supplies throughout Wales.

5. *Aluminium*

Aluminium in drinking water can cause discoloration problems if the aluminium concentration in treated water exceeds 0.1 $mgl.^{-1}$ In the presence of aluminium, iron concentrations which would normally be acceptable may produce pronounced discoloration. Levels of aluminium above 0.3 mgl^{-1} in treated water entering the supply and distribution system generally reflect faults in the treatment processes at works using aluminium sulphate (alum) as a coagulant.

The concentration of aluminium in a particular source water is controlled by pH, the type and concentration of complexing agents that may be present, the oxidation state of mineral components, and the redox potential of the system.[18] Decreases in pH (<5.5) lead to the mobilisation of aluminium from the silicate matrix to the aquatic environment in the labile form. Al^{3+} and aluminium fluoride, hydroxide and sulphate account for most of the liberated aluminium.[5] Acidification may also interfere with the aluminium sulphate coagulation process, which is pH/alkalinity dependent, resulting in high residual aluminium levels in the product water.

The EC 'Drinking Water' Directive MAC of 0.2 mgl^{-1} is based on aesthetic considerations. The WHO guideline value (also 0.2 mgl^{-1}) is essentially a compromise which takes into account that discoloration may occur at this level but that lower levels may be difficult to achieve where alum is used as a coagulant.[18]

Aluminium is one of the most abundant elements in the soil, the majority occurring in the aluminosilicate lattice of particles. The link between acid deposition and health lies in its ability to release large amounts of aluminium into the environment in a more bioavailable form. In addition, episodic changes in raw water quality may interfere with the aluminium sulphate coagulation process resulting in reduced efficiency and increased levels of residual aluminium entering supply.

Aluminium is generally considered to be non-toxic to man. However, the fact that it is not involved in any biological systems, and is normally controlled at very low levels in the body masks its extreme toxicity to the central nervous system. In the 1970's aluminium from tap water used for renal dialysis was identified as the causative agent in dialysis dementia.[19] Point of use water treatment systems such as reverse osmosis are now used to reduce the aluminium level of water used for dialysis to less than 10 μgl^{-1}. Aluminium has been implicated in Alzheimer's disease, a form of pre-senile dementia which affects five to ten percent of those over 65 years of age.[7] A recent epidemiological study carried out by the Medical Research Council[12] has suggested that there is a correlation between the incidence of Alzheimer's disease and aluminium levels in drinking water, but no causal link has yet been established. Perl (pers.comm.) has suggested that aluminium may be toxic to certain high risk groups of the population if mechanisms which exclude or remove it from the central nervous system are faulty, if there is excess exposure to aluminium, if there is a uniquely toxic available form of aluminium, or a combination of these factors.

Should a causal link between aluminium ingestion and Alzheimer's disease be established, the case for stricter control of acidic emission would be strengthened; such a link would also have serious implications for the use of aluminium compounds in potable water treatment processes. It should, however, be noted that even with a concentration of 1.5 mgl^{-1} aluminium (c.f. 'Drinking Water' Directive MAC of 0.2 mgl^{-1}) it has been estimated that drinking water would provide less than 4 per cent of the total daily intake of aluminium.[7]

6. Lead

The main problem of treating low pH, poorly buffered water, whether naturally occurring or exacerbated by acid deposition, is that existing treatment facilities on many small rural works cannot adequately stabilize the pH of the treated water. Although there is little lead in either raw or treated water, failure to stabilize an aggressive water can result in dissolution of lead from communication pipes or lead domestic plumbing. Even where all lead has been removed, high lead levels can result from corrosion of solder joints. In 1974 a national review of lead in drinking water highlighted the problems associated with the supply of plumbosolvent water. The ensuing joint national survey published in 1977 showed that approximately 9.5% of houses in Wales (9% in Britain as a whole) were likely to draw water containing concentrations of lead in excess of 0.1 mgl^{-1}.

The 1980 Lead in Water Action Programme identified those treatment works in Wales (serving some 600,000 resident population) which were supplying plumbosolvent water. Their geographical distribution is plotted (Fig. 9.1), along with those works which received aggressive water (pH <6.5) but where plumbosolvency was not observed, due to either adequate stabilization or the absence of lead from domestic plumbing. In the North and South West of the area more than half the source water was plumbosolvent whereas in the South East of Wales only about 2% was potentially so. Welsh Water's policy is to adjust the pH of the water leaving the treatment works, coupled with ongoing monitoring to assess the effects of such remedial action, rather than to embark on full-scale removal of lead pipework. This remedial work, which will be completed in 1989, has the added benefits of generally inhibiting corrosion throughout the supply and distribution infrastructure as well as minimising dissolution of copper and zinc from the domestic plumbing system.

The EC drinking water standard for lead is an MAC of 0.05 mgl^{-1} in running water, with further action required on properties where a random daytime sample (without prior flushing) frequently, or to an appreciable extent exceeds 0.1 mgl^{-1}. The limit was calculated on the principle that tap water should provide less than 10% of the overall exposure to lead.[13]

The toxicity of lead and the health effects of its ingestion are well documented[18] but recently doubt has been cast upon the level of lead in blood which is considered safe. In the 1960's a level of 60 $\mu g\ dl^{-1}$ of blood in children was considered dangerous enough to warrant medical attention. Whilst this level has been reduced to 25 $\mu g\ dl^{-1}$ it is anticipated that a further reduction to a level as low as 10 $\mu g\ dl^{-1}$ may be necessary to protect foetuses and children.[2] If such a limit is imposed it will require the removal of domestic lead plumbing and any lead pipework connecting the mains to the domestic system in addition to stabilization at the treatment works. In 1981 a survey of Welsh homes estimated that there were 294,000 dwellings with lead

Fig. 9.1. Water-treatment works supplying plumbo solvent and aggressive water.

plumbing.[9] The implications would be of particular significance to private supply owners in areas where the water is likely to be aggressive.

Other elements associated with corrosion of domestic plumbing systems are copper, zinc and cadmium. Copper poisoning is rare due to the astringent taste it imparts to water above 5 mgl^{-1} and its powerful emetic action.[18] Only a small number of individuals who suffer from Wilson's disease are likely to be concerned over copper levels normally encountered, even in aggressive supplies.[14]

Concentrations of cadmium are generally low, rarely exceeding a few micrograms per litre. However, levels at the tap may be much higher if low grade zinc is used in galvanised equipment and storage tanks. Statutory water undertakings actively discourage the use of galvanised fittings in potable water systems. The contribution of solders which contain lead and cadmium should not be overlooked in areas supplied with plumbosolvent waters.[14]

Effects on treatment processes

The chemical form of substances present in untreated source waters, especially surface derived waters, affects the efficiency of their removal by treatment processes. For example, stable soluble complexes of metals with either organic or inorganic ligands may pass through the treatment process unaffected. Subsequent breakdown of such complexes within the supply and distribution system can lead to aesthetic problems at the point of use.

Under certain circumstances some metals form insoluble complexes with humic substances, this being the basis for the use of aluminium and iron salts as coagulants in the production of potable water. Coagulant is added to form an insoluble 'floc' which disrupts the colloidal state of impurities and attracts particles of suspended matter and associated bacteria and viruses, together with dissolved organics and minerals. This prepares the water for filtration. The soft acidic waters of upland Wales which are susceptible to deterioration by further acidic deposition are more readily treated with aluminium than iron based coagulants. The transformation of aluminium sulphate into insoluble aluminium hydroxide floc is a process which is mainly a function of the pH and alkalinity of the water being treated. If there is insufficient proton acceptor present, such as bicarbonate, the pH may fall below 4.5, resulting in failure to produce an acceptable floc. The process of transforming aluminium sulphate into aluminium hydroxide has a by-product of hydroxonium ions which marginally reduces the pH of the aqueous system. The use of chlorine at this stage in the treatment process, either as an oxidant or disinfecting agent, will also depress the pH of the water. It may be necessary to add lime, lime and carbon dioxide, or bicarbonate to raise the pH and alkalinity sufficiently to achieve effective floc formation. Inadequate control of coagulation may result in poor floc formation, under-dosing, over-dosing or floc penetration of the filters, all of which may lead to high aluminium levels in the distributed water.

During the coagulation stage of water treatment the removal of colour from soft acidic water is mainly dependent upon adsorption rather than coagulation and therefore aluminium sulphate must be thoroughly mixed with the untreated source water before pH and alkalinity adjustment produces a floc. This avoids the formation of stable neutral colour complexes, which are both difficult to treat and lead to water quality problems in the supply and distribution system. Coagulation preferentially removes the coloured fraction of humic substances because they are highly substituted with carboxylic groups with which metal ions readily interact.[17] For almost complete colour removal as little as fifty percent of the total organic carbon and chemical oxygen demand may be removed.[1]

Chlorination prior to filtration may be beneficial for colour removal or in aiding coagulation but has the disadvantage of producing trihalomethanes in the treated water leaving the works. These compounds and any organic precursors not removed during treatment may continue to react with free chlorine in the distribution system resulting in water of an unacceptable quality.

In the pH range 5 to 6, aluminium sulphate coagulation and filtration will effectively remove colour, aluminium and iron from the surface water. However, a second filtration stage with the pH in excess of 7.5 is required for the removal of manganese. The release of manganese from the solid to the aqueous phase by acidification may therefore present considerable treatment problems. There are several well established techniques for the removal of manganese at the treatment works but all are difficult to control effectively. Dickson[3] has recently described the successful liming of a Swedish lake to combat manganese problems at source. Since 1975, the lake has been limed every third year and precipitation has occurred on the lake bottom instead of at the consumers' premises. This approach has had the added benefit of successful hatchings of pike and perch immediately after the first addition of lime following about ten years of failure. Similar encouraging results have been observed on a lake serving a water treatment works in mid Wales.[4] It should be noted that autumn or spring 'overturns' could redistribute manganese in the future, or liming prior to water treatment could possibly result in 'fixing' of humic colour, the consequences of which have been described earlier.

Private supplies

The quality of private supplies is the responsibility of the Local Authority's Environmental Health Department and is required to comply with the Drinking Water Directive. No accurate data are available for the number of private supplies in the UK but the most recent estimate of 80,000 private supplies in England and Wales is considered to be low. In Wales approximately 20,000 private supplies have been identified, 15,000 of these being situated in the acid sensitive areas (McDonald, pers. comm.). The majority of these are small supplies serving single properties and the UK Government has granted a ten year relaxation period before compliance with the Drinking Water Directive becomes mandatory. An estimate of the size distribution of private supplies using the four categories defined in the 'Drinking Water' Directive shows that whilst most serve one household only, there are some which are of comparable size to Welsh Water's smaller treatment works (Table 9.4).

Table 9.4. Private supplies – Classification by size of private supplies (see Drinking Water Directive) in comparison of Welsh Water's treatment works.

	Number of supplies	
Population served	Private	Welsh Water
I >500*	565	165
II 50–500	69	49
III <50 (except IV)	568	18
IV 1 property	18404	1

* Whilst most class 1 private supplies serve populations of 500–1000, Welsh Water has several works serving over 100,000 people.

Table 9.5. Private supplies in acid sensitive areas – projected percentage compliance against MAC.

	Percentage compliance against MAC	
	As average value	As maximum value on individual sample
Mn	61.5	16.1
Fe	90.9	32.9
Al	95.2	44.9
pH	32.7	1.4
colour	–	–

The private supplies at most serious risk from acidification are those with little or no treatment. In order to provide an assessment of the likely effects on their quality the compliance of data from the 'Regional Survey of Acid Waters' has been tested against the 'Drinking Water' Directive using both interpretive procedures for MAC (average and individual samples), (Table 9.5). Compliance for pH and manganese is poor. This is most marked when the MAC refers to individual samples, with almost all sources failing on the criterion of pH.

Whilst bacteriological/viral contamination of private supplies is the most serious threat to public health, the effects of an unstabilized water on domestic plumbing is also of major concern. Almost all of the waters monitored in the 1984 survey regularly have pH values below 6.5 with aggressive indices of less than 10 (i.e. highly aggressive). Such waters would require pH correction to prevent plumbosolvency and cuprosolvency. There are many examples of untreated waters having necessitated the complete replacement of domestic plumbing systems within a few years of installation. For the small supply a pH adjusting filter and ultraviolet light disinfection unit should provide adequate treatment.

Cost implications of acidification

Welsh Water Authority

It is difficult to separate the effects of acidification from seasonal or climatological fluctuations in water quality. Consequently, the assessment of additional treatment costs, due solely to acidification, cannot at this stage be quantified.

The most obvious cost of dealing with acidification is in stabilization of the water using either lime or dolomitic filtration media. Those works which the 1980 lead survey identified as requiring further action in relation to plumbosolvency already have lime dosing equipment or dolomitic filtration media installed or planned and the short term effects of acidification should require only additional chemical usage. Current costs of producing water are approximately 0.1 p m^{-3} for lime and 0.4 to 0.8 p m^{-3} for dolomitic filtration media. A simple calculation on increased lime costs necessitated by acidification[8] suggests that for those works within the acid sensitive area (Fig. 9.1) the chemical cost for the 270 Mld^{-1} of water involved would total approximately £3,500 p.a. for each 1 mgl^{-1} decrease in alkalinity of this source water. The Swedish experience has been a drop in alkalinity of 30 mgl^{-1} over twenty years and if acidification on this scale occurred in Wales the yearly cost for stabilisation chemicals alone would rise by £100,000 p.a. (at 1988 prices).

Such a long term deterioration in water quality could require major refurbishment or even works replacement if EC quality standards for drinking water are to be met. Sophisticated treatment processes for automatic pH and coagulation control and colour, iron, manganese and aluminium removal would be required. The recently built Carno works near Ebbw Vale in South Wales is an example of the costly facilities needed to deal with a problematical acidic source water. The former single stage rapid gravity works could not cope with the naturally acidic, highly coloured upland lake water. Facilities at the new site include air flotation, filtration to remove colour and a second filtration stage to remove manganese.

Private supplies

Acidification may have considerable cost implications for private supply owners in acid sensitive areas. Some form of pH correction will be essential to meet the requirements of the EC 'Drinking Water' Directive and to protect public health. The simplest system of pH correction and disinfection is likely to cost several hundred pounds, with larger private supplies requiring more sophisticated and costly treatment facilities. Private supply owners will also have to consider monitoring to ensure that 'Drinking Water' Directive compliance is achieved. Depending on the size of supply this could vary from £50 to £1000 p.a. With 20,000 private supplies such monitoring costs seem likely to be no less than £1–5 million p.a. – although these costs could not be ascribed specifically to acidification.

Conclusions

Water treatment in Wales reflects the predominantly rural nature of the Principality. Whilst there are several major treatment works supplying urban conurbations the majority are small with approximately 50% supplying populations of less than 5,000.

Acidification is generally considered in the light of its environmental impact but it undoubtedly has a significant role to play in water treatment. Although considerable progress has been made with the installation of instrumentation, control and automation (ICA), many smaller works still cannot cope adequately with seasonal and climatic water quality variations. The extent to which acidification is involved in such variations is extremely difficult to assess but it is likely that in the acid sensitive areas of Wales it affects process efficiency and leads to increased operating costs.

Whilst compliance with the EC 'Surface Water' and 'Drinking Water' Directives affords protection of drinking water supplies, there is need for an immediate policy review regarding alkalinity and aluminium. Alkalinity is a key indicator of acidification, it plays an important role in coagulation efficiency and strongly influences water stability. It is not included in the 'Surface Water' Directive and is mentioned only with regard to water softening in the 'Drinking Water' Directive. Aluminium is mobilised by acid deposition and accordingly it should be included in the 'Surface Water' Directive in order that any long term changes are adequately monitored.

In the short term, acidification of source waters which are treated by WWA is most unlikely to cause any adverse health effects to the customer. Nevertheless if current water quality standards are to be maintained in the longer term then considerable capital and revenue costs may be incurred.

The effects of acidification on private supplies are more readily identifiable as many have inadequate treatment or none at all. Approximately 15,000 such supplies are found in the acid sensitive areas and it is essential that remedial action is implemented to reduce the problems associated with corrosion.

References

1. Connor, K. (1986). WRc Processes, Information Sheet 35.
2. Davis, J.M. and Svendsgaard, D.J. (1987). Lead and child development. *Nature*, 329: pp. 297–300.
3. Dickson, W. (1986). Which are the acceptable loadings of atmospheric acids to sensitive waters? *Water Quality Bulletin*, 11: No. 1, pp. 39–43, 61.

4. Donald, A.P. (1988). The effects of lake liming on the raw and treated water quality of upland sources in West Wales. Welsh Water, Scientific Services Report.
5. Driscoll, C.T., Baker J.P., Bisogni, J.J. and Schofield, C.L. (1980). Effects of aluminium speciation on fish in diluted acidified waters. *Nature*, 284: pp. 161–164.
6. Faust, S.D. and Aly, O.M. (1983). Chemistry of Water Treatment, p. 52, Butterworths.
7. Ferry, G. (1987). The genetic roots of dementia. *New Scientist*, 12th March pp. 20–21.
8. Grimvall, A., Cole, C.A., Allard, B. and Sanden, P. (1986). Quality trends of public water supplies in Sweden. *Water Quality Bulletin*, 11: No. 1, pp. 6–11, 58.
9. Grist, I. (1987). Hansard, 10th July.
10. Healey, M. (1987). Research into water quality and health – a Government view. Water Bulletin, No. 253, pp. 8–9.
11. McDonald, M.E. (1985). Acid deposition and drinking water. *Environmental Science and Technology*, 19: No. 9, pp. 772–776.
12. M.R.C. (1988). Medical Research Council Study Unit, Southampton. Unpublished Work. Water Bulletin, No. 313, p.3.
13. Moore, M.R. (1985). Influence of acid rain upon water plumbosolvency. *Environmental Health Perspectives*, 63: pp. 121–126.
14. OECD (1982). Control policies for specific water pollutants, p. 120.
15. Petersen, R.C. Jr., Petersen, L.B.M., Persson, V., Kullberg, A., Hargeby, A. and Paarlberg, A. (1986). Health aspects of humic compounds in acidic environments. *Water Quality Bulletin*, 11: No. 1, pp. 44–49, 62.
16. Reed, J.S. and Henningson, J.C. (1984). Acid precipitation and drinking water supplies. *Journal of the American Waterworks Association*, 76: No. 3, pp. 60–65.
17. Van Breemen, A.N., Nieuwstàd, Th.J. and Van Der Meent-Olieman, G.C. (1979). The fate of fulvic acids during water treatment. *Water Research*, 13: pp. 771–779.
18. WHO (1984). Guidelines for drinking-water quality, Vol. 2, Health criteria and other supporting information.
19. Wing, A.J., Brunner, F.P., Brynger, H., Chantler, C., Donckerwolcke, R.A., Gurland, H.J., Jacobs, C., Kramer, P. and Selwood, N.H. (1980). Dialysis dementia in Europe. *The Lancet*, 26th July, pp. 190–192.
20. WRc (1986). Health aspects of organics in drinking water, TR 231. Water Research Centre.

10. Ecotoxicological studies of acidity in Welsh streams

N.S. WEATHERLEY, C.P. McCAHON, D. PASCOE and
S.J. ORMEROD

Introduction

Faunal distributions in soft waters are often related to total hardness, pH and filterable aluminium (Chaps. 2, 7). These relationships may be explained by one or more of the following hypotheses:

1. Acidity or related aspects of water chemistry exert direct physiological effects which limit species distributions.
2. Species are indirectly influenced by interactions with other organisms which are themselves limited by direct physiological effects.
3. The relationship with water chemistry is spurious, because species are controlled by physiographic effects which correlate with water chemistry.

The purpose of ecotoxicological studies is to examine these hypotheses by identifying the mechanisms through which they operate. This should assist management in determining appropriate remedial action, for example the control of certain water quality parameters.

As in most pollution problems, acid-aluminium toxicity may involve chronic and/or acute stress. Due to the extended exposure periods needed and the more subtle effects involved, chronic stress of low pH has rarely been studied. This paper is concerned with the effects of short term 'acid episodes', conditions often associated with heavy rain or snowmelt, and characterized in streams by increased flow, decrease in pH and increase in aluminium concentrations. In Scandinavia and North America these conditions have been shown toxic to some species of fish and invertebrates.[19, 27] Thus acid episodes may be critical to the maintenance of populations in soft-waters, or in artificially limed waters where episodes often persist.

In Welsh hill-streams, the typical duration of storm episodes of about one day is a suitable time period for experimental simulation. In an attempt to copy natural events as closely as possible, a series of acid episodes was simulated in the field and the biological responses were monitored. It was hoped that this approach would provide more valid results than laboratory tests under artificial conditions. Over the course of three experiments, one per year from 1985 to 1987, the effects of low pH, low pH with high

R.W. Edwards et al. (eds) Acid Waters in Wales. 159–172.

aluminium concentrations (both labile and organically complexed), and lime-treated water were compared.[30, 31, 35, 37, 44, 45]

Experimental protocol

In each experiment a second or third order stream was divided into contiguous zones approximately 100–200m in length. Using the upstream zone as an untreated reference, successive downstream reaches were treated with sulphuric acid, and with sulphuric acid plus aluminium sulphate. In the second experiment, a fourth downstream zone was treated with citric acid, intended to organically complex the aluminium. In the third experiment, the fourth zone was dosed instead with powdered limestone, in order to assess the effectiveness of this technique in mitigating acid and aluminium toxicity. All treatments lasted for 24h, with stream flow, chemical and biological effects being monitored for 24h before, during and after treatment. In 1985 and 1986 the experiment was performed on a circumneutral stream, designated stream 6 in the Llyn Brianne study. Stream 3, used in 1987, was naturally acidic, so test animals were also held in the neutral stream 5 for reference purposes. Chemical data from the three experiments are summarized in Table 10.1. Unless otherwise stated, aluminium concentrations quoted in the text are filterable at 0.45μm. The 1986 experiment coincided with a spate, in which flow increased from 0.2 m^3s^{-1} to a peak of $0.7m^3s^{-1}$, so hydrologically simulating conditions which normally occur during acid episodes.

In addition to observations on the natural benthic fauna (drift and benthic density changes), several species of macroinvertebrates were held in the stream for estimation of survival, aluminium accumulation, haemolymph sodium concentrations, and feeding response. For fish, survival, gill aluminium concentrations, ventilation and blood physiology were monitored.

Acute mortality and feeding response

Of 4 fish and 15 invertebrate species tested, some showed negligible mortality in any of the experimental treatments: *Salvelinus fontinalis* (brook charr), *Chironomus riparius* (true fly), *Ephemerella ignita* (mayfly), *Amphinemura sulcicollis (stonefly), Hydropsyche angustipennis*, *Plectrocnemia conspersa*, *Rhyacophila dorsalis* (caddisflies). Acidification to pH 4.9 (0.05 mg Al l^{-1}) for 24h caused few mortalities amongst invertebrates or fish. More extreme acidification to pH 4.3 (0.05 mg Al l^{-1}) induced mortalities of up to 25% in *Baetis rhodani*, *Ecdyonurus venosus* (mayflies) and *Gammarus pulex* (shrimp), and up to 10% in *Salmo trutta* (brown trout) and *S.salar* (Atlantic salmon).

At pH 4.5–5.0, elevated aluminium levels (0.27–0.67 mg l^{-1}) caused similar invertebrate mortalities to those at pH 4.3 with low aluminium concentrations

Table 10.1. Selected physico-chemical data during experimental treatments. Temperature = overall mean; Ca (mg l^{-1}) = lowest mean of any zone; Al (mg l^{-1}) = filterable at 0.45μm, values in parentheses = ion-exchangeable Al ('labile').

Date	Temp. (°C)	Ca	Reference		Acid		Aluminium			
			pH	Al	pH	Al	pH	Al	pH	Al
5–8/9/85	9.2	2.3	7.0	0.05	4.3	0.05	5.0	0.35		
									Citrate	
5–8/8/86	10.7	3.1	7.2	0.02	4.9	0.05	4.9	0.27	4.9	0.23
				(0.00)		(0.01)		(0.11)		(0.00)
									Lime	
2–5/6/87	9.1	1.8	5.0	0.32	4.5	0.40	4.5	0.9	7.2	0.13

(*B.rhodani*, *Rhithrogena semicolorata* (mayflies); *Dinocras cephalotes*, *Isoperla grammatica* (stoneflies); *Hydropsyche instabilis* (caddisfly)), but much higher salmonid mortalities. At pH 5.0 with 0.35 mg Al l^{-1}, LT50 values were 23.5h for *S.trutta* and 14.7 h for *S.salar* (Fig 10.1). Salmon in their second year, 8 months pre-smolt, were more sensitive than O-group fish (C.P. McCahon & D. Pascoe, unpub. data), and adult *Cottus gobio* (bullhead) had greatly reduced tolerance in the pre-spawning period (C.P. McCahon, unpub. data). The importance of aluminium speciation was indicated by the substantially reduced acute toxicity to both invertebrates and fish when labile aluminium was organically complexed by citrate. However, this treatment may also have reduced oversaturation of aluminium hydroxide, providing an alternative explanation for the reduction in toxicity. Lime treatment reduced the mortality rate of salmon, and virtually eliminated mortality in the other fish species and *G. pulex*.

In addition to interspecific variation in tolerance, there were also differences related to the origin and prior exposure to pollutants of the test animals.[31] Although these phenomena are potentially of great importance, the results here were not consistent enough to provide reliable conclusions. Nevertheless, an observation of related interest was that *G.pulex* infected with the acanthocephalan parasite *Pomphorhynchus laevis* were less tolerant of high aluminium levels at pH 4.5 than uninfected individuals (M. Poulton, pers. comm.).

Sub-lethal effects of acid and aluminium exposure were investigated by measuring the feeding rate of *G.pulex* on leaf discs. Feeding rates were substantially lowered during exposure to pH 4.5 with raised aluminium concentrations; lime treatment failed to counteract this effect (M. Poulton, pers. comm.).

The mortality data summarized here indicate that over 24h hydrogen ions alone become acutely toxic to invertebrates and salmonids when the pH decreases to between 4.3 and 4.9. Within the pH range 4.5–5.0, high aluminium levels are toxic when labile aluminium is present, or when aluminium

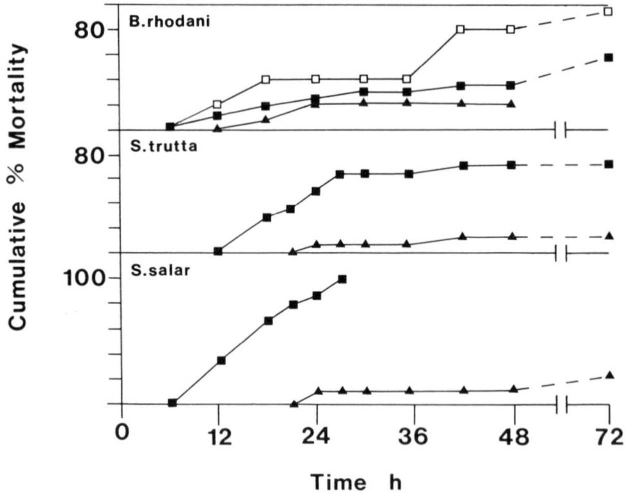

Fig. 10.1. Cumulative percentage mortalities of *B. rhodani*, *S. trutta*, and *S. salar* exposed to pH 4.3 (▲), and pH 5.0 with 0.35 mg Al l^{-1} with (□) or without (■) pre-exposure to 0.05 mg Al l^{-1}.

hydroxide is precipitated. Toxicity varies with species, life stage, previous exposure and presence of additional stressors.

Pathology

Fish toxic responses to acid and aluminium have been associated with the production of excess mucus, especially on gill surfaces.[15] There is also evidence for the accumulation of aluminium in fish gills and the tissues of invertebrates.[20, 42] In the 1985 experiment these phenomena were investigated using staining techniques for mucus and aluminium, and acid digestion of fish gills and invertebrates for aluminium determination.[30]

At pH 4.3, fish gills showed low mucus production and did not stain for aluminium, although concentrations of aluminium in the gills were significantly elevated (Table 10.2). At pH 5.0 with 0.35mg Al l^{-1}, staining revealed much mucus and aluminium on the secondary gill lamellae of both brown trout and salmon. Concentrations of aluminium in the gills were also greatly increased in these conditions (Table 10.2). Although no salmon survived, trout lost most of the aluminium 72h after being returned to circumneutral water low in aluminium. Enhanced mucus production and aluminium accumulation in the gills at pH 5 with elevated aluminium concentrations corresponds with the higher mortality in this treatment than at low pH alone. The absence of correlations between aluminium concentrations in gill tissue and

Table 10.2. Aluminium concentrations ($\mu g\ g^{-1}$ dry wt.) in the gills of fish and whole body of mayflies in each zone ± standard deviation, with numbers of replicates in parentheses.

Zone	Trout	Salmon	Trout post-exp	Salmon post-exp	*Baetis*	*Ecdyonurus*
Ref.	50 (1)	75 (1)	—	—	175 (1)	125 (1)
Acid	1125±175 (3)	700 (1)	800 (1)	550±200 (2)	1200 (1)	500 (1)
Acid + Al	2950±300 (4)	3050±225 (6)	225 (1)	—	1200±100 (2)	3175 (1)

time of death may be explained by a lethal threshold of accumulation, variation in individual response, or by aluminium accumulation not being closely related to the cause of death.

The mayflies, *B.rhodani* and *E.venosus*, showed no mucus production in either treatment and did not stain for aluminium at pH 4.3, although the concentrations of aluminium in the whole body were increased at this pH (Table 10.2). At pH 5 with aluminium, staining showed aluminium over the whole body in both species, and concentrations of aluminium were also greatly increased. However, there was no clear relationship between whole body concentrations of aluminium and cumulative mortality.

These observations suggest a relationship between acute toxicity and the accumulation of aluminium on fish gills. For invertebrates the relationship between toxicity and aluminium accumulation is less clear. At pH 5, toxicity is apparently due to aluminium, whilst at pH 4.3 toxicity is probably caused by hydrogen ions alone.

Ventilation

Gas exchange in invertebrates occurs over much of the cuticle, and the observed aluminium deposition may interfere with this process by physical occlusion of the body surface.[24] In fish, similar effects on the gills are probably exacerbated by the extrusion of mucus onto their surface.

The ventilation response of brown trout was investigated using an aluminium and fish monitor deployed at the experimental site.[37] The monitor holds four fish, recording their ventilation frequency and total muscular power at 1 minute intervals, whilst measuring aluminium concentrations every 20 minutes.

At pH 5.1 (0.002mg Al l^{-1}) fish showed a slight increase in ventilation frequency, accompanied by a 30% loss of power, whilst at pH 4.2 (0.019mg Al l^{-1}) the frequency became very erratic and power fell by 70%. An aluminium concentration of 0.4mg l^{-1} at pH 4.2 caused no change from

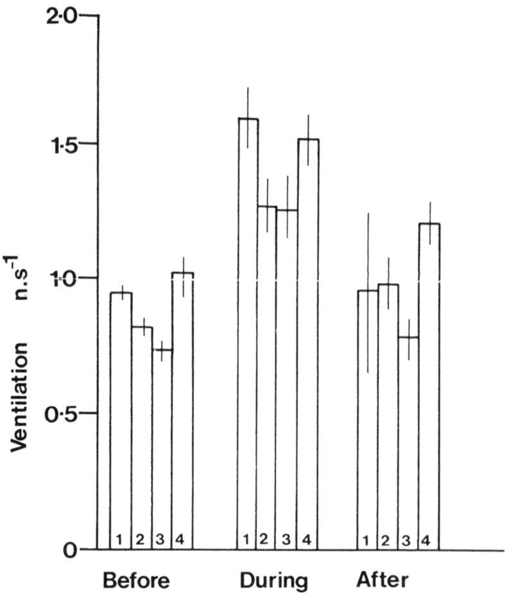

Fig. 10.2. The ventilation frequency (± 1 S.D.) of four brown trout, 24h before, during and after exposure to 0.27 mg Al 1^{-1} at pH 4.9.

pre-exposure ventilation frequencies despite a 70% loss of power. At pH 4.9–5.1 individual fish responded to raised aluminium concentrations by increasing their ventilation frequency and power (Fig. 10.2). In general, ventilation parameters returned to pre-treatment levels within 24h after the pollution episode, corresponding with the observed loss of aluminium from the gills.

A detailed interpretation of the changes in ventilation frequency would require data on other respiratory parameters, such as ventilation volume. In addition, it is not clear whether the power responses were related to ventilation alone or included changes in locomotory activity. However, it seems reasonable to assume that observed changes were signs of stress, and therefore that a sub-lethal response occurred at pH 5.1 (low aluminium) in the absence of mortality. The more severe reaction at pH 4.2 reflected the mortality rate around this pH. High aluminium levels at pH 4.2 partially mitigated the ventilation response, concurring with results of published toxicity tests.[33] The strong, if variable, ventilation response to aluminium around pH 5 is also consistent with the high mortality in these conditions.

Ion regulation and blood physiology

The gills of fish play a major role in ion exchange between the blood and surrounding water. This function is particularly sensitive to environmental

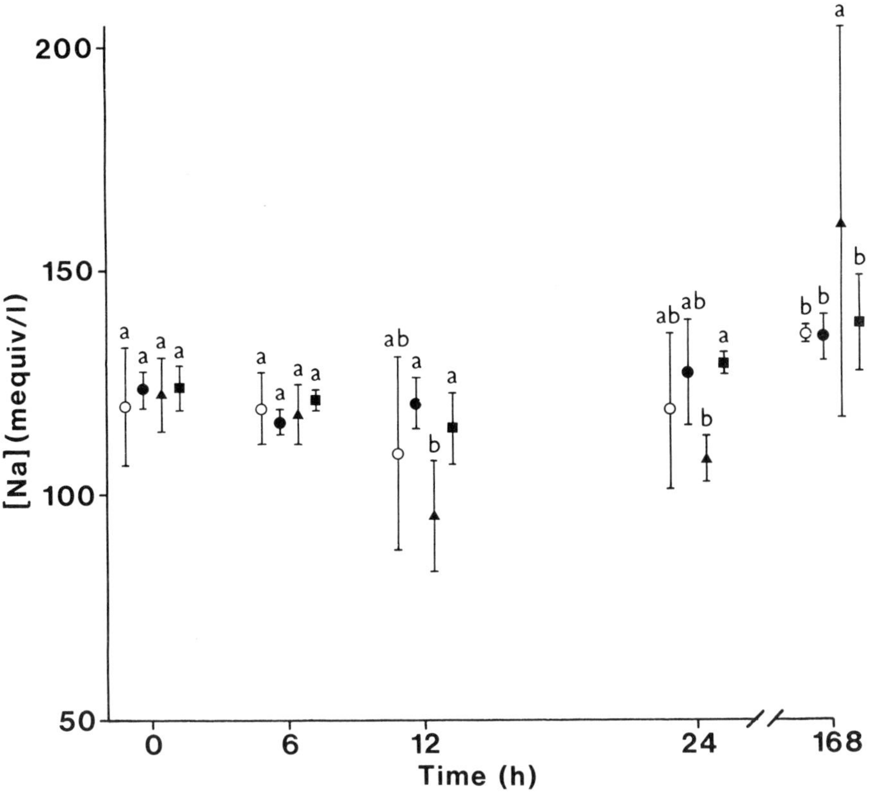

Fig. 10.3. Means and 95% confidence limits of plasma [Na^+] of brown trout exposed to pH 7.2 (○); pH 4.9 (●); 0.27 mg Al l^{-1} at pH 4.9 (▲); 0.27 mg Al l^{-1} with 0.5–0.7 mg citric acid l^{-1} at pH 4.9 (■). Significant differences between zones at the same time (as derived from ANOVAR and Tukeys range tests), occur where individual letters (a, b) are not common to two zones.

stressors[12], and is also likely to be affected by the changes in ventilation parameters outlined above. Sodium regulation in particular has been the focus of much research, plasma [Na^+] being used as a measure of stress. Variations in plasma [Na^+] were monitored using blood samples taken from the caudal vein of brown trout.[37]

After exposure to pH 4.9 (0.01 mg Al l^{-1}) for 12h, plasma [Na^+] was unchanged, but was significantly reduced at pH 4.2 (0.02 mg Al l^{-1}). Plasma [Na^+] decreased within 12h at pH 4.5–5.1 with raised concentrations of aluminium (0.27–0.67 mg l^{-1}) and still exhibited high variance six days after the episode (Fig. 10.3). Addition of citrate eliminated effects on plasma [Na^+], indicating that labile aluminium or precipitating aluminium hydroxide were the toxic forms. Increasing the pH to 7.2 by lime addition also prevented changes in plasma [Na^+], suggesting that precipitation of aluminium hydroxide may not have been important, as this probably occurred in the liming zone. At pH 4.2, 0.4mg Al l^{-1} mitigated the effect of hydrogen ions alone on plasma [Na^+].

Invertebrates face ion regulation problems similar to those of fish, and effects on haemolymph [Na^+] in the crayfish, *Austropotamobius pallipes* were also assessed. After 24h exposure to high aluminium concentrations at pH 4.5–5.0, this species showed a reduction in haemolymph [Na^+] which was alleviated in the liming zone.[45]

In fish, the endocrine system is activated by stress,[10] and in the 1987 experiment, plasma cortisol, thyroxine and glucose concentrations in brown trout were measured (C. Whitehead and J.A. Brown, pers. comm.). High aluminium levels at pH 4.5–5.0 caused hyperglycaemia and increases in plasma cortisol and thyroxine, which were partially mitigated by liming. The cortisol increase is likely to be a factor in the generation of hyperglycaemia. This is a typical response to stress, which mobilizes energy reserves and in this case may help to counteract a fall in plasma osmolarity due to ion loss.[14]

In summary, the effects described above follow a similar pattern to the mortality and ventilation data. It seems probable that stress to ionic and respiratory regulation are both major pathways in the toxicity of acid and aluminium to invertebrates and fish.

Drift and benthic density of invertebrates

The ultimate problem to be resolved by the foregoing studies is whether the species concerned can sustain viable populations when subjected to acid-related stress. This was investigated on the short-term basis of episodes by measuring the downstream loss of indigenous invertebrates (drift), and the differences in their benthic densities before and after treatments.[35, 44]

The drift density (number m^{-3}) of Simuliidae (true flies) increased at pH 4.3 (low Al), but at pH 5 only increased when high aluminium concentrations were present. *B.rhodani* was affected at both pH levels, but most strongly when aluminium levels were elevated at pH 4.9–5.0, showing geometric mean increases in drift density of 5.7–8.4 × (Fig. 10.4). Other taxa (*Dixa puberula*, *Dicranota* sp. (true flies); *Protonemura meyeri* (stonefly); *Ephemerella ignita* (mayfly) only increased in the drift at pH 5 when aluminium levels were high. *B.rhodani* was the only taxon to change significantly in benthic density, decreasing by 64% and 78% at pH 4.9–5.0 with 0.27–0.35 mg Al l^{-1} (Fig. 10.4), and by 89% at pH 4.9 with organically complexed aluminium. In 1986, it was notable that drift densities were not significantly affected by the increased flows of the spate, which would normally accompany an acid episode, but were affected by acid conditions.

Increases in drift density could be the result of mortalities, debilitation or behavioural avoidance. Whatever the mechanism, the populations of *B.rhodani* in the experimental stream would rapidly become extinct if natural episodes approximating the simulated event occurred frequently. Despite the

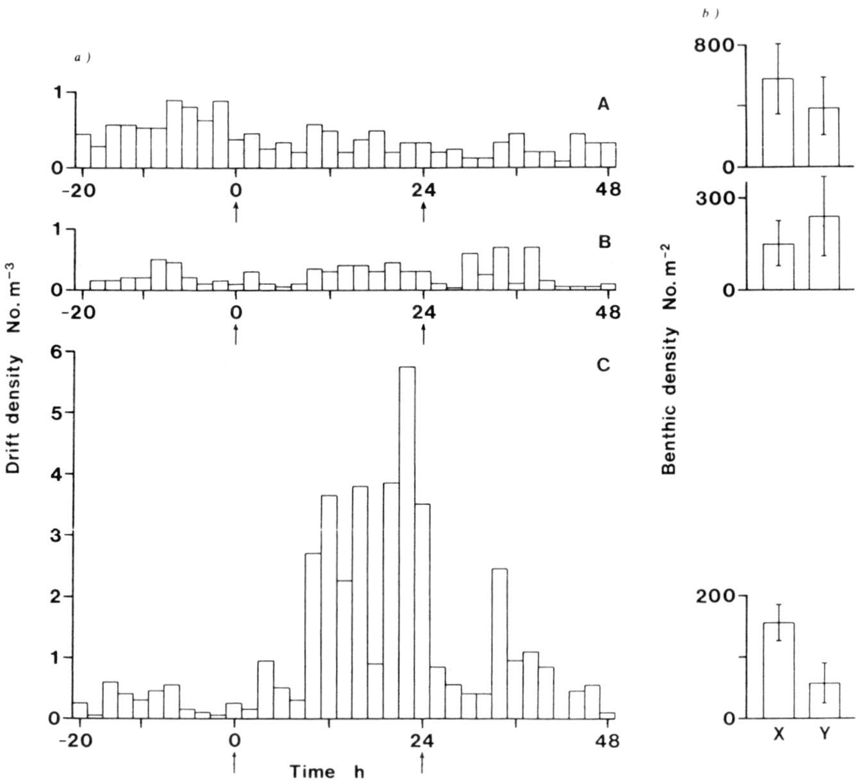

Fig. 10.4. a) Drift density of *B. rhodani* in pH 7.0 zone (A), pH 4.3 zone (B) and pH 5.0 with 0.35 mg Al l^{-1} zone (C). Arrows indicate start (0 h) and end (24 h) of treatments. *b)* Benthic densities (± 1 S.D.) of *B. rhodani* in zones A,B and C before (X) and after (Y) treatments.

absence of mortality, an increased drift of *E.ignita* was demonstrated, which could account for its absence in acidic streams. However, the enhanced drift of *P.meyeri* and *Dicranota* sp. was not consistent with their presence in acidic streams.[43]

Discussion

These studies and several others[23, 47] demonstrate that low pH and high aluminium concentrations are toxic to certain species of fish and invertebrates. It is important to understand these effects so that the responses of fauna to acid-related conditions can be predicted and their effects mitigated.

For fish, hydrogen ion toxicity involves the penetration of H^+ into the blood, with consequent changes in acid-base balance and disruption of oxygen transport.[17] As observed in this study, high aluminium concentrations at around pH 5 exacerbate respiratory stress due to accumulation of the metal coupled with mucus production at the gill surface. These processes presumably lower the diffusion capacity of the gills and impede ventilation.[15, 34] The mucus may aid removal of aluminium, and thus account for its rapid loss when exposed fish are returned to water low in aluminium. As shown here, exposure of fish to elevated hydrogen ion and aluminium concentrations around pH 5, is consistently associated with loss of body Na^+, Cl^- and K^+.[47] This may partly result from hyperventilation with increased blood flow to the gills,[15] and from physical damage to the gill epithelia by hydrogen ions.[48] High hydrogen ion concentrations may also affect sodium exchange by competing with Na^+ for carrier sites, increasing the gradient of H^+ between blood and water (against which blood H^+ is exchanged for Na^+ from the water), and by making the trans-epithelial potential more positive, again hindering H^+–Na^+ exchange.[17, 48] The hyperglycaemia observed in the present experiments was probably a response to help counter the effects of stress.

Although elevated concentrations of aluminium in gill tissue are related to toxicity,[42, 48] the mechanisms are unclear. Aluminium may exacerbate the effects of hydrogen ions by physical blockage and by reducing activity of gill Na-K-ATPase, the enzyme responsible for Na^+ uptake.[48] From theoretical calculations Fivelstad and Leivestad[15] concluded that $Al(OH)^{2+}$ and $Al(OH)_2^+$ were the most toxic species, although speciation at the gill–water interface may well shift from that in the ambient water.[21, 34] Certainly, the toxic role of labile aluminium is well supported,[4, 11, 15] and when this form was eliminated in the present study, by complexation with citrate, there was a reduction in toxicity. However, aluminium additions in our experiment may have resulted in oversaturation and precipitation of aluminium hydroxide, an effect which may also have been reduced by citrate complexation. Moreover, there are clear examples of fish kills resulting from the mixing of acidic aluminium-rich water with circumneutral water, when precipitating aluminium hydroxide was the predominant form,[9] although $Al(OH)_4^-$ may also be toxic at high pH.[28] At about pH 4 when aluminium mitigates H^+ toxicity,[33] Al^{3+} is dominant. This species could compete with H^+ for branchial binding sites, acting in a similar way to Ca^{2+} which reduces membrane permeability to H^+ and Na^+, and hence mitigates acid toxicity.[5, 32]

Consistent with work on Welsh lakes[46] and elsewhere,[40] this study demonstrates that H^+ and Al effects can be alleviated by liming, which raises the pH and concentrations of Ca^{2+}.

The mechanisms of acid-aluminium toxicity to invertebrates may be analogous to those in fish. There are examples of aluminium reducing hydrogen ion toxicity at pH 4–4.5,[21] although there was no evidence for such an effect here, or in other studies.[1, 7] Aluminium is generally most toxic near pH 5,[23] and as in fish, impairment of gas exchange and ion regulation have been suggested as the two main toxic pathways.[24]

Oxygen transport may be disrupted by haemolymph acidosis,[24] and aluminium accumulation could impede gas exchange. These factors could lower respiration efficiency, thus explaining altered oxygen consumption rates amongst invertebrates.[8, 24] However, it is not clear whether the deposition of aluminium compounds over the exoskeleton observed here occurs naturally in acid waters, as no correlations have been found between concentrations of aluminium in stream water and those in indigenous invertebrates.[41] Moreover, Havas[20] recorded decreasing aluminium accumulation at lower pH.

The loss of haemolymph sodium by *A. pallipes* on exposure to low pH with aluminium has been observed in other sensitive crayfish species,[2, 25, 49] and is mainly due to a fall in sodium influx rates.[49] Havas and Likens[21] found that at pH 5, aluminium also caused a reduction in sodium influx in the water flea *Daphnia magna*. The reduction in sodium outflux in this species caused by aluminium at pH 4.5 could account for any ameliorating influence of aluminium at lower pH's, although no evidence for this phenomenon was apparent in the present study. There is limited evidence that aluminium is toxic to invertebrates in the labile form,[6, 38] but also that precipitating aluminium hydroxide is toxic.[21] The present study cannot distinguish between these two modes of action.

Can the toxicity of acid waters explain the observed species distributions? Whilst no detailed extrapolation from our data to the field is possible, species which were sensitive in the tests were generally those whose distribution can be related to pH and aluminium.[43] There were exceptions to this concordance (e.g. the beetle *Elmis aenea*, which showed no drift response but does not occur in acidic streams), but the limitations of the toxicological data mean that direct physiological effects cannot be excluded:

(i) species tolerant of acid episodes may be affected by chronic exposure;[7, 50]
(ii) tolerance can vary widely through the life cycle and during sensitive physiological processes (e.g. smoltification[31, 39]); only a limited range of life stages was tested here;
(iii) there exist a multitude of potential sub-lethal effects – clearly it would not be feasible to investigate all these, but long term measures of effects on fitness would be necessary to exclude them;
(iv) acid stress may become increasingly significant in combination with other forms of stress, as exemplified here for parasitized *G.pulex*.

Interpretation of detrimental physiological effects must also be treated with caution. Sensitive organisms may be able to avoid temporary adverse conditions e.g. by moving to areas of better water quality.[18] The life history timing of a species may prevent sensitive life stages coinciding with acid episodes prevalent in spring or autumn.[19] Moreover, acclimation of a population, or genetic adaptation, may enhance tolerance compared with test populations.[29]

Despite these caveats, important conclusions can be drawn from toxicity data concerning short-term exposure to acid conditions. Acid episodes can rapidly cause acute toxic responses in a number of taxa, ranging from grazing invertebrates to carnivorous salmonid fish. The importance of these effects is supported by field observations of natural episodes.[22, 26, 27] Individuals which survive such events may take over a week to recover[27] (see also this study). These observations are particularly relevant to natural waters which generally have circumneutral pH but are subject to acid episodes, and to limed waters. In limed running waters, acid episodes are often not eliminated,[13] and in limed lakes, influxes of acidic water to the littoral zone frequently occur.[3] Thus liming without the elimination of episodes may not allow full recovery of the fauna.

Whilst there is some evidence that alternative hypotheses may be important in explaining the distribution of some species,[36] direct physiological constraints seem to be crucial in many cases. Future ecotoxicological studies should be directed towards the acquisition of sufficiently comprehensive data to enable the development of models, which might predict population parameters of key species in given environments.[16] This will allow testing of the underlying hypotheses, and ecosystem management to be targeted at specific water quality parameters and values.

References

1. Appelberg, M. (1985). Changes in haemolymph ion concentrations of *Astacus astacus* L. and *Pacifastacus leniusculus* (Dana) after exposure to low pH and aluminium. *Hydrobiologia*, 121: 19–25.
2. Appelberg, M. (1987). Some factors regulating the crayfish *Astacus astacus* L. in acid and neutralized waters. *Annales de la Société Royale Zoologique De Belgique*, 117, supplement 1: 167–179.
3. Allard, M. and Moreau, G. (1985). Short-term effect on the metabolism of lotic benthic communities following experimental acidification. *Canadian Journal of Fisheries and Aquatic Sciences*, 42: 1676–1680.
4. Baker, J.P. and Schofield, C.L. (1982). Aluminium toxicity to fish in acidic waters. *Water, Air and Soil Pollution*, 18: 289–309.
5. Brown, D.J.A. (1983). Effects of calcium and aluminium concentrations on the survival of brown trout (*Salmo trutta*) at low pHs. *Bulletin of Environmental Contamination and Toxicology*, 30: 582–587.
6. Burton, T.M. and Allan, J.W. (1986). Influence of pH, aluminium, and organic matter on stream invertebrates. *Canadian Journal of Fisheries and Aquatic Sciences*, 43: 1285–1289.
7. Burton, T.M., Stanford, R.M. and Allan, J.W. (1985). Acidification effects on stream biota and organic matter processing. *Canadian Journal of Fisheries and Aquatic Sciences*, 42: 669–675.
8. Correa, M., Coler, R., Yin, C.M. and Kaufman, E. (1986). Oxygen consumption and ammonia excretion in the detritivore caddisfly *Limnephilus* sp. exposed to low pH and aluminium. *Hydrobiologia*, 40: 237–241.
9. Dickson, W. (1983). Liming toxicity of aluminium to fish. *Vatten* 4: 400–404.
10. Donaldson, E.M. (1981). The pituitary-interrenal axis as an indicator of stress in fish. *Stress and Fish*, (Ed. A.D. Pickering). Academic Press, London.
11. Driscoll, C.T., Baker, J.P., Bisogni, J.J. and Schofield, C.L. (1980). Effect of aluminium speciation on fish in dilute acidified waters. *Nature, London*, 284: 161–164.
12. Eddy, F.B. (1981). Effects of stress on osmotic and ionic regulation in fish. *Stress and Fish*, (Ed. A.D. Pickering). Academic Press, London.
13. Edman, G. and Fleischer, S. (1980). The River Hogvadsan liming project – a presentation.

Proceedings of an International Conference on the Ecological Impact of Acid Precipitation, pp. 300–301. SNSF Project, Norway.

14. Edwards, D., Brown, J.A. and Whitehead, C. (1987). Endocrine and other physiological indicators of acid stress in the brown trout. *Annales de la Société Royale Zoologique De Belgique*, 117, supplement 1: 331–342.
15. Fivelstad, S. and Leivestad, H. (1984). Aluminium toxicity to Atlantic salmon (*Salmo salar* L.) and brown trout (*Salmo trutta* L.): mortality and physiological response. *Report of the Institute of Freshwater Research, Drottningholm*, 61: 69–77.
16. France, R.L. and LaZerte, B.D. (1987). Empirical hypothesis to explain the restricted distribution of *Hyallela azteca* (Amphipoda) in anthropogenically acidified lakes. *Canadian Journal of Fisheries and Aquatic Sciences*, 44: 1112–1121.
17. Fromm, P.O. (1980). A review of some physiological and toxicological responses of freshwater fish to acid stress. *Environmental Biology of Fishes*, 5: 79–93.
18. Gunn, J.M. (1986). Behaviour and ecology of salmonid fishes exposed to episodic pH depressions. *Environmental Biology of Fishes* 17: 241–252.
19. Hall, R.J., Driscoll, C.T. and Likens, G.E. (1987). Importance of hydrogen ions and aluminium in regulating the structure and function of stream ecosystems: an experimental test. *Freshwater Biology*, 18: 17–43.
20. Havas, M. (1985). Aluminium bioaccumulation and toxicity to *Daphnia magna* in soft water at low pH. *Canadian Journal of Fisheries and Aquatic Sciences*, 42: 1741–1748.
21. Havas, M. and Likens, G.E. (1985). Changes in ^{22}Na influx and outflux in *Daphnia magna* (Strauss) as a function of elevated Al concentrations in soft water at low pH. *Proceedings of the National Academy of Sciences, USA*, 82: 7345–7349.
22. Henriksen, A., Skogheim, O.K. and Rosseland, B.O. (1984). Episodic changes in pH and aluminium speciation kill fish in a Norwegian salmon river. *Vatten*, 40: 255–260.
23. Hermann, J. (1987). Aluminium impact on freshwater invertebrates at low pH : a review. *Speciation of metals in water, sediment and soil systems*. Lecture Notes/Springer-Verlag, in press.
24. Herrmann, J. and Andersson, K.G. (1986). Aluminium impact on respiration of lotic mayflies at low pH. *Water, Air and Soil Pollution*, 30: 703–709.
25. Hollett, L., Berrill, M. and Rowe, L. (1986). Variation in major ion concentration of *Cambarus robustus* and *Orconectes rusticus* following exposure to low pH. *Canadian Journal of Fisheries and Aquatic Sciences*, 43: 2040–2044.
26. Lacroix, G. and Townsend, D. (1987). Responses of juvenile Atlantic salmon to episodic increases in the acidity of some rivers of Nova Scotia, Canada. *Annales de la Société Royale Zoologique De Belgique*, 117, supplement 1: 297–307.
27. Leivestad, H., Hendrey, G., Muniz, I.P. and Snekvik, E. (1976). Effects of acid precipitation on freshwater organisms. *Effects of acid precipitation on forest and freshwater ecosystems in Norway*, pp. 87–111. SNSF Project Research Report, Vol. 6.
28. Leivestad, H., Jensen, E., Kjartansson, H. and Xingfu, L. (1987). Aqueous speciation of aluminium and toxic effects on Atlantic salmon. *Annales de la Société Royale Zoologique De Belgique*, 117, supplement 1: 387–398.
29. Maltby, L., Calow, P., Cosgrove, M. and Pindar, L. (1987). Adaptation to acidification in aquatic invertebrates; speculation and preliminary observations. *Annales de la Société Royale Zoologique De Belgique*, 117, supplement 1: 105–115.
30. McCahon, C.P., Pascoe, D. and McKavanagh, C. (1987). Histochemical observations on the salmonids *Salmo salar* L. and *Salmo trutta* L. and the ephemeropterans *Baetis rhodani* Pict. and *Ecdyonurus venosus* Fabr. following a simulated episode of acidity in an upland stream. *Hydrobiologia*, 153: 3–12.
31. McCahon, C.P. and Pascoe, D. (1989) Short-term experimental acidification of a Welsh stream: toxicity of different forms of aluminium at low pH to fish and invertebrates. *Arch. Environ. Contam. Toxicol.*, 18: 233–243.
32. McWilliams, P.G. (1982). The effects of calcium on sodium fluxes in the brown trout, *Salmo*

trutta, in neutral and acid water. *Journal of Experimental Biology*, 96: 439–442.

33. Muniz, I.P. and Leivestad, H. (1980). Toxic effects of aluminium on the brown trout, *Salmo trutta* L. *Proceedings of an International Conference on the Ecological Impact of Acid Precipitation*, pp. 320–321. SNSF Project, Norway.
34. Neville, C.M. (1985). Physiological response of juvenile rainbow trout, *Salmo gairdneri*, to acid and aluminium – prediction of field responses from laboratory data. *Canadian Journal of Fisheries and Aquatic Sciences*, 42: 2004–2019.
35. Ormerod, S.J., Boole, P., McCahon, C.P., Weatherley, N.S., Pascoe, D. and Edwards, R.W. (1987a). Short-term experimental acidification of a Welsh stream: comparing the biological effects of hydrogen ions and aluminium. *Freshwater Biology*, 17: 341–356.
36. Ormerod, S.J., Wade, K.R. and Gee, A.S. (1987). Macro-flora assemblages in upland Welsh rivers in relation to stream acidity and invertebrate faunas. *Freshwater Biology*, 18: 545–557.
37. Ormerod, S.J., Weatherley, N.S., French, P., Blake, S. and Jones, W.M. (1987b). The physiological response of brown trout *Salmo trutta* to induced episodes of low pH and elevated aluminium in a Welsh hill-stream. *Annales de la Société Royale Zoologique De Belgique*, 117, supplement 1: 435–447.
38. Petersen, R.C.Jr., Petersen, L.M.-M., Petersson, U., Kullberg, A., Hargeby, A. and Paarlberg, A. (1986). Health aspects of humic compounds in acid environments. *Water Quality Bulletin*, 11: 44–49.
39. Rosseland, B.O. and Skogheim, O.K. (1984). A comparative study on salmonid fish species in acid aluminium-rich water II. Physiological stress and mortality of one- and two-year-old fish. *Report of the Institute of Freshwater Research, Drottningholm*, 61: 186–194.
40. Rosseland, B.O., Skogheim, O.K., Abrahamsen, H. and Matzow, D. (1986). Limestone slurry reduces physiological stress and increases survival of Atlantic salmon (*Salmo salar*) in an acidic Norwegian river. *Canadian Journal of Fisheries and Aquatic Sciences*, 43: 1888–1893.
41. Sadler, K. and Lynam, S. (1985). *The mineral content of some freshwater invertebrates in relation to stream pH and calcium concentration*. CEGB Report TPRD/L/2781/N84.
42. Stoner, J.H., Gee, A.S. and Wade, K.R. (1984). The effects of acidification on the ecology of streams in the Upper Tywi catchment in West Wales. *Environmental Pollution (Series A)*, 35: 125–157.
43. Weatherley, N.S. and Ormerod, S.J. (1987). The impact of acidification on macroinvertebrate assemblages in Welsh streams: towards an empirical model. *Environmental Pollution*, 46: 223–240.
44. Weatherley, N.S., Ormerod, S.J., Thomas, S.P. and Edwards, R.W. (1988). The response of macroinvertebrates to experimental episodes of low pH with different forms of aluminium, during a natural spate. *Hydrobiologia*, 169: 225–232.
45. Weatherley, N.S., Thomas S.P. and Ormerod, S.J.. Chemical and biological effects of acid, aluminium and lime additions to a Welsh hill-stream. *Environmental Pollution*, in press.
46. Welsh Water, (1986). *The survival and physiological responses of brown trout stocked in two acidic upland lakes, before and after liming and an assessment of the performance of trout fisheries re-established in them*. Report SW/86/30, 46pp.
47. Wood, C.M. and McDonald, D.G. (1982). Physiological mechanisms of acid toxicity to fish. *Acid Rain/Fisheries*, (Ed. R.E. Johnson), pp. 197–226. Proceedings of an international symposium on acidic rain and fishery impacts on Northeastern North America. American Fisheries Society, Bethesda, Maryland.
48. Wood, C.M. and McDonald, D.G. (1987). The physiology of acid/aluminium stress in trout. *Annales de la Société Royale Zoologique De Belgique*, 117, suppl. 1: 399–410.
49. Wood, C.M. and Rogano, M.S. (1986). Physiological responses to acid stress in crayfish (*Orconectes*): haemolymph ions, acid-base status, and exchanges with the environment. *Canadian Journal of Fisheries and Aquatic Sciences*, 43: 1017–1026.
50. Zimmer, D. (1987). Effects of low pH acclimation on cladocerans: clues to the interaction of physiology and ecology of acid lake zooplankton. *Annales de la Société Royale Zoologique De Belgique*, 117, supplement 1: 139–149.

11. Dissolved inorganic aluminium in acidic stream and soil waters in Wales

C. NEAL, B. REYNOLDS, P.A. STEVENS, M. HORNUNG and S.J. BROWN

Introduction

Acidic soils overlying extensive, base poor bedrock characterise a large part of the Welsh and Scottish uplands, with the consequence that these areas are both acidic and acid sensitive. Stream acidification, with associated loss of fisheries, is being observed on a large scale throughout these areas.[3, 11, 12, 14, 15, 36] High inorganic aluminium concentrations (Al^{3+}, $Al(OH)^{2+}$ etc.) are toxic to fish.[12, 22, 29] Thus the fish losses are associated with waters that are not only acidic but which have anomalously high inorganic aluminium concentrations compared with most natural waters. Acidic deposition and conifer afforestation have both been implicated in the decline in water quality and fisheries status for Welsh and Scottish headwater streams. The relative importance of these two factors is not easily quantified[24, 25, 31, 32] although predictive models are being developed to overcome this shortfall.[6, 7, 8, 38, 39] These models are urgently required to describe the changes in streamwater quality as a function of management strategies in the power and agricultural industries. Contained in these models is the assumption that inorganic aluminium concentrations in the soil and streamwaters are controlled by simple aluminium hydroxide solubility constraints.[23, 26] However, it is now clear that such solubility controls are not appropriate for many of the acidic catchments of concern.[17, 23, 25, 27, 33, 37]

In view of the high toxicity of aluminium to some organisms, the assessment of the impact and the required amelioration of atmospheric pollution and land use change on streamwater quality requires a thorough understanding of inorganic aluminium behaviour. In this chapter this behaviour is explored for the Welsh uplands using specific case studies for the Plynlimon and Beddgelert catchments of central and northern Wales respectively, presently being studied by researchers at the IH and ITE.[18, 19, 25, 30, 31, 34] Comparisons are then made with regional streamwater data, collected by the WWA, and soil water data from the numerous catchment studies in the Llyn Brianne area of South Wales[35] to provide a picture of inorganic aluminium behaviour on a regional scale.

R.W. Edwards et al. (eds) Acid Waters in Wales. 173–188.

Theoretical consideration of processes

The chemistry of inorganic aluminium (Al(inorg)) in acid soil and stream waters may be considered in terms of mineral solubility, ion exchange and water mixing processes, each of which is considered separately below. The assessment of these processes is difficult, owing not only to the variety of reactions taking place but also to the complexity of inorganic aluminium speciation. For example in the pH range 4 to 8 the inorganic aluminium in solution may include significant amounts of any or all of the following species: Al^{3+}, $Al(OH)^{2+}$, $Al(OH)^{+}_{2}$, $Al(OH)^{0}_{3}$, $Al(OH)^{-}_{4}$, together with aluminium-fluoride complexes. The amount of each complex is determined not only by the total aluminium concentration but also by pH and temperature (see Goenaga, Bryant and Williams[13] and Chap. 12). In the following discussion, allowance has been made for all these species in solution, as well as temperature, using the thermodynamic data given by Nordstrom et al.[28]

Solubility controls

The minerals considered as determining stream and soil water aluminium concentrations in acidic and acid sensitive systems are aluminium hydroxide, kaolin, aluminium hydroxy silicate and aluminium hydroxy sulphate,[4, 5, 10, 23, 26, 33] but to simplify analysis only the aluminium hydroxide and aluminium hydroxy sulphate cases are considered here. Owing to the nature of the thermodynamic reactions and the silica concentrations in solution being typically near quartz (SiO_2) saturation, aluminium hydroxy silicate, aluminium silicate and aluminium hydroxide solubility controls are not easily distinguished.[23, 26] In this context the aluminium hydroxide thermodynamic analysis presented should be viewed as covering a spectrum of potential reactions and not a single "type" mineral solubility control.

The solubility reactions and their associated equilibrium constants (K) are:

Al $(OH)_3$ minerals

$$Al(OH)_3 + 3\,H^+ \rightleftharpoons Al^{3+} + 3\,H_2O$$

$$\log_{10}K_{Al(OH)_3} \rightleftharpoons \log_{10} a_{Al^{3+}} - 3\log_{10} a_{H^+}$$

Aluminium Hydroxy Sulphate (Al(OH)So_4)

$$Al\,(OH)SO_4 + H^+ \rightleftharpoons Al^{3+} + SO_4^{2-} + H_2O$$

$$\log_{10}K_{Al(OH)SO_4} \rightleftharpoons \log_{10} a_{Al^{3+}} + \log_{10} a_{SO_4^{2-}} - \log_{10} a_{H^+}$$

Ion exchange

In a system where aluminium, hydrogen and other cations can exchange between soil cation exchangers and soil solution, the exchange reaction is described by the general equation (for ions M and N of charges m+ and n+ respectively, and an adsorbed phase (Ads))

$$n\text{ Ads M} + m\,N^{n+} = m\text{ Ads N} + n\,M^{m+}$$

The equilibrium constant for this reaction ($K_{M,N}$) is given by the equation:

$$K_{M,N} = \frac{(\text{Ads N})^m}{(\text{Ads M})^n} \frac{(a_{M^{m+}})^n}{(a_{N^{n+1}})^m}$$

In the case where the number of ion exchange sites vastly exceeds the number of ions passing through the soil, then for fluctuations in soil water composition the ratio of adsorbed ions remains constant and

$$K_{Al^{3+},H^+} \frac{(\text{Ads Al})}{(\text{Ads H})^3} \text{ constant} = \frac{a_{Al^{3+}}}{(a_{H^+})^3}$$

Water mixing

Water mixing processes may also be important in determining the relationship between Al(inorg) and H ions.[34] Consequently there is a need to model how these two species vary as a function of mixing so that comparisons can be made with field observation. Unfortunately such a step is difficult since during the mixing process H^+ varies non-conservatively due to bicarbonate buffering in solution. As a first step in resolving the effects of mixing, a basic calculation is made to allow soil and groundwaters to mix and degas CO_2 in the stream. To do this it is assumed that pCO_2 in the soil/groundwater and the stream is 25 and 2 times the atmospheric value respectively, and that bicarbonate principally controls the acid buffering. The model is calibrated such that the soil and groundwater endmembers in solution have a composition which on degassing gives the observed endmember stream composition.

The Plynlimon and Beddgelert study areas

The work described in detail is based on two areas of upland Wales; the headwater catchments of the rivers Wye and Severn in mid Wales and Beddgelert forest in north Wales (Table 11.1). The bedrock geology at Plynlimon consists of lower Palaeozoic mudstones, shales and grits containing occasional calcite veins. The Beddgelert site is underlain by Ordovician slates cut by bands of intrusive dolerite. Locally derived glacial and post-glacial drift deposits are present in both areas. A mosaic of upland soils occurs at the sites consisting of peats, stagnopodzols, brown earths and stagnogleys. Stagnopodzols[1] are the predominant soil type and are typically 70 cm thick with an organic-rich O horizon of between 3 and 10 cm thickness. This is underlain by a leached, greyish, heavily textured E horizon between 10 and 20 cm thick. Below this is about 40 cm of fine textured podzolic B horizon merging into a stoney C horizon. At the forest sites a layer of spruce needle litter has developed (L horizon).

In the headwater catchments of the Wye (Cyff, Gwy and Nant Iago; Table 11.1) the vegetation consists of an acid grassland dominated by *Nardus* and

Table 11.1. Summary features of the Plynlimon and Beddgelert catchments.

	PLYNLIMON					BEDDGELERT	
	Moorland			Forest		Forest	
	1	2	3	4	5	6	7
Catchment area (ha)	390	308	105	347	335	1.4	4.7
Altitude range (m)	380–730	355–690	380–630	350–690	339–738	306–420	306–420
Hill top peat (%)	37	14	30	52	33	5	0
Improved moorland (%)	7	39	27	0	0	0	0
Forest (%) cover	1	0	3	48	77	88	80
Rainfall (mm)	2245	2242	2272	2215	2409	2880	2280

1, Afon Gwy; 2, Afon Cyff; 3, Nant Iago; 4, Afon Hafren; 5, Afon Hore; 6, Beddgelert spruce plantation; 7 Beddgelert harvested spruce plantation forest.

Agrostis species with *Eriophorum* sp. on the peats. Parts of the upper Wye have been agriculturally improved by the periodic addition of lime and compound fertilizers over the last forty years.[19] The headwater catchments of the river Severn (Hore and Hafren; Table 11.1) were planted, predominantly with Sitka spruce (*Picea sitchensis*), in three phases between 1937 and 1964. The Beddgelert study site was planted with Sitka spruce in 1936.

Soil water from the L and O horizons was collected using zero-tension tray lysimeters. In the mineral horizons (E, B, C) porous ceramic cup soil water samplers were used following laboratory pre-treatment and a three month period of field equilibration.[16, 20] Three sets of trays and porous cups were used at the grassland and forest sites at Plynlimon and four sets at the Beddgelert forest site. Streamwater samples were collected weekly and the soil water samples at fortnightly intervals.

In the laboratory, samples from Plynlimon were filtered through 0.45 μm pore size membrane filters, and samples from Beddgelert through GF/F glass fibre filters, although pH was determined on an unfiltered subsample. Total filtrable aluminium concentrations for Plynlimon streamwaters were determined by ICPOES, following evaporative concentration. For the soil waters from Plynlimon and soil and streamwaters from Beddgelert, inorganic aluminium (Al(inorg)) was determined as labile monomeric aluminium using the speciation technique of Driscoll;[9] Al(inorg) for Plynlimon streamwaters was calculated assuming that organic aluminium was constant at 0.8 μMol Al l^{-1}. Concentrations of SO_4 were determined by ion chromatography and Si and total F by conventional auto-analyser methods.

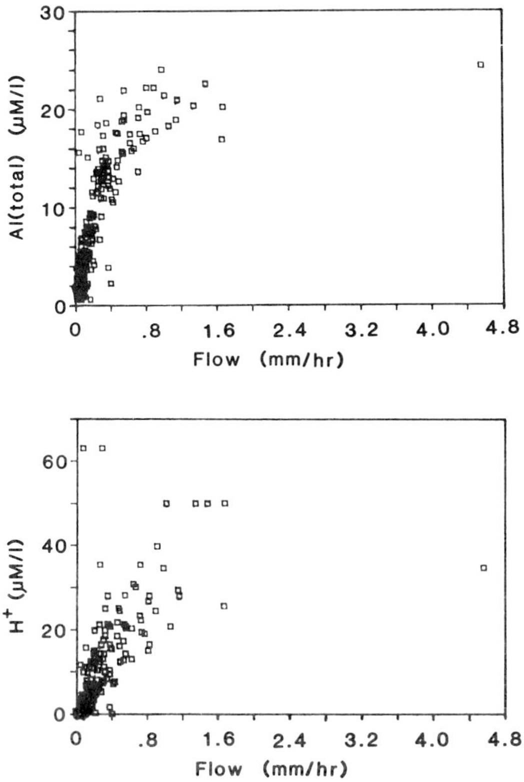

Fig. 11.1. (a) Al(total) and (b) H^+ relationships with flow for Plynlimon forest streamwaters.

H^+ are observed. At both localities the Al(inorg) and H^+ trends are distinct for each horizon (Fig. 11.4). The large concentration ranges observed highlight the dynamic nature of the hydrological and chemical reactions occurring. Analysis of the stability relationships reveals that all the waters are undersaturated with respect to $Al(OH)_3$ except in the lowest horizons (B and C) where they are approximately saturated with respect to the most crystalline form of $Al(OH)_3$ (Fig. 11.4). The degree of undersaturation increases markedly in the surface soil horizons. In addition the soil waters are undersaturated with respect to $Al(OH)SO_4$, indicating that this mineral does not control Al(inorg) concentrations. Indeed these results suggest that this mineral is not present within the soils.

Ion exchange materials are known to be present in significant amounts within the soils. Empirically, ion exchange reactions between Al^{3+} and H^+ can be described by:

$$K = \frac{a_{Al^{3+}}}{(a_{H^+})^x}$$

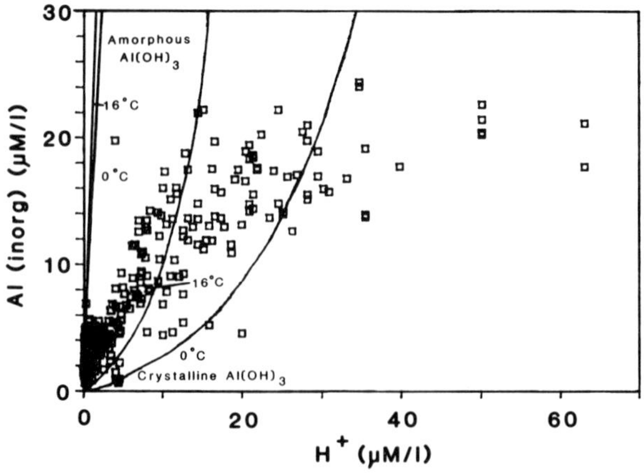

Fig. 11.2. H^+–Al(inorg) relationships for Plynlimon forest streamwaters.

where K and x are constants with x having a value of 3. Detailed investigation of the relationships between H^+ and Al^{3+} using thermodynamic equilibrium diagrams reveals a relationship which approaches cubic in the upper soils (O and E) with different equilibrium constants for each of the horizons. For the lower horizons, (B and C) the cubic relationship breaks down and the power term varies between 1 and 3. Thus, in the upper soils an ion exchange equation may be applicable but in the lower horizons more complex relationships are required. Reasons for this discrepancy in the lower horizons include:

1. the cation exchange capacity is not large compared to the throughput of ions in the soil water (ie. Ads Al^{3+}/Ads$^3H^+$ is not constant);
2. the cation exchange capacity varies as a function of the H^+ concentration and/or the ionic strength of the solution;
3. the selectivity coefficient ($k_{Al,H}$) varies as a function of the H^+ concentration in solution;
4. binding of different aluminium species occurs on the exchange surface.

The alternatives may be interdependent or dependently related but they are consistent with views presented in the literature for podzolic and acidic soil.[2, 21]

The ion exchange reaction, throughout the profile, may however be viewed in another way. In the organic soils reactions between H^+ and ions other than positively charged aluminium species occur. Thus fluctuations in H^+ concentration are determined by fluctuations of ionic strength in the organic soil, aluminium fluctuations being determined by weathering/cycling reactions. In the lower mineral soils aluminium exchanges with cations other than H^+, H^+ being partially consumed by weathering (dissolution) reactions. Thus by this alternative approach there is no direct causal link between aluminium and H^+ concentrations by ion exchange reactions.

Presently it is unclear which if any of these reactions is appropriate. However ion exchange processes are important within the catchments owing

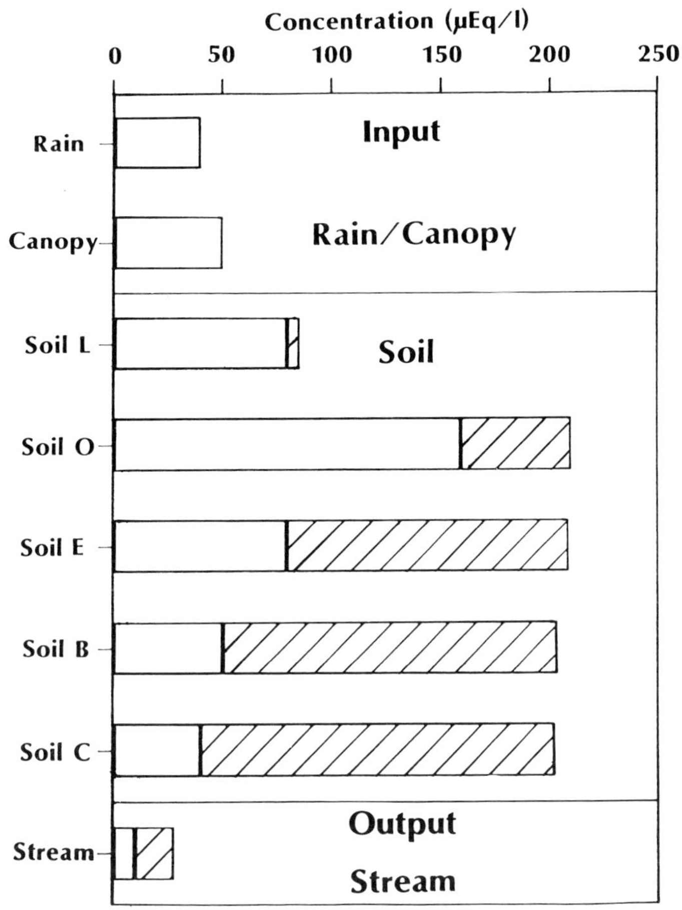

Fig. 11.3. Mean Al(inorg) and mean H^+ concentrations in the Plynlimon forest catchment. (Shaded and unshaded areas represents Al(inorg) and H^+ concentrations respectively.)

to the large exchangeable cation store. This is seen on a broad front by comparing aluminium concentrations in the soil and streamwaters. The concentrations vary in the order semi-natural grassland < plantation forest < recently harvested plantation forest. This is the same order as the anion concentration in solution; an increase in the anion concentration should result in ions of higher charge being preferentially released to solution.[32]

Soil – stream-water relationships

Stream waters at baseflow are all less acidic than the corresponding soil waters (Tables 11.2, 11.3). This indicates that the baseflow waters are derived from within the groundwater zones not sampled in these surveys. Mean streamwater Al(inorg) and H^+ concentrations are lower than those in the

Table 11.3. Inorganic aluminium concentration (μMol l^{-1}) ranges for Plynlimon and Beddgelert soil waters.

		HORIZON				
		L	O	E	B	C
Plynlimon						
Grassland	Range	–	0.4–6	4.4–23	3.3–15	9–39
podzol	Avg.	–	2.1	10.5	8.8	21.0
Forest	Range	–	2.6–6	9–42	21–64	15–94
podzol	Avg.	–	11.9	25.4	46.4	60.3
Forest	Range	–	3.0–17	26–92	36–95	39–74
gley	Avg.	–	7.5	53.6	64.7	54.9
Beddgelert						
Forest	Range	0.1–3.7	0.1–44	4.1–27	2.2–54	25–60
podzol	Avg.	0.6	4.6	14.3	33.5	40.5
		pH for Plynlimon and Beddgelert soil waters				
Plynlimon						
Grassland	Range	–	3.9–4.6	4.3–4.7	4.4–4.7	4.7–5.0
podzol	Avg.	–	4.3	4.4	4.5	4.8
Forest	Range	–	3.7–4.7	3.9–4.5	4.2–4.8	4.5–5.0
podzol	Avg.	–	4.1	4.1	4.4	4.7
Forest	Range	–	3.7–4.4	3.8–4.2	4.2–4.6	4.4–4.7
gley	Avg.	–	4.0	4.0	4.3	4.5
Beddgelert						
Forest	Range	3.5–4.3	3.7–4.5	3.9–4.3	4.4–4.8	4.4–4.7
podzol	Avg.	4.0	4.0	4.1	4.5	4.5

soils (Fig. 11.3). These factors suggest that mixing of soil and groundwaters are important in determining stream Al(inorg) and H^+ chemistry.

The results for simulated two component mixing (Fig. 11.5), in which 'groundwater' is mixed with surface soil waters, show two features. Firstly, if Al(inorg) is conserved in solution during mixing, the model over predicts Al(inorg) concentrations at low hydrogen ion concentrations. At higher hydrogen ion concentrations (>10 μMol 1^{-1}) a much better fit is observed. The poor fit at low hydrogen ion concentrations results from a decrease in hydrogen ion concentration in the stream due to bicarbonate buffering and CO_2 degassing, but without a concomittant change in Al(inorg) concentrations. Therefore, a precipitation reaction is required to explain the change in Al(inorg) concentration. From the previous discussion, a simple precipitation in the stream of $Al(OH)_3$ or an aluminium hydroxy silicate does not occur. However, the application of a two component mixing model, involving groundwater and upper soil water, where Al(inorg) is allowed to precipitate as microcrystalline Gibbsite or an aluminium hydroxy silicate analogue in the soil, but not in the stream, gives a good fit (Fig. 11.5). As mentioned previously, the uncertainty in the thermodynamic relationships would permit several mineral phases to give the same result. Given the dynamic nature of

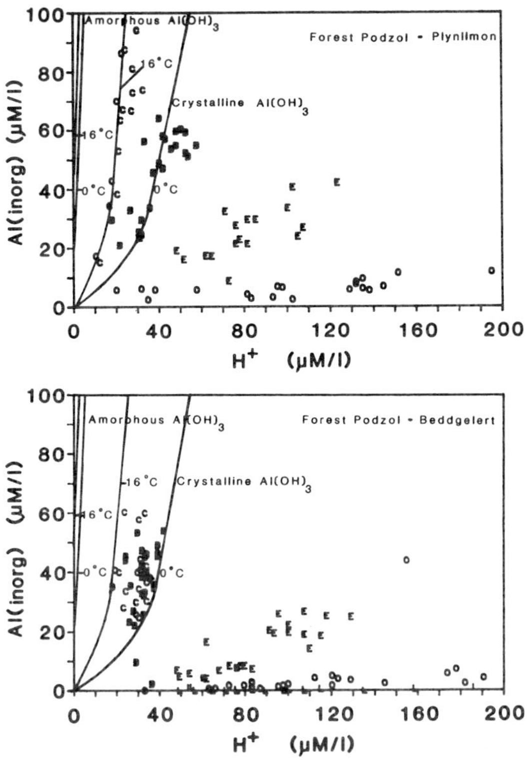

Fig. 11.4. H^+–Al(inorg) relationships for Plynlimon and Beddgelert forest soil waters.

the streams, nucleation of Al(inorg) precipitates is unlikely to proceed to equilibrium. However, within the soil, where a large number of nucleation centres are available, equilibrium precipitation might be approached more closely. Hence, the two component mixing model gives a reasonable description of Al(inorg) regulation in the stream.

Aluminium–hydrogen ion relationships; a regional view

The Plynlimon and Beddgelert streamwater results are reflected on a regional scale by the extensive data of WWA. Thus for the average Al(inorg) and H^+ concentration for each stream, thermodynamic analysis indicates neither $Al(OH)_3$, aluminium hydroxy silicate nor $Al(OH)SO_4$ solubility controls. However, a two-component mixing model might well be considered to hold (Fig. 11.6). For individual catchments the composition of each mixing end-member may be different; this is seen when data are plotted for all rather than mean values.

The Plynlimon and Beddgelert soil water results are also reflected on a

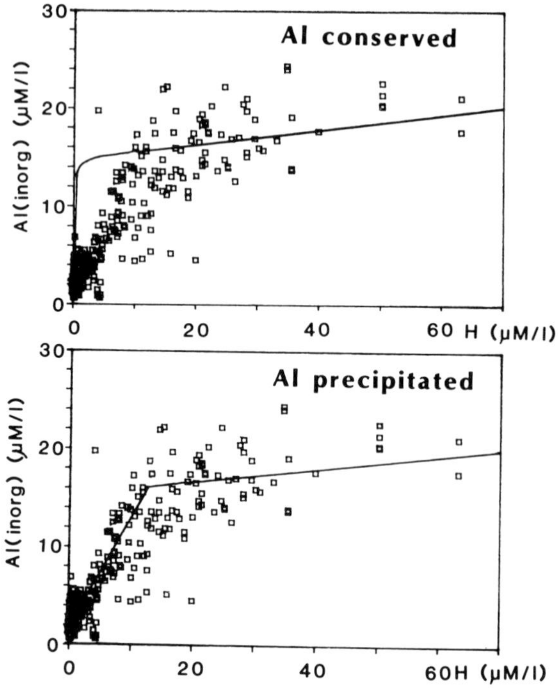

Fig. 11.5. Two component mixing models for Plynlimon forest streams.

regional scale when comparisons are made with the data collected as part of the Llyn Brianne study. These data cover a wide range of upland soils supporting different vegetation and landuse types. The highest H^+ and lowest Al(inorg) concentrations are observed in the upper (O, E) horizons while the lowest H^+ and the highest Al(inorg) concentrations are observed in the lower (B,C) horizons (Fig. 11.7). However, for the Llyn Brianne data and linear behaviour observed in the earlier plot of Al(inorg) against H^+ is not observed except in the O and A horizons. This simply reflects that various different forms of land use and soil types have been monitored with the consequence that the H^+ and Al(inorg) concentrations will be determined by factors additional to those for the semi-natural grassland and forested catchments of Plynlimon and Beddgelert. Indeed, if limed soils as well as different soil types had been included in the Plynlimon data analysis, very similar results would have been found to those for Llyn Brianne. As in the previous cases $Al(OH)_3$ and $Al(OH)SO_4$ solubility controls are not operative.

Conclusions

Large temporal and spatial variations in aluminium concentration occur in Welsh streamwaters. Hydrological and land use controls primarily account for these variations. Aluminium and H^+ are highly correlated so that the

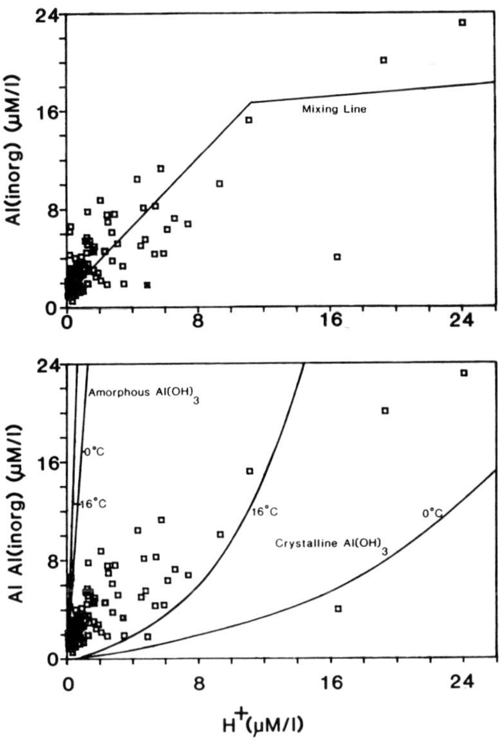

Fig. 11.6. H^+–Al(inorg) relationships for 131 Welsh streams.

most acidic streams have the highest aluminium concentration. The temporal variation in both aluminium and H^+ concentration can be explained in terms of the mixing of waters from the upper soil (L, O, A horizons) and groundwater source, each endmember composition being determined by the bedrock type/reactivity and the land use history of the soils. Variations in stream and soil water aluminium concentrations in the order semi-natural moorland < conifer forested moorland < recently harvested forest can be explained by ion exchange reactions related to changes in the anion concentrations passing through the system, and weathering. In no case can the streamwater aluminium concentration be related simply to $Al(OH)_3$ and $Al(OH)SO_4$ solubility controls. The same is also true for the soil waters.

Within the modelling context, the results presented here are of importance when considering regional predictions of stream acidification on both a short (hourly) and long (decade) time scale. The major point made in this paper is that the classic $Al(OH)_3$ and $Al(OH)SO_4$ solubility relationships assumed to hold within all the models are inappropriate. Additionally, owing to the nature of the ion exchange reactions, simple cubic relationships may not be adequate for some of the soils of concern. Consequently there is a need to re-evaluate how aluminium mobilisation is depicted in the currently used models.

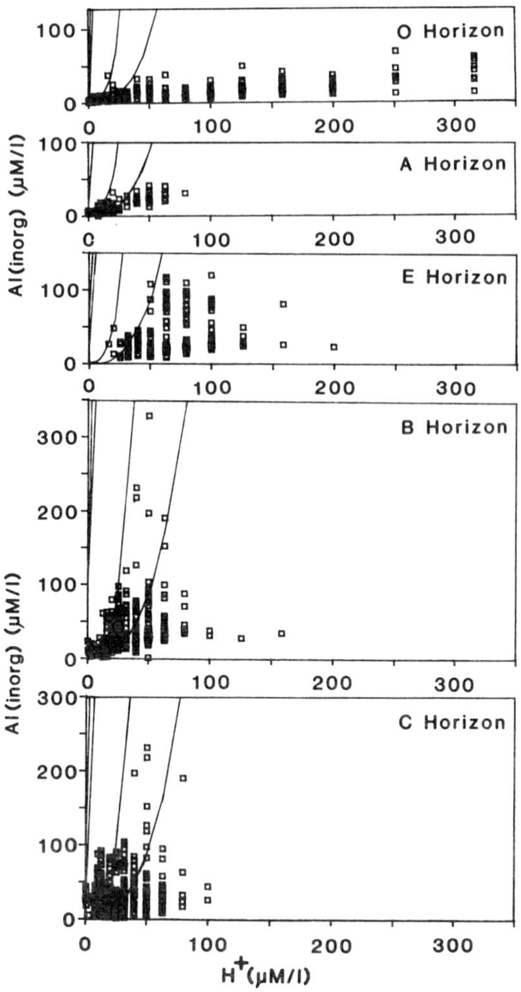

Fig. 11.7. H^+–Al(inorg) relationships in soil water of different horizons for a spectrum of upland soil under different vegetation (moorland and forest) and land use in the Llyn Brianne area of mid-Wales.

References

1. Avery, B.W. (1980). Soil Classification for England and Wales. Soil Survey Tech. Monogr. No. 14 Soil Survey, Harpenden.
2. Bache, B.W., and Sharp, G.S. (1976). Soluble polymeric hydroxy-aluminium ions in acid soils. *Journ. Soil Sci.*, 27: 167–174.
3. Battarbee, R.W., Flower, R.J., Stevenson, A.C. and Rippey, B. (1985). Lake acidification in Galloway: A palaeoecological test of competing hypotheses. *Nature*, 314 (6009), 350–353.
4. Van Breemen, N. (1973). Calculations of ionic activities in natural waters. *Geochim. et Cosmochim. Acta*, 37: 101–107.
5. Van Breemen, N. (1976). Genesis and solution chemistry of acid sulphate soils in Thailand. Wageningen, Centre for agricultural publishing and documentation.
6. Christophersen, N., Seip, H.M. and Wright, R.F. (1982). A model for steamwater chemistry at Birkenes, Norway. *Water Resour. Res.*, 18: 977–996.
7. Christophersen, N., Rustad, S., and Seip, H.M. (1984). Modelling streamweter chemistry with snowmelt. *Phil. trans. of Royal Soc.* (London), B305, 427–439.
8. Cosby, B.J., Wright, R.F., Hornberger, G.M. and Galloway, J.N. (1985). Modelling the effects of acid deposition: assessment of a lumped parameter model of soil water and streamwater chemistry. *Water Resour. Res.*, 21: 51–63.
9. Driscoll, C.T. (1980). Chemical characteristics of some dilute acidified lakes and streams in the adirondack region of New York State. Ph.D. thesis, Cornell Univ., Ithaca, N.Y.
10. Eriksson, E. (1981). Aluminium in groundwater; possible solution equilibria. *Nordic Hydrology*, 12: 43–50.
11. Flower, R.J. and Batterbee, R.W. (1983). Diatom evidence for recent acidification of two Scottish lochs. *Nature*, 305(5930), 130133.
12. Gee, A.S. and Stoner, J.H. (1984). The effects of seasonal and episodic variations in water quality on the ecology of upland waters in Wales. Institute of Water Pollution Control/Institute of Water Engineers and Scientists joint meeting on acid rain, 40 pp.
13. Goenaga, X, Bryant, R. and Williams, D.J.A. (1987). Influence of sorption processes on Al determination in natural waters. *Anal. Chem.*, 59: 2673–2676.
14. Harriman, R. and Morrison, B. (1981). Forestry, fisheries and acid rain in Scotland. *Scott. For.*, 36: 89–95.
15. Harriman, R. and Wells, D.E. (1985). Causes and effects of surface water acidification in Scotland. *Water Pollut. Control*, 84: 215–224.
16. Hetsch, W., Beese, F. and Ulrich, B. (1979). Die Beeintlussung der Bodenlosung durch Saugkerzen aus Ni-Sintermetall & Keramik Z. *Pflanzenernachr. Bodenkd.*, 142: 29–38.
17. Hooper, R.P. and Shoemaker, C.A. (1985). Aluminium mobilisation in an acidic headwater stream: temporal variation and mineral dissolution disequilibria. *Science*, 229: 463–465.
18. Hornung, M., Reynolds, B. and Hatton, A.A. (1985). Land management, geological and soil effects on streamwater chemistry in upland mid-Wales. *App. Geog.*, 5: 71–80.
19. Hornung, M., Stevens, P.A. and Reynolds, B. (1986). The impact of pasture improvement on the soil solution chemistry of some stagnopodzols in mid-Wales. *Soil Use Manag.*, 2: 18–26.
20. Hughes, S., and Reynolds, B. (1988). Cation exchange properties of porous ceramic cups: implications for field use. *Plant and Soil*, 109: 144–144.
21. Lee, R., Bache, B.W., Wilson, M.J. and Sharp, G.S. (1985). Aluminium release in relation to the determination of cation exchange capacity of some podzolized New Zealand soils. *Journ. Soil Sci.*, 36: 239–253.
22. Muniz, I.P. and Leivestad, H. (1980). Toxic effects of aluminium on the brown trout *Salmo trutta* L. In: Drablos, D. and Tollan, A. (eds.): Ecological Impact of acid precipitation, 320–321, SNSF-project.
23. Neal, C. (1988). Aluminium solubility relationships in acid waters; a practical example of the need for a radical reappraisal. *Journ. Hydrol.* 104: 141–159.
24. Neal, C., Whitehead, P.G., Neal, R. and Cosby, J. (1986a). Modelling the effects of acidic

deposition and conifer afforestation on stream acidity in the British Uplands. *Journ. Hydrol.*, 86: 15–26.
25. Neal, C., Smith, C.J., Walls, J. and Dunn, C.S. (1986b). Major, minor and trace element mobility in the acidic upland forested catchment of the upper River Severn, Mid Wales. *Q.J. Geol. Soc.*, 143: 635–648.
26. Neal, C., Skeffington, R.A., Williams, R. and Roberts, D.J. (1987). Aluminium solubility controls in acid waters: The need for a reappraisal. *Earth and Planetary Sciences Letters*, 86: 105–112.
27. Nordstrom, D.K. and Ball, J.W. (1986). The geochemical behaviour of aluminium in acidified surface waters. *Science*, 232: 54–56.
28. Nordstrom, D.K., Valentine, S.D., Ball, J.W., Plummer, L.N. and Jones, B.F. (1984). Partial compilation and revision of basic data in the WATEQF programs. USGS, Wate-Resources Investigation Report, 84–4186, 1–19.
29. Overrein, L.N., Seip, H.M. and Tollan, A. (1981). Acid precipitation on forest and fish. Final report of the SNSF project 1972–1980, 1–175.
30. Reynolds, B., Hornung, M. and Hughes, S. (1983). Some factors controlling variations in chemistry of an upland stream in mid-Wales. *Cambria*, 10: 130–145.
31. Reynolds, B., Neal, C., Hornung, M. and Stevens, P.A. (1986). Baseflow buffering of streamwater acidity in five mid-Wales catchments. *J. Hydrol.*, 87: 167–185.
32. Reynolds, B., Neal, C., Hornung, M., Hughes, S. and Stevens, P.A. (1988). Impact of afforestation on the soil solution chemistry of stagnopodzols in mid-Wales. *Water Air and Soil Pollution*, 38: 55–70.
33. Seip, H.M., Muller, L. and Naas, A (1984). Aluminium speciation: comparison of two spectrophotometric analytical methods and observed concentrations in some acidic aquatic systems in southern Norway. *Water Air and Soil Pollution*, 23: 81–95.
34. Stevens, P.A. (1987). Throughfall chemistry beneath Sitka spruce of four ages in Beddgelert Forest, North Wales, UK. *Plant and Soil*, 101: 291–294.
35. Stoner, J.H. (1987). Llyn Brianne Acid Waters Project. 1st tech. report, Welsh Water Authority, 1–91.
36. Stoner, J.H., Gee, A.S. and Wade, K.R. (1984). The effects of acidification on the ecology of streams in the Upper Tywi catchment in West Wales. *Environment Pollution (Series A)*, 35: 125–157.
37. Sullivan, T.J., Christophersen, N., Muniz, I.P., Seip, H.M. and Sullivan, P.D. (1986). Aqueous aluminium chemistry response to episodic increases in discharge. *Nature*, 323: 324–326.
38. Whitehead, P.G., Neale, R., and Neal, C. (1985). Predicting the effects of acid deposition on water quality. Inst. of Hydrol. report (contract ENV–829–UK(H)).
39. Whitehead, P.G., Neal, C. and Neale, R. (1986). Modelling the effects of hydrological changes on streamwater acidity. *Journ. Hydrol.*, 84: 353–364.

12. Determination of aluminium speciation in acid waters

X. GOENAGA and D.J.A. WILLIAMS

Introduction

The determination of the speciation of aluminium in acidified freshwaters is an important factor in the interpretation, prediction and possible modification of aluminium toxicity and mobilisation in acidified catchments.

Aluminium has a very complex chemistry at low pH and it can be present in many forms, eg. as the free aquo-ion, as inorganic and organic complexes, polymerized, adsorbed onto solid phases and as part of the mineral structure of suspended particles. The speciation of aluminium in an aqueous system consists of determining the concentration of each of the physico-chemical forms present. Recently, several experimental aluminium speciation schemes have been proposed,[1] which only measure aluminium fractions containing a mixture of species with similar behaviour. The use and limitations of such aluminium fractionation schemes have been discussed elsewhere.[6, 8, 9, 12] Work is also being done on the effect of sample collection and storage[2] and to assess differences in speciation derived from the use of different schemes.[1]

The prediction of the concentration of individual species can only be achieved if laboratory fractionations are complemented with mathematical models of thermodynamic equilibria.[13, 17] Models available in the literature[17, 22] only apply to the inorganic monomeric aluminium fraction. However, significant amounts of aluminium in acidified surface waters can be associated with aquatic humic substances (HS) (referred to as Al_o) and accurate knowledge of this fraction is important. Analytical schemes available for the determination of Al_o are tedious and are subject to errors arising from the quantitative separation of Al_o from other aluminium species. Therefore, a mathematical model that predicts the concentration of organically complexed aluminium as well as inorganic aluminium species in solution is desirable.

This chapter identifies a suitable strategy for the determination of important aluminium fractions, and gives particular attention to the development of a mathematical model for predicting the concentration of inorganic and organic monomeric aluminium species in equilibrium. The use of the speciation scheme is illustrated using freshwater samples from the Llyn Brianne

R.W. Edwards et al. (eds) Acid Waters in Wales. 189–201.

catchment. Model predictions of Al_o are compared with experimental determinations.

Chemical characterisation of samples

The equilibrium model for aluminium requires that the concentration of potential complexing ligands and other physico-chemical factors that can influence aluminium speciation are known. Therefore, it is important that the chemistry of freshwater samples under study is well characterized.

Freshwater samples were collected from the surface of the Llyn Brianne reservoir and from three different catchment streams (1, 2, 9). Details of sample collection and chemical analysis are given elsewhere.[12, 13]

Average and extreme values of the main chemical parameters measured in weekly spot samples from the three streams and the reservoir (only average) are shown in Table 12.1. These measurements were carried out in <0.45µm filtrates, and therefore the values shown here may not correspond to truly dissolved concentrations of the elements being analysed. Most iron and significant amounts of aluminium detected in samples from these locations are actually associated with particles >0.015µm, while manganese and other species are mainly dissolved.[12]

The concentration of suspended solids observed in spot samples is low ($<4 mgl^{-1}$), but during storm episodes their concentration can increase 100 fold.[13] This increase in suspended solids concentration is accompanied by a change in the mean particle volume distribution (from $d_p<1\mu m$ to $d_p > 1\mu m$), but it is not clear whether the same conclusion applies to the particle surface distribution.[12] The concentration of suspended solids is the water quality parameter of stream water samples from Llyn Brianne which suffers the largest variation during storm episodes.

The suspended solids collected from stream 2 during a storm episode are mainly micas and quartz type minerals.[14] The measurement of electrokinetic properties (Malvern Zeta Sizer II) of these particles (pH=3.5−10, $I=10^{-3}M$), before and after oxidation of organic matter with H_2O_2, suggested that they are coated with HS. Adsorbed HS may actually control their surface properties, including adsorption of metals. Over the range of pH of most samples from Llyn Brianne (pH = 4.5 to 5) their electrokinetic charge ranges from −0.1 to −0.2 μCcm^{-2}. Model particles of similar mineralogy and electrokinetic properties (illite and quartz) were found[14] to adsorb aluminium over the pH range of the freshwater samples studied here, and it is likely that their adsorption capacity is increased by the presence of HS.[7] Thus, it is not surprising that elevated levels of adsorbed aluminium were detected in samples with high suspended solids concentrations.[13]

Table 12.1. Average and extreme water chemistry of catchments 1, 2 and 9, and average water chemistry at the surface of the reservoir. Data were obtained from weekly spot samples collected in 1984–1985 and filtered through 0.45μm membrane filters before analysis.

		1		2		9		Reservoir
		Mean	Range	Mean	Range	Mean	Range	Mean
pH		4.9	4.3–7.0	4.6	4.3–5.0	5.3	4.1–6.2	5.3
cond	(μScm^{-1})	53	39–70	48	42–56	35	24–53	48
NH_4^+	(μeql^{-1})	1.5	1.4–5.0	1.4	1.4–1.5	1.5	1.4–2.4	2.5
NO_3^-	"	10.4	7.1–28.6	12.9	7.1–28.6	14.8	7.1–10.0	14.3
SO_4^{2-}	"	154	98–260	131	40–168	95	63–132	63
Cl^-	"	247	112–366	213	141–592	146	77–226	196
Na^+	"	203	135–434	174	134–220	139	78–228	177
K^+	"	5.3	0.8–25.6	3.2	0.5–8.2	5.2	1.8–22.0	6.4
Ca^{2+}	"	59	13–129	53	35–75	43	29–73	40
Mg^{2+}	"	60	24–102	46	35–60	54	34–95	37
Cu^{2+}	(μgl^{-1})	3.4	1–19	3.6	2–9	2.3	1–4	–
Pb^{2+}	"	3.9	2–18	16	2–33	3.4	2–8	–
Al_f	"	385	30–851	430	270–680	120	22–380	169
Fe_f	"	68	3–243	60	20–174	96	3–1160	–
Mn_f	"	99	8–170	80	8–125	64	4–250	–
F^-	"	14	–	14.5	–	11.5		15
SiO_2	(mgl^{-1})	1.4	0.7–1.8	1.6	1.2–1.8	0.9	0.5–1.8	0.9
HS	"	1.0	0.2–5.3	0.9	0.2–4.1	0.8	0.2–2.7	–
ss	"	1.7	1.0–4.0	1.7	1.0–4.0	2.1	1.0–5.0	3.5

Aluminium speciation scheme

The scheme used for the determination of aluminium speciation is shown in Fig. 12.1. The total concentration of monomeric aluminium (Al'_m) in the sample is measured with pyrocatechol violet (PCV).[12] Then the sample is filtered through a 0.015μm filter (Nuclepore) and the concentration of monomeric aluminium in the filtrate (Al_m) is measured. Al_{ad} is the difference between Al'_m and Al_m. When the mathematical model developed herein (SPECIAL, see below) accounts for Al-HS interactions then the concentrations of all the species that constitute Al_m (including Al_o) are given by the model. When SPECIAL is used without considering Al-HS interactions the filtrate is passed through an ion exchange resin (Amberlite IR-120) and Al_o is measured in the column effluent. The concentration of inorganic monomeric aluminium (Al_i) is the difference between Al_m and Al_o and the concentration of the species that constitute Al_i can be obtained with SPECIAL.

The use of PCV for the determination of aluminium, the effect of filtration on sample composition and the use of ion exchange resins for the separation between Al_o and Al_i have been discussed elsewhere.[12] This scheme uses a very small pore size membrane filter (d_p=0.015μm) to separate colloidal and adsorbed aluminium before determining the concentration of species in

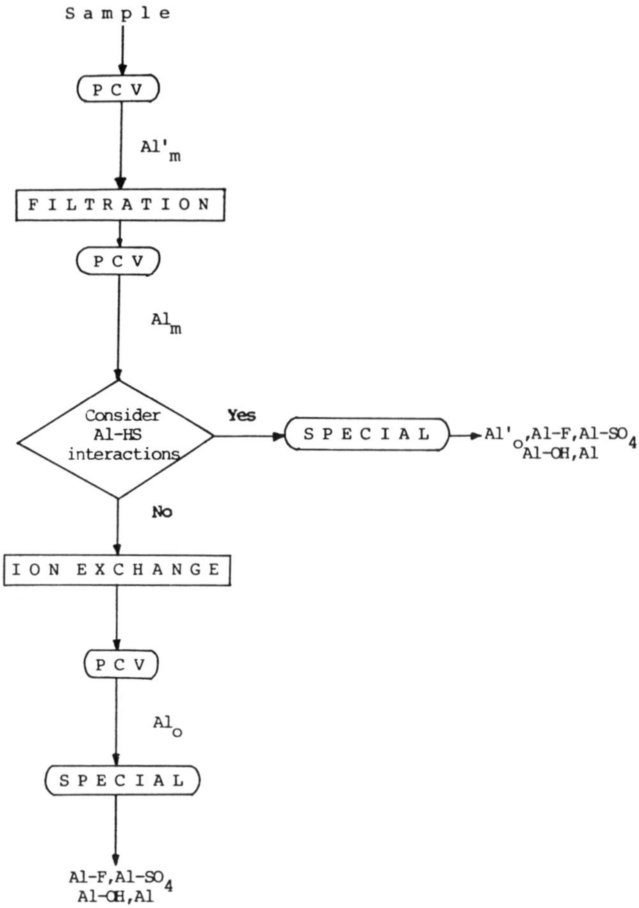

Fig. 12.1. Aluminium speciation scheme for acid waters. The filter pore diameter is 0.015 μm.

solution, and it also allows estimation of Al_{ad}. A feature of the scheme is that Al_o can be either predicted or measured experimentally.

Mathematical model (special)

Aluminium complexation reactions in acid waters

The chemical composition of acidic freshwater samples from the various locations shows (Table 12.1) that OH^-, F^-, SO_4^{2-} and HS are the ligands capable of complexing aluminium. Both fluoride and hydroxyl can give rise to several complexes with aluminium ($AlL_n^{(3-n)+}$) which become more significant as the concentrations of OH^- and F^- increase.

Table 12.2. Set of inorganic reactions and thermodynamic data (25°C) used by SPECIAL.

Inorganic reactions	Log K	Source	ΔH ($Kcalmol^{-1}$)	Source
$Al^{3+} + H_2O = AlOH^{2+} + H^+$	–4.97	[3]	11.90	[3]
$Al^{3+} + 2H_2O = Al(OH)_2^+ + 2H^+$	–9.90	"	22.00	"
$Al^{3+} + 4H_2O = Al(OH)_4^- + 4H^+$	–23.0	"	44.06	[4]
$Al^{3+} + SO_4^{2-} = AlSO_4^+$	3.2	[15]	2.29	[16]
$Al^{3+} + 2SO_4^{2} = Al_2(SO_4)_3^-$	5.2	"	3.07	"
$Al^{3+} + F^- = AlF^{2+}$	7.10	[20]	0.7	[25]
$Al^{3+} + 2F^- = AlF_2^+$	11.98	"	1.1	"
$Al^{3+} + 3F^- = AlF_3^\circ$	15.83	"	2.0	"
$Al^{3+} + 4F^- = AlF_4^-$	18.53	"	2.2	"
$Al^{3+} + 5F^- = AlF_5^{2-}$	20.20	"	2.2	"
$Al^{3+} + 6F^- = AlF_6^{3-}$	20.67	"	1.8	"

Table 12.3. Thermodynamic data (25°C) for Al^- HS interactions used by SPECIAL N_1 and N_2 expressed in $eqgC^{-1}$.

		pH				ΔH
Al-HS interactions		3.5	4.0	4.5	5.0	
$Al^{3+} + HS_1 = AlHS_1$	$LogK'_1$	7.06	7.43	7.45	7.70	3.9
	$-LogN_1$	3.58	3.48	3.03	2.88	
$Al^{3+} + HS_2 = AlHS_2$	$LogK'_2$	5.74	5.48	4.78	4.90	3.9
	$-LogN_2$	3.48	3.18	2.78	3.48	

The cations that can compete with aluminium for HS over the pH range 3.5–5.5 are Ca^{2+}, Mg^{2+}, Mn^{2+}, Cu^{2+}, Pb^{2+}, Fe^{2+} and Fe^{3+}. The concentration of iron (<0.015μm) is below the detection limit (<30μgl^{-1}), and the concentration of Cu^{2+} and Pb^{2+} in freshwaters from the Llyn Brianne catchment (Table 12.1) is too low to have a significant effect on aluminium speciation.[23] Complexes of Ca^{2+}, Mg^{2+} and Mn^{2+} with HS are too weak to affect significantly aluminium complexation by HS.[5]

The set of reactions and their corresponding thermodynamic data which comprise SPECIAL are listed in Tables 12.2 and 12.3. Conditional equilibrium constants for HS-Al reactions have been determined.[14] Two site models[28] were used to represent Al–HS reactions over the pH range 3.5 to 5, i.e. HS are considered to be made up of two types of complexing ligands (HS_1 and HS_2) capable of complexing aluminium:

$$HS_1 + Al^{3+} \rightleftharpoons AlHS_1 \quad (1)$$
$$HS_2 + Al^{3+} \rightleftharpoons AlHS_2 \quad (2)$$

The parameters of these two reactions are the concentration of each site type (N_1 & N_2) and two conditional stability constants (K'_1 & K'_2). As these parameters were only determined at pH 5.0, 4.5, 4.0 and 3.5, the values of the parameters at any pH in the range 3.5–5 were obtained by interpolation. The enthalpies of reaction at 25°C for eqns. (1) & (2) were assumed to be the same as that determined by Plankey and Patterson.[21]

Solution of the law of mass action and mass balance equations

The problem of calculating the equilibrium speciation of a system where many simultaneous reactions occur consists of minimizing the Gibbs free energy of the system subject to the constraint of mass balance.[29] This is usually done using the equilibrium constant approach,[19] where the set of non-linear laws of mass action for basic components and derived species are incorporated into the set of linear mass balance equations for the basic components, and solved iteratively for the correct equilibrium composition.

The generalized form of the non-linear law of mass action equations that have to be solved is

$$K_n = \{C_n\}^{m_n} / \prod_i \{M_i\}^{m_{in}} \qquad (3)$$

where, K_n is the stability constant for the formation of the complex species (eg. $AlOH^{2-}$, AlF^{2-}, etc.) whose activity is $\{C_n\}$, m_n is the stoichiometric coeffecient of the complex C_n (>0), $\{M_i\}$ is the activity of the i-th species and m_{in} are stoichiometric coefficients (<0 for reactants and >0 for products).

The mass balance equation for each basic component (eg. Al, SO_4, F) can be written as

$$[AC_i] = [A_i] + \sum_n m_n[C_n] \qquad (4)$$

where, AC_i is the concentration of the i-th basic component, $[A_i]$ is the concentration of the free ionic species and C_n is obtained from eqn. (3).

Solution of the set of non-linear equations

The numerical techniques that can be used to solve the set of non linear equations arising from the stability constant approach to equilibrium speciation have been reviewed by Zeleznik and Gordon[31] and van Zeggeren and Storey.[30] These techniques are mainly Newton-Raphson, successive approximations and the so called nested iterations method. Herein, one of the successive approximations methods[11] was chosen. This method is particularly suited for:

1. equilibrium systems which do not consider equilibrium with solid phases in calculating equilibrium speciation in solution and
2. systems where the set of non-linear simultaneous equations to be solved is not excessively large.

The iteration algorithm by which a further approximation to the solution is obtained from the last calculated solution is expressed as[11]

$$\{A_i\}^{k+1} = \{A_i\}^k \left[\frac{AT_i}{AC_i^k}\right]^{1/p_i} \tag{5}$$

where A_i^{k+1} and A_i^k are successive approximations to the equilibrium concentration of the i-th reactant, AT_i and AC_i are the analytically measured and the numerically calculated (from eqn. (4)) concentrations of each basic component respectively, and the parameter p_i is the largest stoichiometric coefficient of the i-th reactant in order to secure convergence.

Activity and temperature corrections

In the limit of infinite dilution, all ionic activities approach ionic concentrations, and activity coefficients approach unity. Single ion activities are calculated with the expression: $\{A_i\} = \gamma_i[A_i]$, where $\{A_i\}$ is the activity of ionic species A_i, $[A_i]$ is its concentration and γ_i is its activity coefficient. γ_i were evaluated here using the extended Debye-Hückel equation (valid at $I<10^{-1}M$).[26]

The stability constant of each reaction, K_n, is, in general, dependent on temperature and independent of pressure (for $P <$ ca. 100atm).[27] The variation of K_n with temperature over narrow ranges of temperature can be represented by the van't Hoff eqn.

$$\text{Log } K = \text{Log } K^\circ - \frac{\Delta H^\circ}{2\,3R}\left[\frac{1}{T} - \frac{1}{T^\circ}\right] \tag{6}$$

where, K° is the stability constant at the reference temperature T° (25°), K is the stability constant at temperature T, R is the gas constant and ΔH° is the enthalpy of reaction at 25°C.

Implementation of SPECIAL

The algorithm used to perform the various calculations embraced by SPECIAL is illustrated in Fig. 12.2. The input data are pH, m_{in}, m_n, K_n, ΔH°_n, AT_i and the ion size parameters (a) of the Debye-Hückel equation. A computer code in FORTRAN F77 implemented SPECIAL on a VAX 8700 computer.

Results and discussion

Analytical determinations

A particular advantage of this speciation scheme (Fig. 12.1) is that colloidal material is separated before determining Al_i and Al_o, so that possible interference by Al_{ad} in their determination is avoided.[12] This becomes particularly

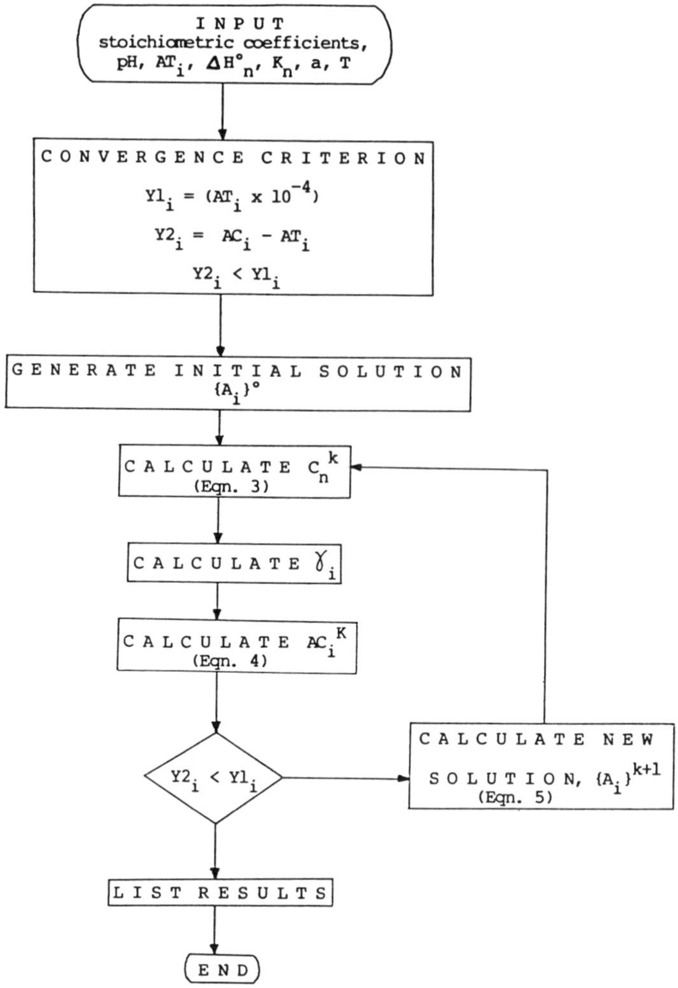

Fig. 12.2. Flow diagram for the calculation of equilibrium speciation using SPECIAL.

important with samples containing high suspended solids (ss) concentrations occurring during storm and snowmelt episodes. For example, spot samples (ss < ca. $5 mgl^{-1}$) contained Al_{ad} $<20 \mu gl^{-1}$ and in samples from episodes (ss$>20 mgl^{-1}$) Al_{ad} was 50–200μgl^{-1}1.[13] A significant proportion (ca.40–60%) of the adsorbed aluminium is associated with coarse particles ($d_p>0.4 \mu m$), and therefore, standard sample filtration through 0.4μm pore size filters will reduce the level of interference. However, the concentration of adsorbed aluminium associated with the colloidal fraction (0.015-0.4μm) may be important (<ca.100μgl^{-1}) and could induce some error in the determination of Al_i and Al_o. This error probably is not important if only parameterization of toxicity using Al_i is needed, but could be significant if a detailed toxicological analysis is to be carried out.

Since filtration through 0.015μm pore size filters is a slow process, the volumes of sample to be filtered were minimized by, (i) using small volumes of sample (3.5ml) for each PCV analysis, and (ii) using a small ion exchange resin packing (1ml).[12]

By conditioning the ion exchange resin with electrolyte solutions of the same ionic strength and pH as the samples treated, the pH variation of the ion exchanged samples was pH<0.2, which minimized decomplexation and exchange of labile Al-HS complexes. Similarly, residence times of the samples in the column were optimized (t_R=0.5-1min) for samples collected from the Llyn Brianne catchment in order to reduce exchange of labile Al-HS complexes while all inorganic monomeric aluminium forms are exchanged.

Although the presence of polymeric aluminium forms in acidified freshwater samples has not been ascertained, they might be present[24] and, if they are, PCV is likely to detect them. As these forms of aluminium are also exchangeable they would be measured as a part of Al_i. On the other hand, PCV might not be a strong enough Al-complexing agent to detect the more stable Al-HS complexes.

Determination of inorganic monomeric aluminium species using SPECIAL

SPECIAL was validated using an equilibrium system whose solution has been reported.[10] The system is the set of four simultaneous hydrolysis reactions of aluminium for the formation of $AlOH^{2+}$, $Al(OH)_2^+$, $Al(OH)_3^\circ$ and $Al(OH)_4^-$, with stability constants taken from May et al.[18] The aluminium speciation of this system expressed as percentage of the total aluminium concentration, between pH 4 and 8, obtained using SPECIAL was compared with the speciation obtained by French[10] (Fig. 12.3). Both graphs show the same trends for all species. The agreement between the predictions of both models at every pH is within a few percent for all aluminium species.

The concentration of the species that constitute Al_i in samples from the Llyn Brianne catchment with different chemistry has been obtained using SPECIAL. A detailed account of these results is given elsewhere.[13] The concentration of fluoride complexes and hydroxy-complexes of aluminium are of the same order as Al_o and constitute most of the Al_i in spot samples with pH>5.0, but the concentrations of Al^{3+} and sulphate complexes are very small. In episode samples (pH<ca. 4.7) Al^{3+} is the predominant species, followed by hydroxo-complexes, with flouride and sulphate complexes being negligible.

Determination of organically complexed aluminium

The concentration of organically complexed aluminium and the concentration of other species in Al_m (Fig. 12.1) in samples from the Llyn Brianne catchment were predicted using SPECIAL. It should be noted that conditional stability constants for Al-HS interactions at pH>5.0 are not available, therefore, the concentration of organically complexed aluminium for samples with

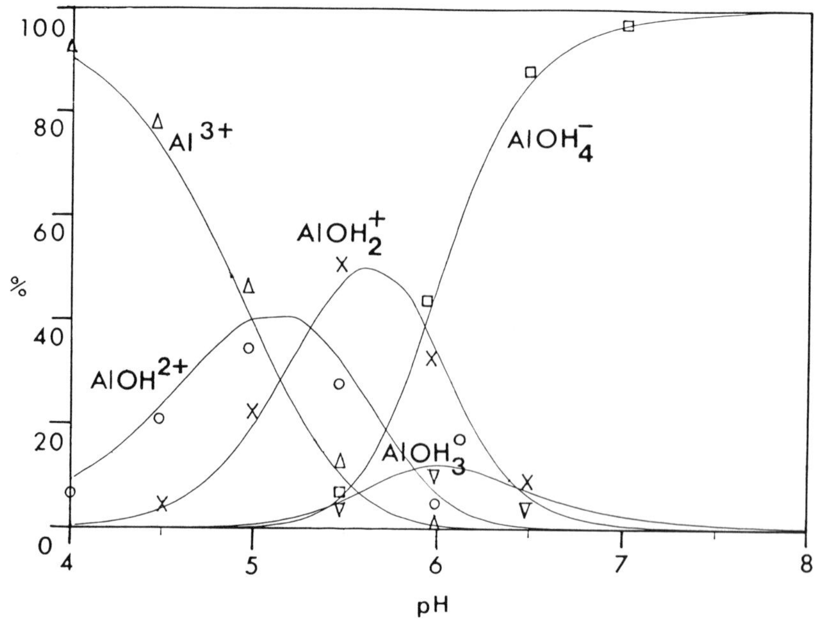

Fig. 12.3. Aluminium speciation as a function of pH in a system considering only aluminium hydrolysis (thermodynamic data from May et al. 1979): French 1985 (discrete distributions) and calculated using SPECIAL (continuous distributions).

pH>5.0 could not be predicted; samples with pH<4.3 were not available.

Predicted values of organically complexed aluminium (Al'_o) are in reasonable agreement with values measured experimentally (Al_o) (Fig. 12.4). This is important considering that the conditional stability constants for Al-HS reactions were determined for a HS sample which might differ from the HS present in other samples. The line with slope 1 that passes through the origin shows the case when predictions and measurements coincide. The set of points indicates that for a given concentration of Al^{3+}, HS and pH, the concentration of organically complexed Al per unit mass of HS predicted is slightly larger than that measured experimentally for the low metal loading range (Al_o<1.0mmoll^{-1}), and it becomes smaller than the experimental one as metal loading increases. This set of points can be fitted by linear regression to a straight line with slope 0.6.

Conclusions

1. A more accurate experimental determination of monomeric aluminium forms (Al_m, Al_o and Al_i) is achieved by removing Al_{ad} associated with particles >0.015μm by filtration before colorimetric determinations and sample fractionation are carried out. If suspended solids are not removed

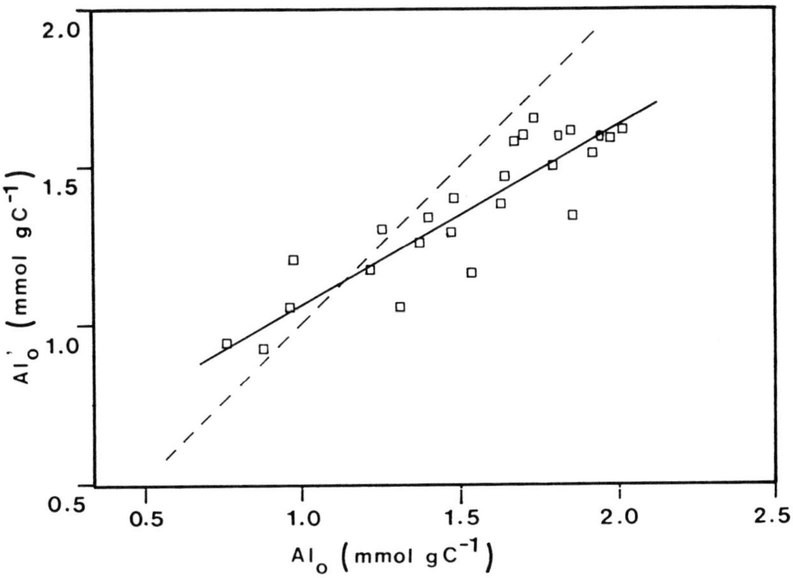

Fig. 12.4. Concentration of organically complexed aluminium in freshwaters from the Llyn Brianne catchment: (Al_0), measured experimentally, and (Al'_0), predicted using SPECIAL.

at all or only particles >0.45μm are removed by filtration, interference by Al_{ad} can produce an overestimation of Al_m, Al_o and Al_i.

2. Al_{ad} is negligible (<20μgl^{-1}) for spot samples with ss <5mgl^{-1} collected from the Llyn Brianne catchment, but it is significant (<200μgl^{-1}) for episode samples with high ss concentrations (>20mgl^{-1}). Between 40 and 60% of the adsorbed aluminium was found to be associated with >0.4μm particles.
3. The high concentrations of suspended solids found in the streams during episodes may act as important sources or sinks of mobile aluminium, depending on the physico-chemical conditions and composition of the aqueous solution and therefore, they may control to some extent the aluminium chemistry of these samples. The extent to which adsorbed aluminium is biologically available is unknown, but it is likely to be less active than inorganic monomeric aluminium.
4. The mathematical model (SPECIAL) developed for thermodynamic equilibrium of aluminium in acid waters includes Al–HS interactions as well as inorganic reactions. This model can be used to predict the concentration of organically complexed aluminium, together with the concentration of inorganic monomeric aluminium species. Model predictions compare favourably with experimental determinations of Al_o in samples from the Llyn Brianne catchment. It is envisaged that if its application is validated using a wider range of samples from various locations, model predictions can substitute the tedious and errorsome experimental determinations of Al_o.

References

1. Amankwah S.A. (1987). Evaluation of analytical procedures for aluminum speciation studies. Symposium on Environmental Chemistry of Aluminum. 194th National Meeting (New Orleans), American Chemical Society, Washington.
2. Arent, L.J. and Lewis, T.E. (1987). The stability of aluminum species in natural field audit samples: possible application as a quality control standard. Symposium on Environmental Chemistry of Aluminum. 194th National Meeting (New Orleans), American Chemical Society, Washington.
3. Baes, C.F. and Mesmer, R.E. (1976). The Hydrolysis of Cations. Wiley, New York.
4. Ball, J.W., Nordstrom, D.K. and Jenne, E.A. (1980). Additional and revised data for WATEQ2, computerized model for trace and major element speciation in mineral equilibria of natural waters. Geological Survey Water Resources Investigations, Menlo Park, California.
5. Buffle, J. and Altmann, R.S. (1987). Interpretation of metal complexation by heterogeneous complexants. In: Aquatic Surface Chemistry: Chemical Processes at the Particle Water Interface (ed. W. Stumm), John Wiley & Sons, New York.
6. Campbell, P.G.C., Bisson, M. Bougie, R., Tessier, A. and Villeneuve, J.P. (1983). Speciation of aluminum in acidic freshwaters. *Analytical Chemistry*, 55: 2246–2252.
7. Davis, J.A. and Leckie, J.O. (1978). Effect of adsorbed complexing ligands on trace metal uptake by hydrous oxides. *Environmental Science and Technology*, 12: 1309–1315.
8. Driscoll, C.T. (1984). A procedure for the fractionation of aqueous aluminum in dilute acidic waters. *Int. J. Environ. Anal. Chem.*, 16: 267–285.
9. Florence, T.M. (1982). The speciation of trace metals in waters. *Talanta*, 29: 345–365.
10. French, P. (1985). Thermodynamic modelling of aluminium speciation and solubility in natural waters. Technical report, 1041-M, Water Research Centre.
11. Ginzburg, G. (1976). Calculation of all equilibrium concentrations in a system of competing complexation. *Talanta*, 23: 149–152.
12. Goenaga, X., Bryant, R. and Williams, D.J.A. (1987). Influence of sorption processes on aluminum determinations in acidic waters. Anal. *Chem.*, 59: 2673–2678.
13. Goenaga, X. and Williams, D.J.A. (1988). Aluminium speciation in surface waters from a Welsh upland area. *Environmental Pollution*, 52: 131–149.
14. Goenaga, X. (1988). Aluminium Speciation in Acid Waters. PhD thesis, University of Wales, U.K..
15. Hem, J.D. (1968). Graphical methods for studies of aqueous aluminum hydroxide, fluoride and sulphate complexes, U.S. Geological Survey Water Supply Paper 1827-B.
16. Izatt, R.M., Eatough, D., Christensen, J.J. and Bartholomew, C.H. (1969). Calorimetrically determined Log K, ΔH° and ΔS° values for the interaction of sulphate ion with several bi- and tri-valent metal ions. *J. Chem. Soc.*, A: 47–53.
17. Johnson, N.M., Driscoll, C.T., Eaton, J.S., Likens, G.E. and McDowell, W.H. (1981). Acid rain, dissolved aluminum and chemical weathering at the Hubbard Brook experimental forest, New Hampshire. *Geochim. Cosmochim. Acta*, 45: 1421–1437.
18. May, H.M. Helmke, P.A. and Jackson, M.L. (1979). Gibbsite solubility and thermodynamic properties of hydroxy-aluminum ions in aqueous solutions at 25°C. *Geochim. Cosmochim. Acta*, 43: 861–868.
19. Nordstrom, D.K. et al. (19 authors) (1979). A comparison of computerized chemical models for equilibrium calculations in aqueous systems. In: Chemical Modeling in Aqueous Systems (ed. E.A. Jenne), Symposium Series no. 93. American Chemical Society, Washington.
20. Pitter, P. (1985). Forms of occurence of fluorine in drinking water. *Water Research*, 19: 281–284.
21. Plankey, B.H. and Patterson, H.H. (1987). Kinetics of aluminum – fulvic acid complexation in acidic waters. *Environmental Science and Technology*, 21: 595–601.
22. Schecher, W.D. and Driscoll, C.T. (1987). An evaluation of uncertainty associated with equilibrium calculations. *Water Resources Research*, 23: 525–534.

23. Schnitzer, M. and Khan, S.U. (1972). Humic Substances in the Environment. Marcel Dekker, New York.
24. Smith, R.W. and Hem, J.D. (1972). Effect of aging on aluminum hydroxide complexes in dilute aqueous solutions. Geological Survey Water-Supply Paper 1827-D, Washington.
25. Smith, R.W. and Martell, A.E. (1976). Critical Stability Constants. Plenum Press, New York.
26. Stumm, W. and Morgan, J.J. (1981). Aquatic Chemistry. An Introduction Emphasizing Chemical Equilibria. John Wiley & Sons, London.
27. Truesdell, A.H. and Jones, B.F. (1973). WATEQ a computer program for calculating chemical equilibria of natural waters, NTIS Technical Report PB2–20464, Springfield, VA77.
28. Turner, D.R., Varey, M.S., Whitfield, M., Mantoura, R.F.C. and Riley, J.P. (1986). Electrochemical studies of copper and lead complexation by fulvic acid. I. Potentiometric measurements and a critical comparison of metal binding models. *Geochim. Cosmochim. Acta*, 50, 289–297.
29. Westall, J., Zachary, J.L. and Morel, F.M.M. (1976). MINEQL a computer program for the calculation of chemical equilibrium composition of aqueous systems. Technical Note no. 18, Dept. of Civil Eng. MIT, Cambridge, MA.
30. van Zeggeren, F. and Storey S.H. (1970). The Computational Aspects of Chemical Equilibria. Cambridge University Press, London.
31. Zeleznik, F.J. and Gordon, S. (1968). Calculation of complex chemical equilibria. *Industrial and Engineering Chemistry*, 60: 27–57.

13. Catchment characteristics and basin hydrology: their effects on streamwater acidity

S.C. BIRD, R.P.D. WALSH and I.G. LITTLEWOOD

Introduction

Hydrological factors are important in accounting for the spatial and temporal variations in streamwater acidity and water chemistry in upland Wales. However, the hydrological character of a catchment is itself highly dependent on other environmental variables, notably basin morphology, climatic factors, vegetation and land use, solid and drift geology, soils and Man. The hydrological system provides the means by which these variables act to influence streamwater acidity, and Fig. 13.1 depicts the more important hydrological influences on streamwater acidity and their links to other environmental variables. Since geology, land use and soils are dealt with in separate chapters, attention here is focussed on climate, basin morphometry, and hydrology and their influences on streamwater acidity in upland Wales. First, climatic factors and their role in contributing to problems of streamwater acidity are described. Then, using the Llyn Brianne study catchments as examples, the effects of varying catchment morphology on stream responses to rainfall and on acidity are examined. Third, the key role played by runoff processes and the runoff responses of streams and how they reflect differences in topography, soils and land use is examined. Finally, the nature of the dynamic relationship between streamflow and acidity for some catchments at Llyn Brianne is described.

Climate and weather

Long-term climatic records of upland sites in Central Wales are few. There are two automatic weather stations (AWS), Trawsnant (378m altitude) in the Llyn Brianne area established in 1985, and the meteorological station at Moel Cynedd (358m altitude) in the Plynlimon area some 40 km north of Llyn Brianne established in 1968.

Mean annual rainfall in the Llyn Brianne area ranges from around 2000mm at Trawsnant AWS to 1800mm in the Camddwr tributary catchment and is a little lower than in the Plynlimon area, where rainfall averaged 2192mm over

R.W. Edwards et al. (eds) Acid Waters in Wales. 203–221.

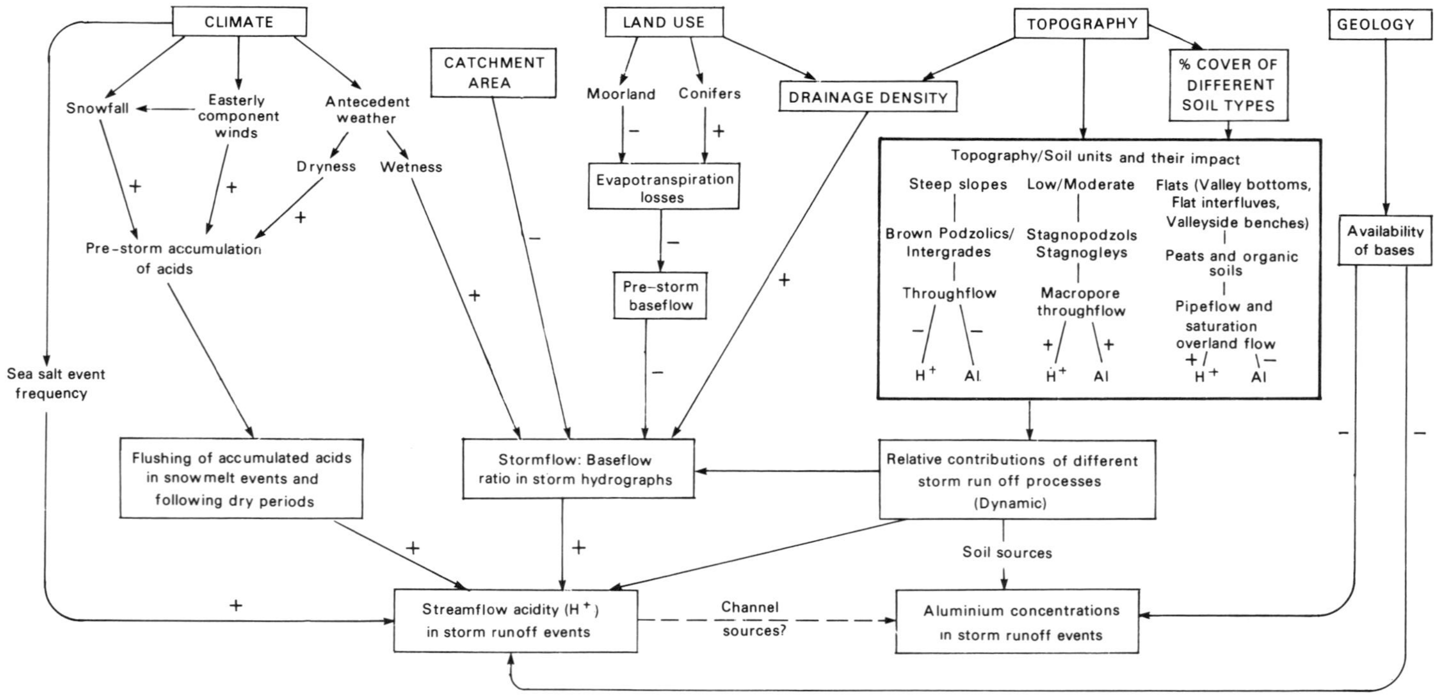

Fig. 13.1. A conceptual model of hydrological factors influencing streamwater acidity and aluminium levels in storm runoff events in upland Wales. (Direct relationships between variables are indicated by +; inverse relationships by −).

the period 1968–77 at Moel Cynedd and around 2500mm at higher altitudes in the upper Severn and upper Wye catchments. However, the two areas are broadly similar in their rainfall characteristics. Both exhibit the marked autumn and early winter rainfall maximum, characteristic of western and upland Wales.[23] Fig. 13.2a shows that, over the 10-year period 1968–77, monthly mean rainfall at Moel Cynedd, Plynlimon, ranged from 284mm in the peak month of November down to 117mm in June;[26] over the rather longer period 1961–85 at Nant y Maen in the north of the Camddwr catchment in the Llyn Brianne area, monthly means ranged from 210mm in December to 99mm in June. Rainfall frequency is very high (Fig. 13.2b) averaging 198.5 days per year at both Trawsnant AWS (Llyn Brianne) and Moel Cynedd (Plynlimon) during the period 1985–87, with a longer-term mean at Moel Cynedd for 1968–75 of 193 days per year.[25] High rainfall frequency is one of the main reasons for the high interception losses in forested areas of central Wales.[27]

The relatively high frequency of large falls of rain in upland Wales (Fig. 13.2e) is also likely to be an important factor in determining the frequency and magnitude of acid runoff episodes. During the period 1985–87 at Trawsnant, there were 70 days per annum with at least 10mm, 18 days with at least 25mm and 5 days with at least 50mm, with peak frequencies during autumn and early winter when the most active depressions tend to occur. This means that acid runoff episodes which, as will be seen later, are typically associated with storm peaks of streamflow, are more frequent in Central Wales than in areas with less frequent heavy rainstorms.

Both rainfall amounts and intensity-frequency characteristics vary considerably both regionally and locally within upland Wales (Fig. 13.2e). For example, within the Llyn Brianne area, rainfall inputs in the southern catchments draining directly into Llyn Brianne (streams 1, 2, 3, 4, 5 and 6) are some 15% higher than in the more northerly catchments (streams 7, 8, 9, 10 and 11), despite the higher altitude of the latter (see Chap. 1; Fig. 1.1). This is reflected in the higher peakflows and baseflows of the southern catchments. Differences in rainfall are greatest in autumn and early winter when orographic enhancement of rainfall is most effective in the southern catchments, which have greater local relief and are more exposed to the southwesterly winds.

Figs. 13.2c and 2d show snowfall frequency and the number of days with snow lying at Moel Cynedd for 1968–77[26] and 1985–87. Snowfall frequency is low (37 days per annum) compared with northern England and Scottish upland locations, despite the higher total precipitation of Moel Cynedd; the contrast in lying snow frequency (30 days at Moel Cynedd, compared with 42 at Malham in the North Pennines, and 65 at Glenmore Lodge in Scotland) is just as striking. Because of the correlation between snowmelt and high acidity events, this is highly significant and indicates that acidity problems in Wales are not as severe as they would be if snowfall were more frequent.

In addition, frost frequency is high, averaging 95 days per year at Moel

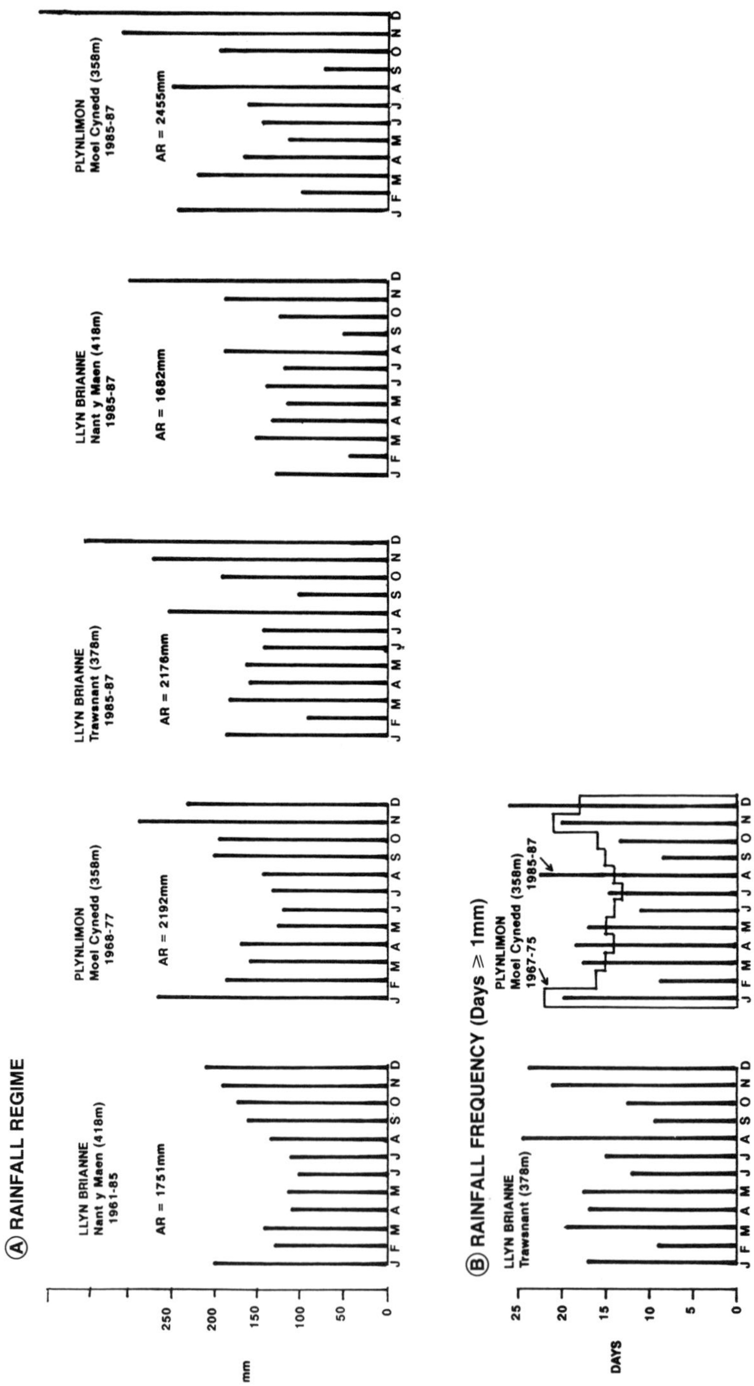

(A) RAINFALL REGIME
LLYN BRIANNE Nant y Maen (418m) 1961-85
AR = 1751mm
PLYNLIMON Moel Cynedd (358m) 1968-77
AR = 2192mm
LLYN BRIANNE Trawsnant (378m) 1985-87
AR = 2176mm
LLYN BRIANNE Nant y Maen (418m) 1985-87
AR = 1682mm
PLYNLIMON Moel Cynedd (358m) 1985-87
AR = 2455mm
mm
250
200
150
100
50
0
J F M A M J J A S O N D
(B) RAINFALL FREQUENCY (Days ≥ 1mm)
LLYN BRIANNE Trawsnant (378m)
PLYNLIMON Moel Cynedd (358m)
1967-75
1985-87
DAYS
25
20
15
10
5
0

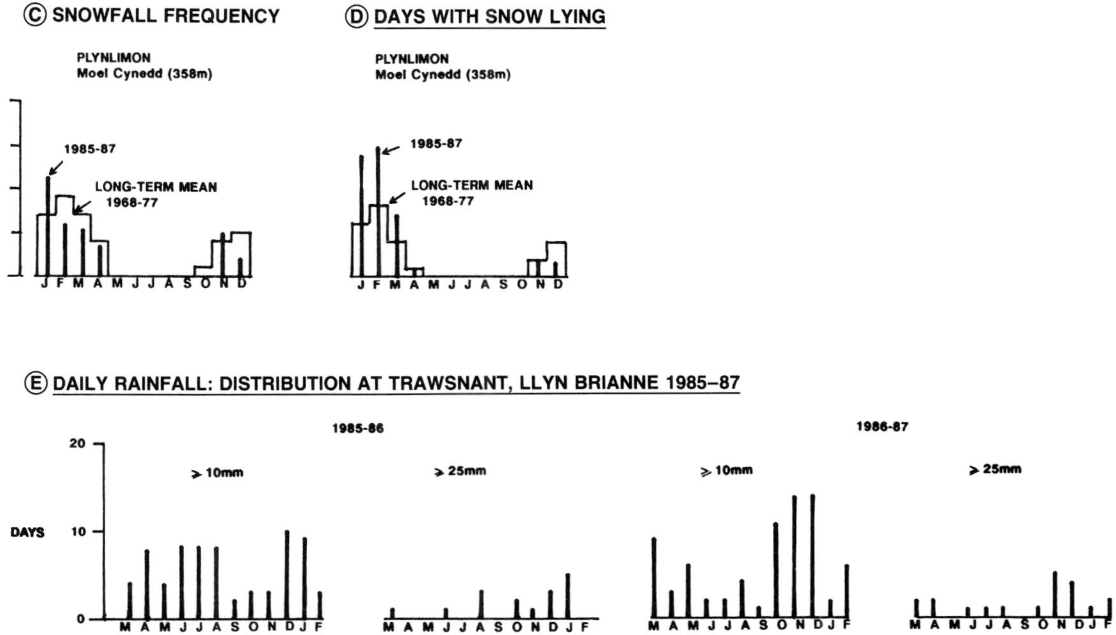

Fig. 13.2. Rainfall and snowfall characteristics at Llyn Brianne and Plynlimon.

Cynedd,[26] and, in 1983–87, around 90 days per year at Trawsnant at Llyn Brianne. Soil temperature at 30 cm depth, however, rarely dips below freezing, probably reflecting the high water content of the soil and the insulating properties of the thick grass vegetation cover.

Wind direction is an important variable as regards deposition of atmospheric gases and aerosols. The location of upland Wales to the west of the main industrial areas of Great Britain is significant with respect to the amount and time of occurrence of wet, dry and occult acid deposition. In contrast with the Pennines and the Grampians, the prevalent westerly and south-westerly winds of Great Britain are relatively unpolluted in Wales, though sea-salt events are more common because of wind trajectories from the Atlantic. The polluting airstreams for Llyn Brianne are south-easterly, easterly and north-easterly winds that have passed over the industrial areas of South Wales, the Midlands and North West England respectively (see Chap. 4). Winds with an easterly component (0–180° E of N) occurred at Llyn Brianne during the period 1985–87 for about 35% of the time. During the two year period March 1985 to February 1987, there were seven months when the frequency of winds with an easterly component was 50% or greater; notable months with winds having a dominantly easterly component were October 1985 (75%), February 1986 (100%) and January 1987 (84%), the latter two also being cold, snow-bound months.

Catchment topography and morphometry

The morphology of a drainage basin influences both the total volume and time distribution of its discharge,[3, 21] with implications for streamwater acidity levels. Fig. 13.1 indicates three of the most important morphological factors influencing hydrology and streamwater acidity: drainage basin area, drainage density and topography.

Streamflow pH generally increases in a downstream direction as catchment area increases. In part this is due to a downstream increase in the proportion and temporal reliability of groundwater-derived baseflow, which because of its long residence time and contact with bedrock, glacial and periglacial deposits, tends to be of relatively high pH. At Llyn Brianne a second reason is that downstream sections of catchments tend to be incised within steep valley-side slopes characterised by relatively permeable mineral soils. These promote throughflow of comparatively high pH. In contrast headwater sections tend to be dominated by high altitude peat-covered interfluves that generate pipeflow and overland flow of low pH. The relative importance of the latter thus tends to decline with increasing catchment size.

Drainage density and steep topography both tend to increase the speed and magnitude of storm responses to rainfall. As storm runoff tends to be of low pH, the more peaked the storm hydrograph, the greater the storm runoff/baseflow ratio, and hence the more acid streamwater at peakflow will be (Fig.

Table 13.1. Some morphometric and topographic characteristics of selected catchments at Llyn Brianne.

Basin	Area km^2	Basin relief m	Mean slope	Drainage density km^{-1} km^{-2}
1	2.525	197	0.171	1.72
2	1.053	147	0.121	1.20
3	0.643	111	0.129	2.08
4	0.328	118	na	2.34
5	0.680	216	0.254	2.84
6	0.659	197	0.197	1.72
8	0.836	141	0.203	1.40
9	0.491	110	0.132	1.71
10	0.342	145	0.168	2.05
11	0.717	122	0.173	1.16

13.1). The influence of topography, however, is not simply one of mean slope.

The extent and distribution within a catchment of easily saturated low slope areas that act as important source areas of storm runoff are also important. Furthermore, there are strong interrelationships between such factors as slope angle and slope profile position and soil type; moderate and steep slopes tend to be more permeable and have less easily saturated soil types than those of lower slope.

Although in general in upland Wales, topography is steep and drainage densities are high by British standards, there is considerable local variation which in turn has an impact on streamwater acidity. At Llyn Brianne, the study catchments were subjected to detailed morphometric analysis by Boon,[4] who identified some important constrasts (Table 13.1). Although all catchments are small, afforested catchments 1 and 2 are substantially larger, and catchment 4 and moorland catchment 10 are substantially smaller than the mean. Whereas the larger catchments have reliable baseflows in summer, smaller catchments tend to dry up in long dry spells. Drainage densities (as derived from the blue-line network of O.S. 1:25,000 maps) range from 1.16 to 2.84 km^{-1} km^{-2}; true drainage densities in the afforested areas, however, are much higher than the values given in Table 13.1, which does not include the afforestation drainage ditch network. Mean slopes range from 0.12 to 0.25 and basin relief from 110 to 216m. Moorland catchment 5, which has the highest drainage density, highest mean slope and greatest basin relief, not surprisingly experiences the greatest storm response to rainfall of all the Llyn Brianne catchments.

More exact quantification of catchment topography for hydrological purposes can be partially achieved using map digitizing and computer surface fitting techniques.[13] Such techniques have recently led to the development of hydrological models of catchments[1, 2] that take into account dynamic contributing areas of overland flow and throughflow.[15] As a preliminary application of one such model 'TOPMODEL', described in Chap. 16, most of the Llyn Brianne basins have been digitized yielding the catchment surface slope information presented in Fig. 13.3 and Table 13.1. The cumulative slope frequency curves shown in Fig. 13.3 clearly demonstrate differences in the areal extent of land of different slope between the various catchments. The steep slopes of moorland catchment 5 (65% of the land with slope angles exceeding 0.2) and the large expanses (>10%) of low slope peaty interfluve and valley bottom land in afforested catchments 1, 2 and 3 clearly contrast.

Catchment hydrology

General features and the impact of conifer afforestation

Catchments in upland Wales are characterised by high and rather flashy runoff. This reflects the high rainfall amounts and intensities, the high relief and the relatively impermeable nature of the geology, surficial deposits and soils of the region. There are, however, important variations between catchments related to contrasts in land use and topography. Long-term catchment studies at Plynlimon have demonstrated that evapotranspiration losses under conifer forest are more than twice as high as under moorland; this has been attributed to the higher interception losses produced by a denser and aerodynamically rougher forest canopy.[6, 26] Thus the annual runoff and baseflows of afforested streams are significantly reduced compared with those of moorland streams. As a consequence, the proportion of well-buffered, high pH baseflow in the streamflow of afforested catchments during storm peaks is reduced, and this is an important contributory reason for the lower pH levels of afforested streams (Fig. 13.1). Effects of afforestation on the size of the flood peaks are more complex. The tendency for the greatly expanded drainage network produced by drainage ditching to increase flood peaks in early years following afforestation[29, 30] becomes subordinate, as a forest becomes more mature, to the effects of soil drying caused by increased interception and transpiration.[16]

Runoff process

Newson[28] has stressed the complexity of runoff process hydrology and hydrochemistry in catchments in upland Wales. He points out that the assumption, often made in acid runoff modelling studies,[7] that runoff pathways can be simply divided into 'shallow/rapid/unbuffered' and 'deep/slow/well buffered' types, characterised by contrasting and easily prescribed chemical characteristics, does not apply in upland Wales. Here impermeable peats, peaty podzols and peaty gleys are of great spatial significance and pipeflow tends to be an important runoff process.

In upland Wales five broad categories of runoff process, each of contrasting

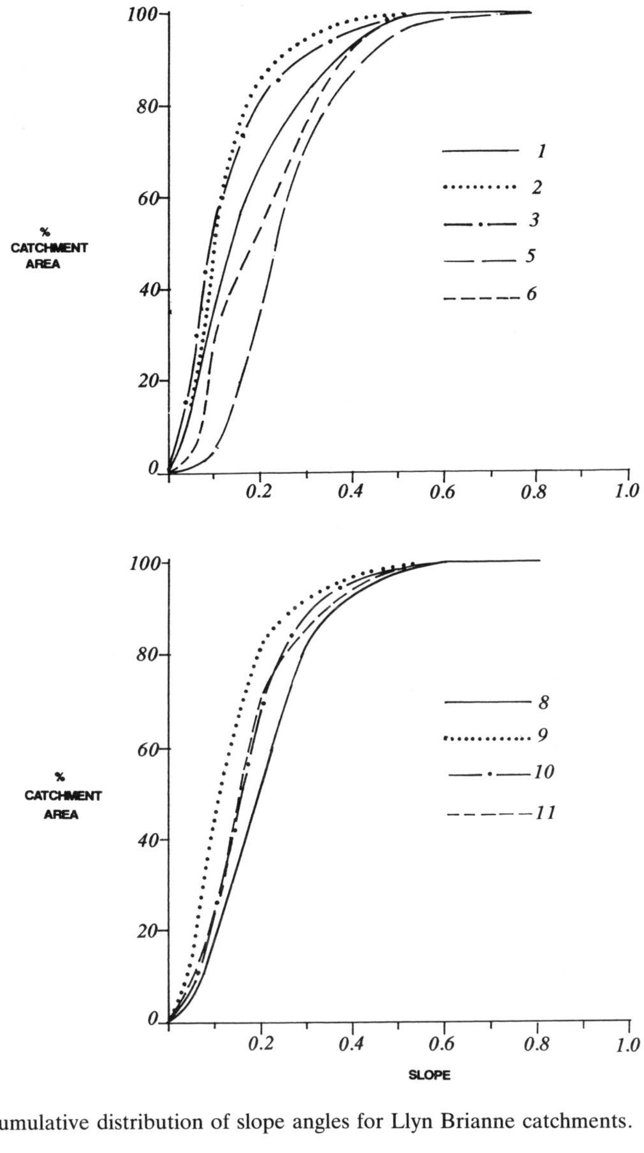

Fig. 13.3. Cumulative distribution of slope angles for Llyn Brianne catchments.

chemical and hydrological significance, are important: (1) shallow throughflow (often partly via macropores), (2) pipeflow, (3) saturation overland flow, (4) deep throughflow at the soil/rock or soil/till interface and (5) groundwater (though this is often difficult in practice to distinguish from deep throughflow). These runoff processes and their spatial distribution within catchments vary not only temporally with antecedent moisture conditions (and hence weather and season), but also with differences between catchments in their topography, soil, types vegetation and surficial deposits.[14] Different slope/

vegetation/soil units tend to be characterised by contrasting runoff processes. The first three are of greatest significance as regards generating storm runoff and enhanced streamwater acidity and these are considered in greater detail.

Throughflow tends to be important only on well-drained, permeable soil types (notably brown podzolics, intergrades and stagnopodzols) generally associated with steeper slopes in the catchments. Throughflow velocities are slow compared either with overland flow[24] or pipeflow fed by overland flow, and catchments dominated by throughflow tend to be characterised by smaller peak discharges, but high recession limb flows as throughflow reaches the streams. However, throughflow responses will be faster on higher angle than on lower angle slopes and also where bracken vegetation encourages macropore flow through root holes. Throughflow in shallow mineral soils overlying parent rock, tends to be relatively well-buffered, siginificantly less acid and contains lower aluminium levels than saturation overland flow and peat pipeflow, which have had little or no contact with mineral soil or bedrock. However, throughflow from more thoroughly weathered soil types tends to be of low pH and high in aluminium because of the base-poor nature of these soils.

Soil pipes of various types are important over much of upland Wales.[11, 18] However, pipes differ greatly and can play contrasting roles in runoff generation.[19] At Llyn Brianne, deep, large pipe systems are confined to the deep peatlands of extensive flattish interfluves (particularly in catchments 1, 2, 9 and 11): these pipes tend to flow seasonally. Shallow, ephemeral pipes are common on relatively gentle (<15°) slopes where peaty soils cover glacial or soliflucted till or parent rock. Pipes are also found in ill-drained valley-side benches and in peat-covered valley bottoms. In each case, pipes are associated with less permeable soils than those where throughflow is important.

Most work suggests that shallow pipes with connections to saturation overland flow source areas located upslope can be important in generating runoff for the falling limb of storm hydrographs, but do not respond quickly enough to generate the flood peak itself. Deeper pipes, particularly in deep peat areas, tend to be important baseflow contributors.

Saturation overland flow (mostly comprising rainfall falling on areas of saturated soil in catchments) generally produces the storm peak of runoff hydrographs in humid vegetated areas.[10] At Llyn Brianne, the most important locations of easily saturated soils are valley bottom wetlands, slope hollows and benches, around channel heads and (during and following very wet antecedent weather) on peat-covered ridge-top flats.

The spatial extent of different types of runoff process, reflecting differences in topography and soil type, varies significantly from catchment to catchment at Llyn Brianne, with implications both for stream hydrology and stream hydrochemistry. Catchments with high proportions of valley bottom or interfluve areas of low slope, usually characterised by more impermeable peats and stagnogleys, tend to be dominated by saturation overland flow and pipeflow in storm hydrographs; whereas catchments with little easily saturated land, and characterised by moderate to steep slopes with well-drained soils, tend to be dominated by throughflow.

Stream hydrographs

As a consequence of the high relief and predominantly impermeable soils and geology of the region, streamflow responses to rainfall in upland Wales tend to be flashy and baseflow levels tend to be comparatively low. The resultant high stormflow/baseflow ratios are an important contributory reason for the high acidity of many streams in the region in storm runoff events. The high frequency of rainstorms (and hence storm runoff events also) means that acid episodes are also frequent.

However, the exact nature of storm runoff responses to rainstorms varies from catchment to catchment with basin characteristics. An example of local contrasts in streamflow response is given in Fig. 13.4 which depicts the streamflow hydrographs of six catchments in the Llyn Brianne area over a 60 hour period, including a multimodal rainfall event. All flows are expressed in per unit area terms (1^{-1} km^{-2} s^{-1}) to facilitate intercatchment comparisons. Three features of the hydrographs are of particular significance: the magnitude of the baseflow (streamflow prior to the rainstorm); the magnitude and timing of the flow peak; and the characteristics of the falling limb.

Pre-storm baseflow levels ranged from 40 to 95 1^{-1} km^{-2} s^{-1}. Afforested catchments, because of their greater evapotranspirational losses, tend to have lower baseflows than moorland catchments,[16] as seen in Fig. 13.4. However, factors other than land use also influence baseflow levels, as the afforested catchment 3 and the moorland catchment 10 exhibit anomalously high and low baseflow levels respectively. The former's high baseflow may be related to its exceptionally high percentage (around 20%) of low slope (<0.05) peaty interfluve land (Fig. 13.3), which generates substantial baseflow from deep pipe systems in the peat. The low baseflow of moorland catchment 10 may simply reflect a small catchment area (0.35 km^2) and high drainage density. In both cases catchment morphometry and related factors significantly influence baseflow.

The peak flows resulting from the rainstorm exhibit a wide range between catchments, though the timing of the peaks (around 1 to 1.5 hours after rainfall peaks) was uniform and suggests a common dominant runoff process source, namely saturation overland flow. Significantly higher peak flows occurred in moorland catchment 5 (750 1^{-1} km^{-2} s^{-1}), by far the steepest catchment (Fig. 13.3), and in two of the afforested catchments, catchment 3 and catchment 2 (650 and 635 1^{-1} km^{-2} s^{-1} respectively), where the very high drainage densities resulting from forest ditches increase responses to rainfall. The relatively subdued responses of the two moorland catchments 10 and 11 probably reflect their low drainage densities and comparative lack of easily

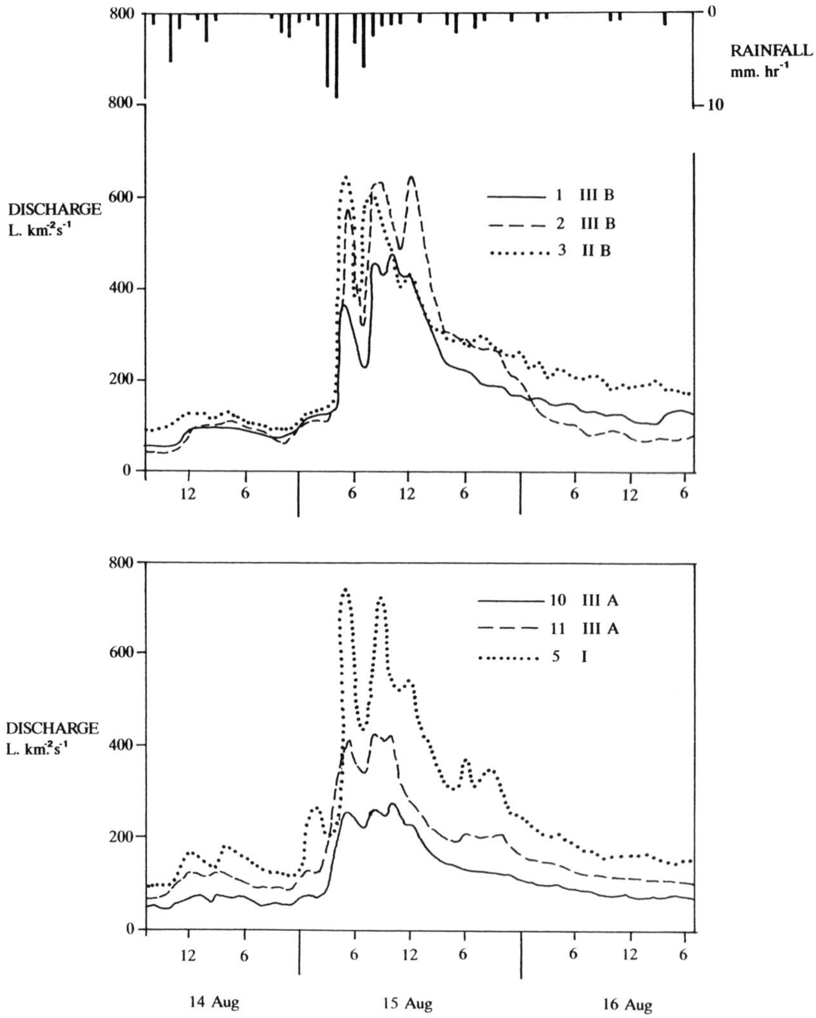

Fig. 13.4. Stream hydrographs and rainfall at Llyn Brianne, 14–16 August 1985.

saturated low slope land (Fig. 13.3). The low response of afforested catchment 1 is more difficult to explain, but may partly reflect drier pre-storm soils and partly a dampening effect of its large catchment area.

The considerable difference in falling limb characteristics between catchments at Llyn Brianne (Fig. 13.4) cannot be attributed to land use. The very steep, but stepped, decline to pre-storm baseflow levels in catchment 2 contrasts with the longer duration falling limbs of the other two afforested catchments. Differences in magnitude of falling limb flows also occur between the moorland streams.

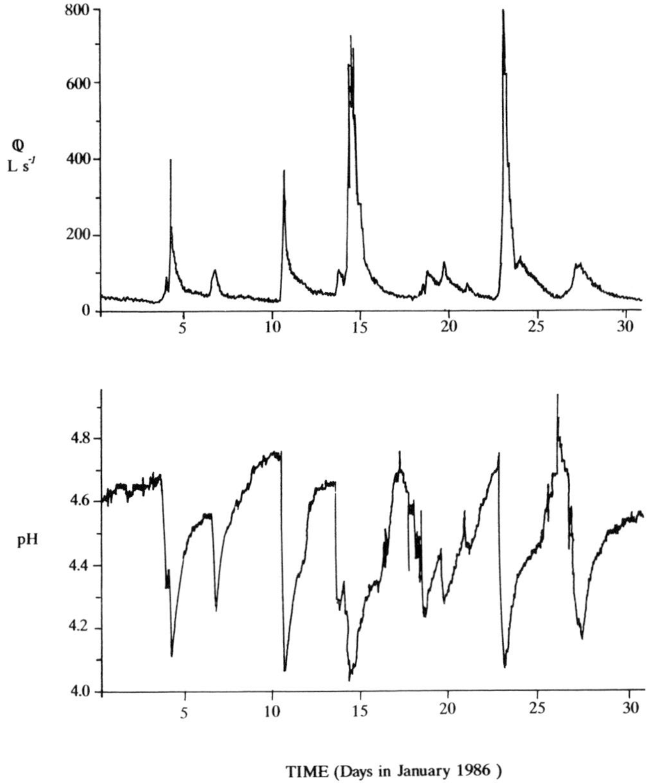

Fig. 13.5. Streamflow (Q) and streamwater pH for afforested catchment 2 at Llyn Brianne in August 1985.

Streamflow and pH dynamics at Llyn Brianne

The typical pH response of upland streams during runoff events is for pH to decrease and increase as streamflow increases and decreases respectively. Fig. 13.5 shows the example of afforested catchment 2 at Llyn Brianne during August 1985; such patterns have also been noted at Plynlimon,[28] in North West England,[8] and in the Loch Dee area of the South West Scotland.[5] The decrease in pH with each increase in streamflow (and vice versa) results largely from changes in the relative contribution of storm runoff at lower pH and baseflow runoff at higher pH (i.e. the stormflow/baseflow ratio of Fig. 13.1). For catchments at Llynn Brianne, like those at Plynlimon,[28] overall linear relationships exist between pH and the logarithm of flow ($\log_e$ Q), as the scatter plot for catchment 1 during October 1985 demonstrates (Fig. 13.6). Best-fit straight-line equations relating pH and $\log_e$ Q, however, vary from stream to stream with differences in geology, soils, morphometry and vegetation,[28] associated variations in runoff processes and mean rates of atmospheric deposition.

There is considerable scatter about any straight-line plot of pH and log Q, but the arrows in Fig. 13.6b indicate that a large proportion of the scatter is accounted for by hysteresis. Hystereses types A-B and C in Fig. 13.6b are common features of stream acidity behaviour at Llyn Brianne[22] and elsewhere.[9, 17] Several alternative hypotheses can be proposed to explain these hysteresis patterns.

The first possible explanation for lower pH on the falling, rather than on the rising, limb involves a systematic change in the proportions of 'old' pre-storm water (of higher pH) and 'new' rainstorm water (of lower pH) in the runoff processes contributing to the storm peak. Although generally stormflows in upland basins are produced by rapid runoff of rainfall (or snowmelt) along surface and near-surface hydrological pathways, some of the storm water particularly in the early part of the storm hydrograph may comprise pre-storm water stored in the catchment as soil moisture or in surface pools. This pre-storm water, which because of its long residence time and contact with soil and vegetation may be of signficantly higher pH than the incoming rainfall (or snowmelt), may be pushed out by, and mixed with, the incoming rainfall. As the storm progresses however, the proportion of lower pH 'new' water in storm runoff progressively increases as the supply of higher pH pre-storm water becomes exhausted. Thus, as indicated in Hysteresis type A-B (Figure 13.6b), the pH of the falling limb tends to be lower than that of the rising limb and in moderate storms the trough in pH may come after the hydrograph peak. In large events, however, particularly after wet antecedent conditions, the storm runoff at peak flow may already comprise 100% new water and hence streamwater pH approaches that of rainfall (or snowmelt) and the minimum pH more coincides with peak streamflow, producing Hysteresis type C (Figure 13.6). Clearly, these are rather simple explanations based solely on hydrology; many chemical processes also operate within catchments which modify the hydrologically driven variations in stream pH. The first rain after a long dry spell may also flush any accumulation of atmospherically deposited material, or the products of oxidation in dry soils, into the stream and cause streamwater pH to decrease to a lower level than usual for a given peakflow. Similarly, lower than normal streamwater pH can occur as a result of a heavy deposition of sea-salt.

A second possibility is that Hysteresis type A-B may be caused by changes in the relative contributions of different runoff processes during storms. For example, stream water during the rising limb of a hydrograph may be contributed by saturation overland flow. The early falling limb of flows may then be characterised by an increasing relative contribution of water which reaches the stream channel by somewhat slower runoff processes of contrasting pH such as macropore throughflow in stagnopodzols and peat pipeflow.

A third possibility is that the hysteresis may be due to variation in the arrival times of waters from different hydrological source areas: peatland

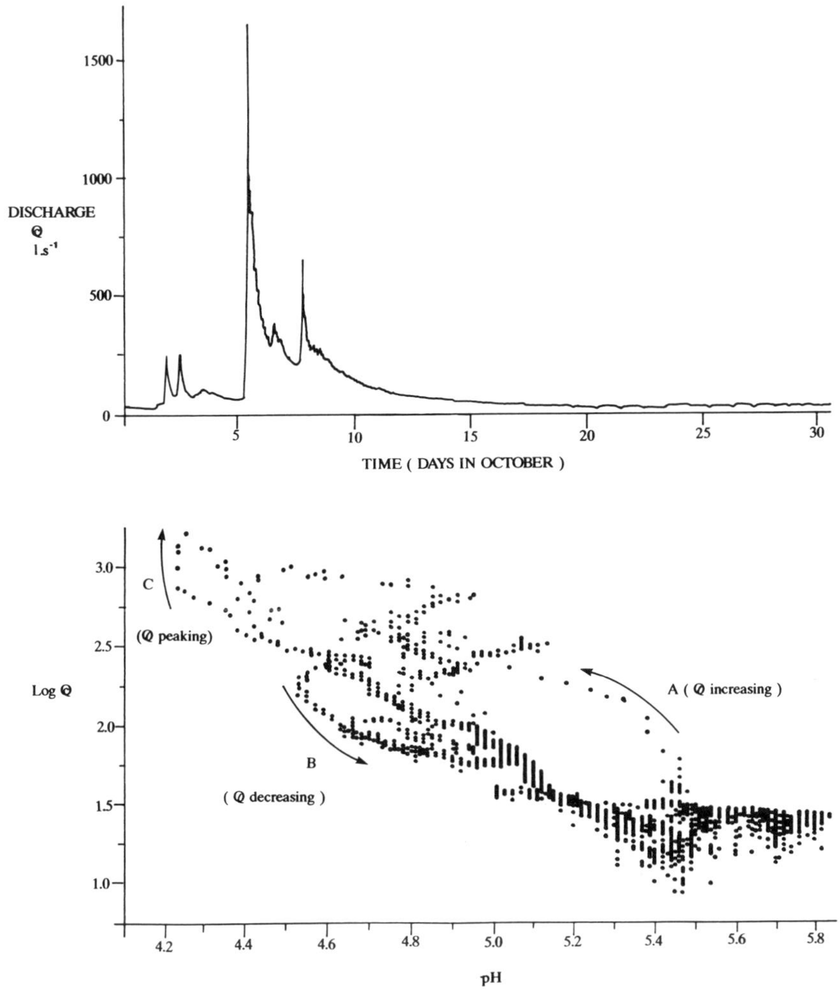

Fig. 13.6. Streamwater pH and discharge at afforested catchment 1 at Llyn Brianne in October 1985: a) discharge during October 1985, b) log flow/pH hysteresis.

sources of low pH are often relatively remote from a catchment outlet and require time to saturate and exceed their surface retention capacity. Indeed, it is likely that all three of these possible explanations may apply to varying degrees in different catchments in the Welsh uplands. Furthermore they are by no means mutually exclusive.

However, during periods of low flow when the stream comprises water which has been resident in the catchment for a relatively long time, chemical processes are more important than hydrological ones. Thus the basic hydrochemical status of a stream is governed by its geology, soils, vegetation and type and rate of atmospheric deposition.

Whilst linear regression analysis of flow and pH may be adequate for an initial assessment of pH response to streamflow, time series analysis techniques are more appropriate for describing the dynamic behaviour of stream acidity.[33, 34] An analysis of data from several of the Llyn Brianne catchments[22] has shown that simple equations can be derived to describe the stream H^+ responses during moderately large runoff events when Hysteresis type C occurs in addition to Hysteresis type A-B as shown in Fig. 13.6b.

The pattern of variation in H^+ with flow in the Llyn Brianne area is remarkably uniform, irrespective of land use. However, there are occasions when H^+ increases disproportionately to the increase in flow. Fig. 13.7 shows the river stage, electrical conductivity and pH recorded at 15-minute intervals for catchment 1 during January 1986. It can be seen that the minimum pH corresponding to a high peak flow on the 9th January was about pH 4.5. However, during subsequent smaller runoff events, pH decreased to less than 4.5. Fig. 13.7 also shows that the electrical conductivity of the stream at the same time decreased by several units ($\mu s\ cm^{-2}$) as streamflow increased to a relatively high level, but during subsequent events conductivity increased by 10 $\mu s\ cm^{-2}$ or more when flow increased only moderately. A similar pattern of stream conductivity changes was recorded over the same period at most of the other Llyn Brianne catchments, and is probably due to variable responses to streamflow of ionic species other than H^+.

An increase in conductivity can be due to flushing from the catchment of accumulations of solutes following dry periods,[32] but this process is thought not to have caused the streamwater ionic enrichment (Fig. 13.7) because effective flushing would have occurred during the high runoff event on the 9th January. The most likely cause of the ionic enrichment which occurred in most streams in the area from the 14th January onwards was a high atmospheric deposition of sea-salts which originated from the surface of a stormy Atlantic during that period. Sodium can be absorbed temporarily onto ion exchange sites in the soil so displacing other cations into solution,[12] and in the base-poor soils typical of upland Wales many of these displaced cations will be H^+. Extra H^+ generated in this way accompany the extra sea-salt chloride ions as they travel conservatively towards the stream. The result is a lower minimum stream pH than if high atmospheric deposition of sea-salts had not occurred.

Some acid runoff events in South West Scotland have also been attributed to atmospheric deposition of sea-salts,[20] but such events occur only once or twice a year in that region.[12] Chloride is the major anion in rainfall in the Llyn Brianne area[31] and sea-salts are of episodic importance as a contributory cause of low stream pH more frequently than in South West Scotland.

Finally, an example of a lower than normal pH caused by flushing of an accumulation of dry-deposited solutes occurred at Llyn Brianne during a rain-on-snow event during early March 1986. This is reported in more detail in Chap. 15.

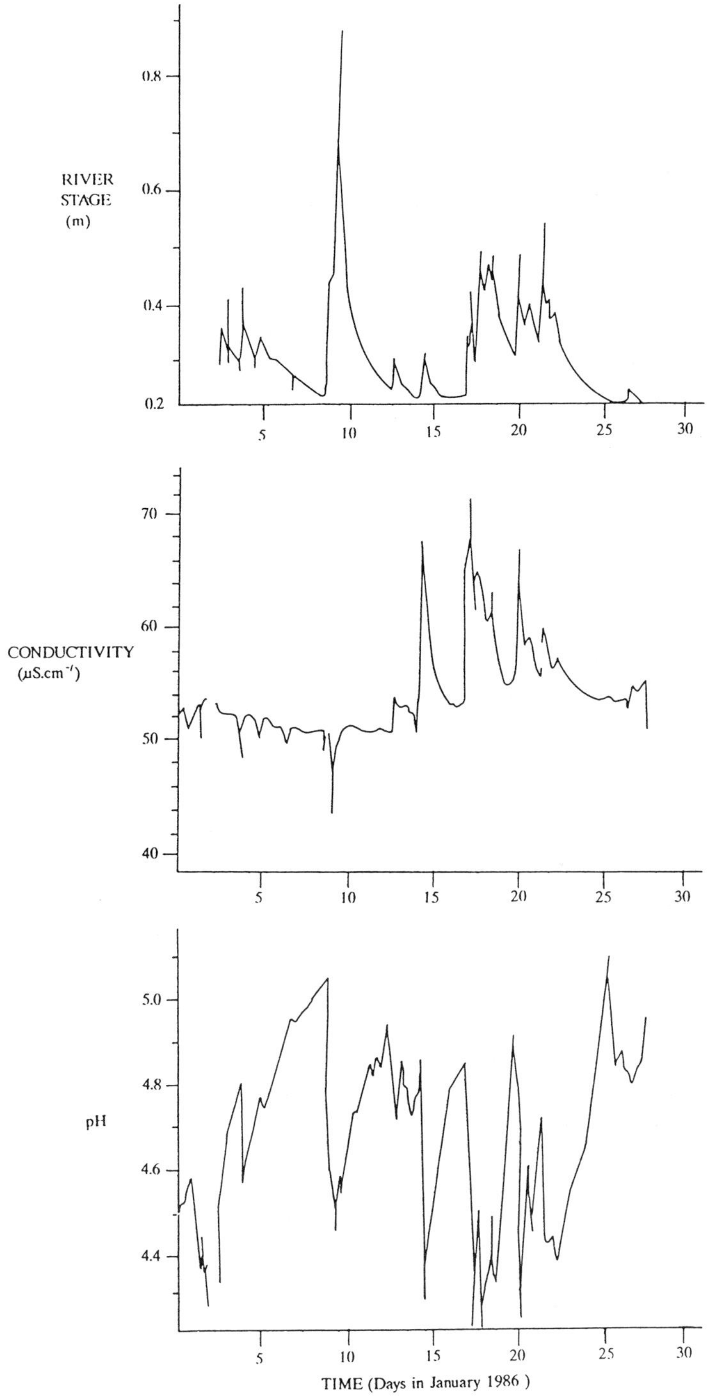

Fig. 13.7. River stage (a), conductivity (b) and pH (c) in afforested catchment 1 at Llyn Brianne in January 1986.

Conclusions

Current research suggests that there are strong interrelationships between topography, soil type and vegetation on the one hand, and runoff processes and their hydrochemistry on the other.[9, 28] Hence, a clear knowledge of catchment hydrology is crucial to the understanding of acidification problems in upland Wales. Comparison of catchments on the basis of morphometry offers a good initial indication of any likely differences in catchment hydrological response. Catchment morphometry, allied with the differences in land use, vegetation cover, soils, geology and climate, largely control the volume and timing of individual catchment runoff responses. They also determine the complex hydrological pathways taken by the runoff and hence the range of soil and rock environments encountered. Subject to the quality of the bulk precipitation inputs, hydrological factors are clearly important in determining variations in the hydrochemical response of streams.

References

1. Beven, K.J. and Kirkby, M.J. (1979). A physically based, variable contributing area model of basin hydrology, *Hydrological Sciences Bulletin*, 24 (1): 43–69.
2. Beven, K.J., Kirkby, M.J., Schofield, N. and Tagg, A.S. (1984). Testing a physically based flood forecasting model (TOPMODEL) for 3 UK catchments, *J. Hydrol.*, 69: 119–143.
3. Beven, K.J. and Wood, E.F. (1983). Catchment geomorpology and the dynamics of runoff contributing areas, *J Hydrol.*, 65: 139–158.
4. Boon, R. (1986). *Sediment studies at Llyn Brianne.* Unpublished report to Catchment Characteristics Sub-Group, Llyn Brianne Acid Waters Study, 6pp.
5. Burns, J.C. and Lees, F.M. (1987). *A review of continuous monitoring of streamflow pH at Loch Dee 1981 to 1987*, Loch Dee Project Working Paper 4/87, Solway River Purification Board.
6. Calder, I.R. and Newson, M.D. (1979). Land use and upland water resources in Britain – a strategic look, *Wat. Resour. Bull.*, 15 (6): 1628–1639.
7. Christopherson, N., Seip, H.M. and Wright, R.F. (1982) A model of streamwater chemistry at Birkenes, Norway. *Wat Resour Res*, 18 (4): 97–996.
8. Crawshaw, D.H., Prigg, R.F., Diamon, M., Robinson, J.F. and Cragg-Hine, D. (1986) *Episodic variations in stream water chemistry associated with acid rainfall and run-off and the effect on aquatic ecosystems, with particular reference to fish populations of N.W. England*, Final Report to European Commission Contract No. ENV-867-UK(H), North West Water Authority Report.
9. Cresser, M.S. and Edwards, A. (1987). *Acidification of Freshwaters*, Cambridge University Press, 136pp.
10. Dunne, T. and Black, R.D. (1970). Partial area contributions to storm runoff in a small New England watershed, *Wat. Resour. Res.*, 6 (5): 1296–1311.
11. Gilman, K. and Newson, M.D. (1980). *Soil pipes and pipeflow, a hydrological study in upland Wales*, Geo Books, Norwich, 1st edition.
12. Harriman, R. and Wells, D.E. (1985). Causes and effects of surface water acidification in Scotland. *Wat. Poll. Control*, 84: 215–224.
13. Heerdegen, R.G. and Beran, M.A. (1982). Quantifying source areas through land surface curvature and shape, *J Hydrol.*, 57: 359–373.
14. Hewlett, J.D. and Hibbert, A.R. (1967). *Factors affecting the response of small watersheds to*

Recording Chinese mitten crabs

Report your sightings by email or phone

Please tell us **where** and **when** you saw a Chinese mitten crab, your **contact details** and, if possible:

- was it male or female?
- take a photograph:
  - take photo from above
  - use a 50p coin to show scale
  - try to show claws clearly

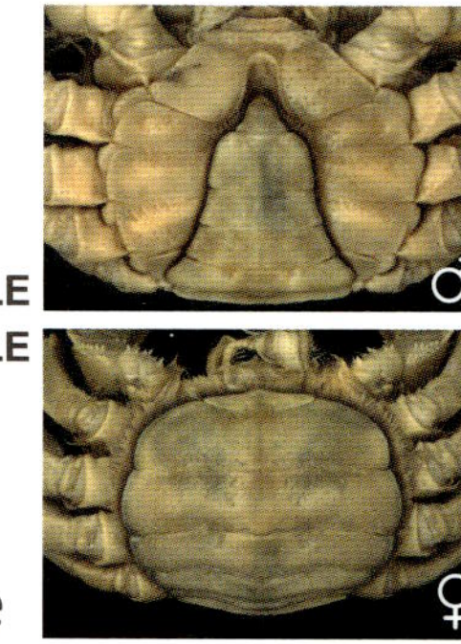

You can also report sightings by text or picture message to 07806 938789 or online at

www.mittencrabs.org.uk

Have you seen a Chinese mitten crab?

Chinese mitten crab
(*Eriocheir sinensis*)

- Grey-green to dark brown crab
- Long walking legs
- Squarish body up to 86 mm across
- Dense brown 'fur' on the white-tipped claws
- Habitat: rivers, brackish water estuaries, rarely along the marine inshore coast
- 4+4+4 pattern of teeth around front of shell
- Juveniles may lack 'fur' on claws

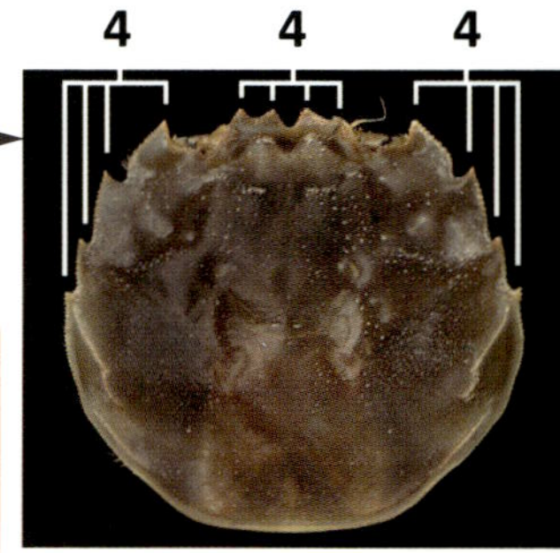

Not to be confused with:

Common shore crab
(*Carcinus maenas*)

- Lacks 'fur' on claws
- Shell shape more triangular
- Found on the sea shore and in estuaries but never in freshwater rivers

The broad-clawed porcelain crab
(*Porcellana platycheles*)

- 2 long antennae
- Only 3 pairs of walking legs apparent
- Found on the sea shore but never in freshwater rivers

Powerful claws

Handle with care

precipitation in humid areas, In: W.E. Sopper and H.W. Lull (eds), Proc. Int. Symp. on Forest Hydrology, Pergammon Press, Oxford, 275–290.
15. Hewlett, J.D. and Nutter, W.L. (1970). The varying source area of streamflow from upland basins, Proc. Symp. on Interdisciplinary Aspects of Watershed Management. *American Society of Civil Engineers*, 65–83.
16. Hornung, M. and Newson, M.D. (1986). Upland afforestation: influences on stream hydrology and chemistry, *Soil Use and Management*, 2, 2, June 1986, 61–65.
17. Jenkins, A., Whitehead, P.G. and Whitcombe, A. (1987). Hydrochemical responses in the Allt A Mharcaidh, *Royal Society Surface Water Acidification Programme Mid-term Review Conference*, Bergen, 22–26 June 1987.
18. Jones, J.A. (1971). Soil piping and stream channel initiation, *Wat. Resour. Res.*, 7: 602–610.
19. Jones, J.A. (1979). Extending the Hewlett model of stream runoff generation, *Area*, 11: 110–114.
20. Langan, S. (1987). Episodic acidification of streams at Loch Dee, SW Scotland, *Transactions of the Royal Society of Edinburgh: Earth Sciences*, 78: 393–397.
21. Langbein, W.B. (1947). Topographic characteristics of drainage basins, *U.S. Geol. Survey Water Supply Paper* 968c: 125–157.
22. Littlewood, I.G. (1987). *Streamflow – pH dynamics in small moorland and conifer afforested catchments in the Upper Tywi Valley, Wales*, Paper presented at the National Hydrology Symposium, University of Hull, 14 – 16th September 1987.
23. Mayes, J. (1987). Rainfall changes in Wales over the last hundred years; regional contrasts and some implications., *Cambria*, 14 (1): 57–70.
24. Mosley, M.P. (1982). Subsurface flow velocities through selected forest soils, South Island, New Zealand, *J. Hydrol.* 55: 65–92.
25. Newson, A.J. (1976). *Some aspects of the rainfall of Plynlimon, mid Wales.* Institute of Hydrology Report 34, Wallingford.
26. Newson, M.D. (1979). The results of ten years experimental study on Plynlimon, mid – Wales and their importance to the water industry, *J. Inst. Wat. Engrs. and Sci.*, 33: 321–333.
27. Newson, M.D. (1980). Water balance at selected sites. In Doornkamp, J.C., Gregory, K.J. and Burn, A.S. (eds). *Atlas of Drought in Britain, 1975–6*, Trans. Inst. Br. Geogr., London, 37–38.
28. Newson, M.D. (1986). Slope and channel runoff processes in upland catchments: interfaces between precipitation and streamflow acidity, *Cambria*, 13 (2): 197–212.
29. Robinson, M. (1980). *The effect of pre-afforestation drainage on the streamflow and water quality of a small upland catchment.* Institute of Hydrology Report no. 73.
30. Robinson, M. (1986). Changes in catchment runoff following drainage and afforestation, *J Hydrol.*, 86: 71–84.
31. Stoner, J.H. and Gee, A.S. (1985). Effects of forestry on water quality and fish in Welsh rivers and lakes. *J. Inst. Wat. Engrs. and Sci.*, 39: 27–54.
32. Waling, D.E. and Foster, I.D.L. (1975). Variations in the natural chemical concentration of river water during flood flows, and the lag effect: some further comments, *J. Hydrol.*, 26: 237–244.
33. Whitehead, P.G., Neal, C. and Neale, R. (1986a). Modelling the effects of hydrological changes on stream water acidity, *J. Hydrol.*, 84: 353–364.
34. Whitehead, P.G., Neal, C., Seden-Perriton, S., Christopherson, N. and Neale, R. (1986b). A time series approach to modelling stream acidity, *J. Hydrol.*, 85: 281–303.

14. Water quality changes from input to stream

M. HORNUNG, B. REYNOLDS, P.A. STEVENS and S. HUGHES

Introduction

Only a small part of the precipitation deposited on to terrestrial systems falls directly into surface waters, the bulk falling on to vegetated land surfaces. As the precipitation moves through the vegetation canopy its chemistry is modified, as a result of interactions with plant surfaces, before reaching the ground as throughfall and stemflow. The throughfall and stemflow waters form the input to soils where they are further modified by soil-solute interactions. Soil waters may enter streams directly as throughflow, surface flow or return flow or they may form the input to groundwaters in underlying drift deposits and bedrock. In the drift and bedrock the chemistry of percolating waters is influenced by solute-rock mineral interactions before forming the groundwater input to surface waters.

This chapter uses data from a series of studies in upland Wales to explore the changes in chemistry as precipitation moves through the vegetation, soil and bedrock compartments of terrestrial ecosystems. The three main study sites are Beddgelert Forest in north Wales, the Plynlimon experimental catchments, and the Llyn Brianne experimental catchments in central Wales (see Chap. 1). Together, the sites cover a range of vegatation types/land uses, soils and some variation in bedrock geology. Details of sites and methodologies are described elsewhere.[8, 18, 21, 22]

Precipitation-vegetation canopy interactions

The chemistry of precipitation is modified as it passes through vegetation canopies by 4 main processes:

a. concentration of solutes in precipitation due to evaporation from the canopy;
b. solution, and wash-off of dry deposited material, aerosols and occult deposition;

R.W. Edwards et al. (eds) Acid Waters in Wales. 223–240.

c. leaching of materials from vegetation as a result of ion exchange and solution of plant products;
d. uptake of ions from precipitation by foliage or canopy living organisms.

The impact of these processes varies with atmospheric concentrations of gaseous materials, site type, the physiology and nutrient cycling strategies of the vegetation.[15, 23]

Data from Beddgelert, Plynlimon and Llyn Brianne show that concentrations of almost all solutes are higher in throughfall than in bulk precipitation (Figs. 14.1 and 14.2). The exceptions are ammonia and nitrate which show reductions in concentrations in the *Nardus-Festuca* throughfall, indicating uptake in the canopy. The largest increases are shown by potassium, sodium, magnesium, chloride and sulphate. The potassium is mainly derived from canopy leaching while the increase in sodium, magnesium, chloride and sulphate largely results from solution of dry deposits and wash-off of aerosols and occult deposition. Some increase in concentration will result from evaporation of precipitation from the canopy but this will be small compared with the effects of leaching and wash-off. The data show significant differences in throughfall chemistry between the vegetation types (Fig. 14.1). Thus, solute concentrations are greater in the throughfall from the spruce canopies than from the *Nardus* and *Festuca* at Plynlimon. Atmospheric inputs as particulate, aerosol and occult deposition are greater on to the rougher tree canopies than on to the smoother grass canopies.[7] Evaporation losses are also greater from the tree canopies than from grassy vegetation; losses from conifer canopies in exposed sites are generally equivalent to 20–30% of incoming precipitation. Results from Plynlimon also show a variation in throughfall chemistry with species of conifer. Throughfall from larch was generally more acid than that from spruce of a similar age and growing on similar soils (Fig. 14.1): atmospheric deposition on to the larch canopy seems to be greater than on to the spruce canopy.[19]

The acidity of throughfall from the grassland canopy at Plynlimon was similar to or less acid than bulk precipitation. Canopy leaching apparently buffers part of the acidity in precipitation, dry and occult deposition. Throughfall from the oak, spruce and *Molinia* canopies showed marked seasonal variations in acidity (Fig. 14.3). In general, the oak throughfall was less acid than precipitation while the trees were in leaf but similar to, or more acid than precipitation in winter. Spruce throughfall at both Plynlimon and Llyn Brianne was more acid than precipitation during late winter and early spring, but similar to that of precipitation, or less acid, during the rest of the year. The increase in acidity during late winter reflects an increased incidence of easterly winds, and dry deposition of SO_2 and NO_x during that period; in summer, canopy leaching in both spruce and oak buffers input acidity to a greater or lesser extent. These results parallel those reported elsewhere in Britain.[13] Throughfall below *Molinia* at Llyn Brianne was more acid than bulk precipitation during the growing season but less acid during the senescent

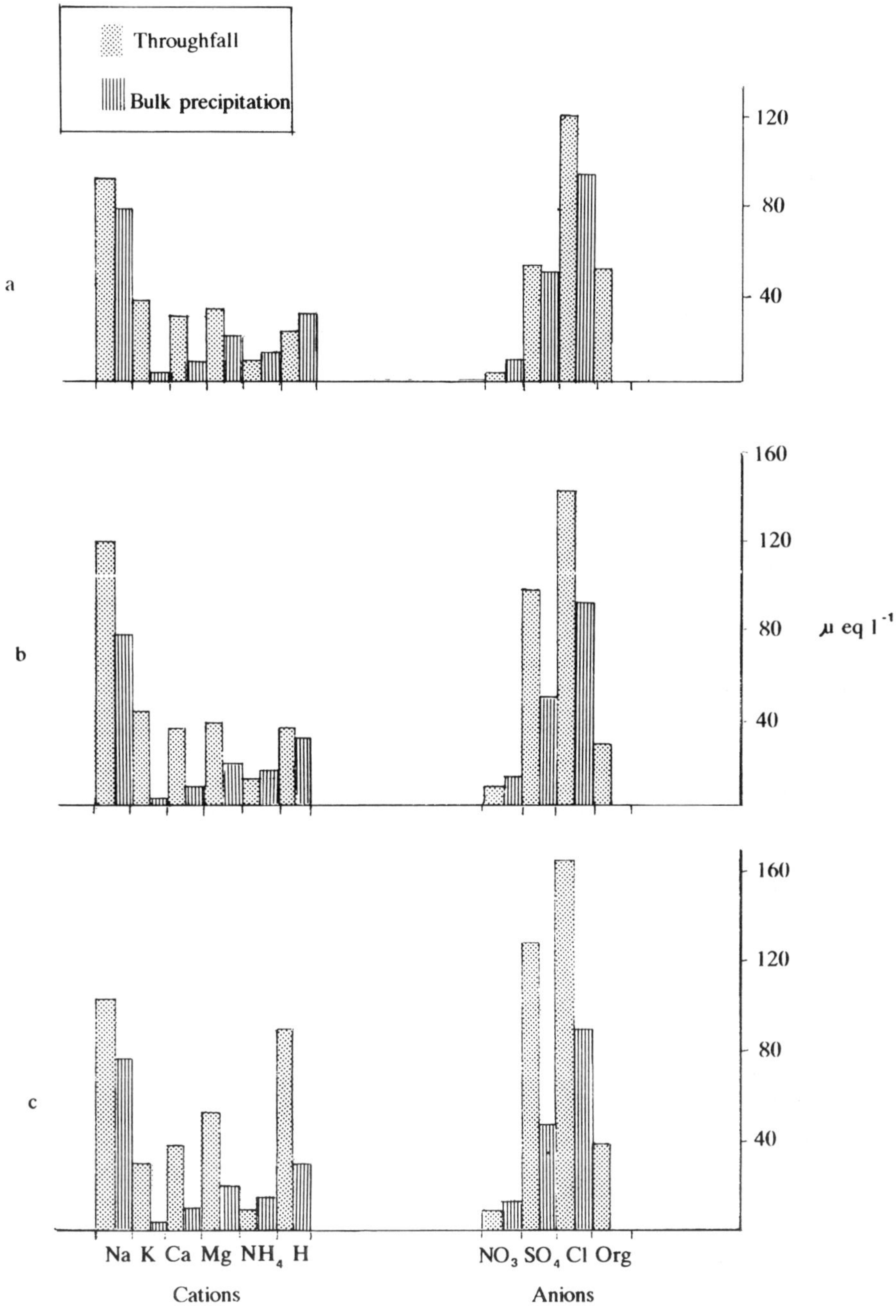

Fig. 14.1. Mean solute concentrations in throughfall from (a) *Nardus-Festuca* grassland, (b) Sitka spruce, (c) and Japanese larch at Plynlimon.

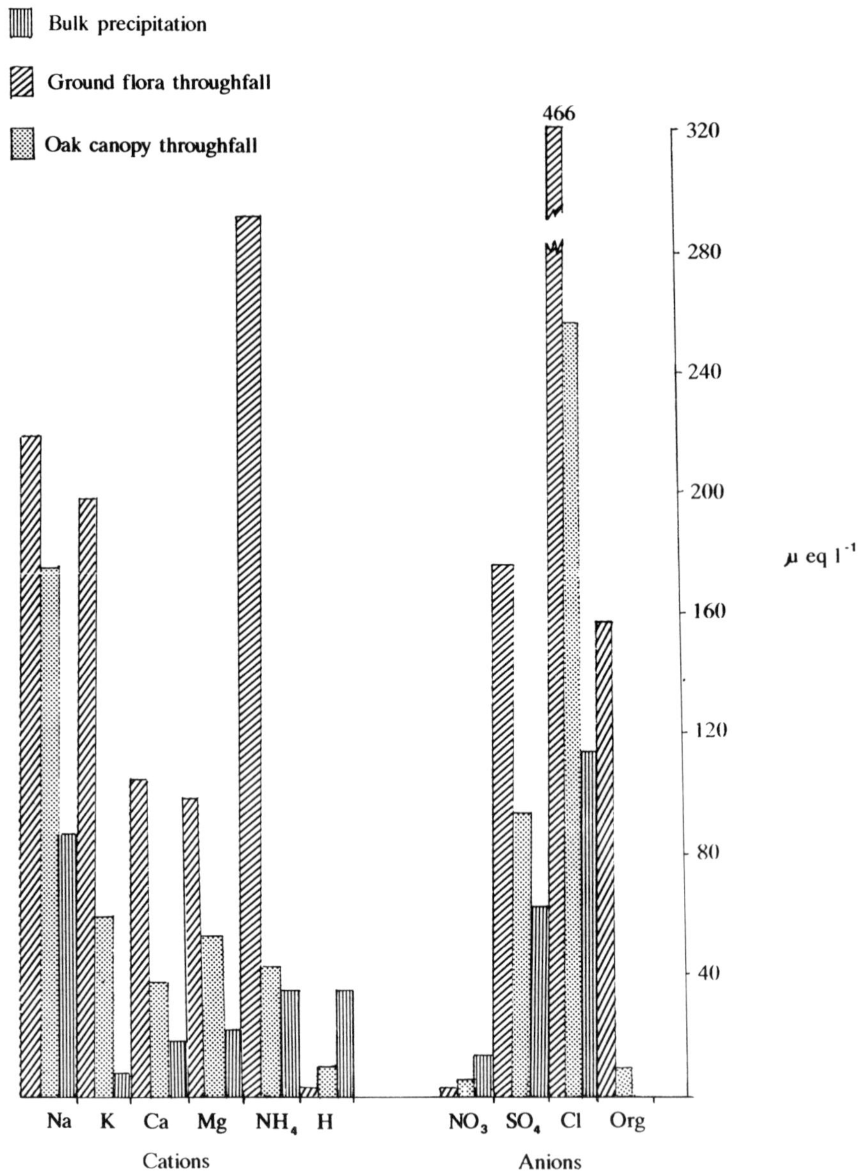

Fig. 14.2. Mean solute concentrations in throughfall from oak and from a grass-bracken understorey.

period (Fig. 14.3). In summer there is apparently significant canopy retentions of ions while in autumn and winter, decomposition of the dead material buffers input acidity.

Data from Beddgelert Forest also show a marked variation in throughfall chemistry, including acidity, of Sitka spruce with age. Throughfall was consis-

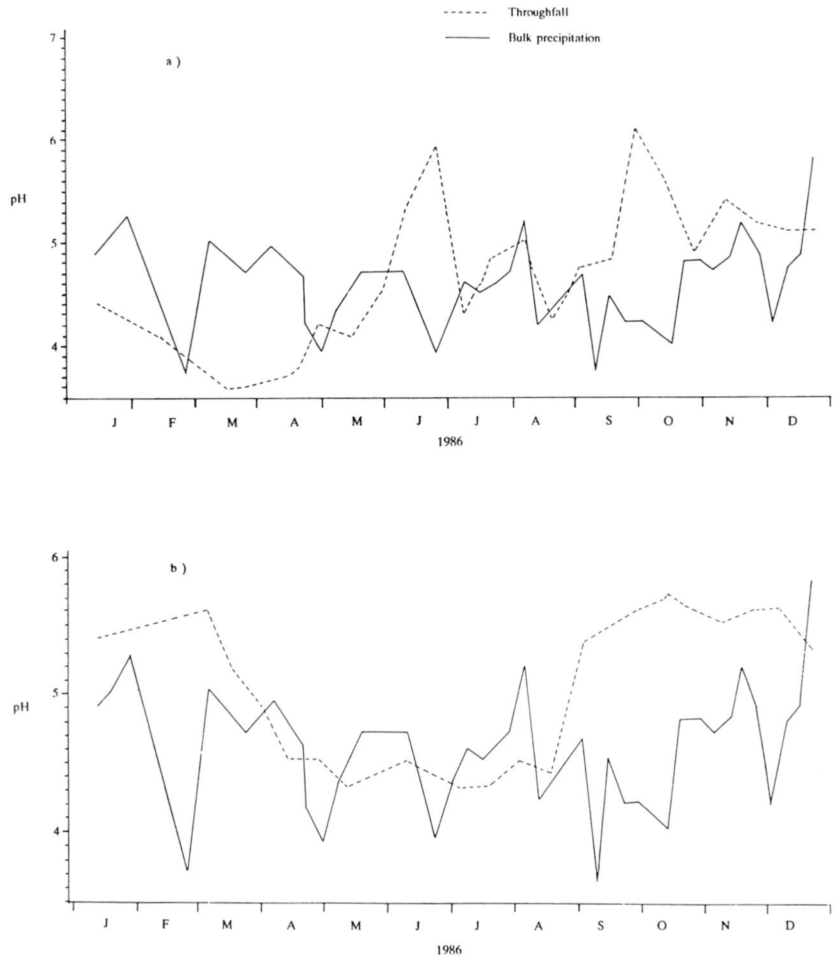

Fig. 14.3. Temporal variations in pH of bulk precipitation and throughfall from (a) Sitka spruce and (b) *Molinia*.

tently more acid than bulk precipitation below 55-year-old Sitka spruce but consistently less acid below 12- and 25-year-old spruce.[21] Dry and occult deposition of acidic ions are probably greater on to the canopy of the older crop while canopy leaching of base cations appears greater from the younger trees.

In deciduous woodlands the throughfall from the tree canopy passes through the canopy of a groundflora before reaching the soil. The concentrations of most solutes are increased further during contact with the groundflora canopy but acidity is generally reduced (Fig. 14.2). The buffering of the acidity is mainly due to canopy leaching of base cations but in the example shown in Fig. 14.2 the production of ammonium during decomposition of bracken fronds is also important.

Table 14.1. Volume weighted mean annual concentrations in bulk precipitation, throughfall and stemflow of 55-year-old Sitka spruce in Beddgelert Forest, north Wales (μeq l^{-1}).

	Bulk precipitation	Throughfall	Stemflow
pH	4.50	4.07	3.62
H	36	88	240
Na	131	422	1392
K	4	37	94
Ca	15	41	159
Mg	30	100	355
NH_4	18	32	44
NO_3	14	44	84
SO_4	74	162	483
Cl	145	477	1760

A proportion of the water reaching the ground below trees does so as stemflow. This water is always more acid and contains significantly higher solute concentrations than throughfall (Table 14.1). There are, however, marked differences in stemflow chemistry between species; in the Llyn Brianne area, stemflow from Sitka spruce is more acid than that from oak, mean annual pH being 4.5 and 4.9 respectively. The proportion of the water, and hence solute flux, to the ground as stemflow varies with tree age but it is always smaller than throughfall. Under 55-year-old Sitka spruce in Beddgelert Forest, stemflow formed 9% of the water flux but 19% of the sulphate flux. Stemflow is input to a small area around the base of the tree and some authors have suggested it may move rapidly through soil, along cracks, and be important in generating acidic stream events.

The acidity of throughfall and stemflow is not necessarily the most important factor influencing the solute interactions with underlying soils and acidification of surface waters.[14] The concentration and flux of mobile acid anions constitute the major factor driving leaching and acidification in soils, including mobilisation of aluminium.[16] The concentrations and fluxes of sulphate and chloride increase in the following pattern in upland Wales: coniferous trees > hardwood trees >grass communities. This seems to be mainly due to the greater atmospheric deposition on to trees, particularly the conifer canopies. Thus, the change of land use from moorland to forest can result in an increase in the input of mobile acid anions and acidity to soils in upland Wales. Occult deposition on to exposed conifer canopies at high altitude may be an important factor causing the increased inputs of mobile anions.[5, 24] Concentrations of both pollutant and non-pollutant ions are greater in mist and fog than in rainfall.

The increase in mobile acid ions through afforestation is difficult to quantify. Catchment studies at Plynlimon suggest that sulphate and chloride outputs from afforested catchments are 150–250% of those from adjacent grassland catchments.[9] The difference is thought to be mainly due to the greater atmospheric deposition on to forest canopies. There may, however, be a significant release of sulphate from the forest soils as a result of drying, and oxidation of organic soil horizons.

Solute-soil interactions: soil water chemistry

The solute chemistry of waters held within, and moving through soils, is influenced by a large number of soil-solute reactions which include: ion exchange, mineral solution and hydrolysis, oxidation-reduction reactions, anion sorption-desorption, mobilisation and degradation of organic acids, plant uptake and release.[16] The impact of the various processes in any given soil will depend on its biological, chemical and mineralogical characteristics and on plant-soil interactions. The characteristics will vary between soils but also between horizons in a given soil.

1. Vertical variations in soil solution chemistry

The impact of the soil-water interactions on water chemistry will be illustrated using data on soil water chemistry from stagnopodzols. These soils are characteristically acid, with a low base saturation (Table 14.2), no free carbonates and few readily weatherable silicates; they usually have an organic, or peaty surface horizon of up to 40 cm thickness. The stagnopodzols occur naturally under acid grasslands, *Molinia* and shrub heath communities, and have been extensively afforested. The solution chemistry of one such soil below an acid grassland in mid Wales is given in Table 14.3.

Waters from the 0 horizon are more acid than the throughfall input, but the pH of the soil waters increases with soil depth to 4.7. Sodium is the dominant cation in the soil solution, accounting for about 50% of the cationic charge, and this reflects the dominance of marine-derived ions in the precipitation inputs. Potassium concentrations in the soil solution are much lower than in throughfall, reflecting plant uptake; calcium uptake probably explains the parallel reduction in calcium concentrations. Calcium and magnesium concentrations are very low in the soil waters (14–40 $\mu eq\ l^{-1}$), and show little variation with depth within the soils; this suggests little mobilisation of these elements from sources within the soils, i.e. from exchange surfaces or by mineral weathering. In contrast to the trends in base cation concentrations, there is a significant increase in concentrations of aluminium and silica in the mineral soil horizons. Aluminium concentrations range between 40 and 70 $\mu eq\ l^{-1}$, accounting for about 15% of the total cationic charge.

Chloride is the dominant anion in waters throughout the soil profile, accounting for 60–65% of the inorganic anion sum and reflecting the influence of marine-derived ions on precipitation input chemistry. Sulphate accounts for about 25% of the inorganic anion charge; this is greater than would be expected in an unpolluted area.[3, 11]

Increased sulphate inputs to acid soils in polluted areas of north America

Table 14.2. pH, cation exchange capacity (CEC) (μeq.100 g^{-1}) and base saturation (%) of stagnopodzols from the Plynlimon and Llyn Brianne sites.

Horizon	pH	CEC	Base saturation
	Plynlimon		
O	3.67	10.26	35
E	3.75	9.42	5
Bs	4.20	4.83	7
C	4.48	1.37	28
	Llyn Brianne		
O	3.64	14.3	27
E	3.94	10.86	8
Bs	4.23	6.38	8
C	4.35	4.53	8

and Scandinavia have been linked to mobilisation of aluminium into soil solution and drainage waters. The relatively high aluminium levels noted above may reflect the enhanced sulphate input from pollutant sources, as suggested for forest sites in the north east of the USA by Cronan and Schofield.[4] The increase in water acidity in the surface soil, compared with throughfall, is probably due to solution and deprotonation of organic acids plus mobilisation of H^+ ions by ion exchange. The marked increase in aluminium and silica concentrations in the mineral horizons will reflect release from weathering and, in the case of aluminium, ion exchange. The speciation of the aluminium in solution varies between soil horizons. Organically complexed aluminium is more important, and may dominate in the organic 0 horizon while labile, monomeric aluminium[6] dominates in the mineral horizons (Table 14.4). The inorganic, monomeric form is that which is toxic to aquatic biota in acidic, low calcium waters (Chap. 10). Reynolds et al.[20] have, however, argued that, in the mineral horizons of these soils, ion exchange is the proximate mechanism controlling aluminium mobilisation.

The release of aluminium will be balanced by retention of H^+ ions and this explains part of the increase in solution pH with depth. Degradation of organic acids, as solutions move through the soils, probably also contributes to the increase in pH. The data suggest little retention of sulphate, by adsorption, in these soils; at some sites this is a major mechanism reducing water acidity and buffering acidic atmospheric deposition.[2, 10] The sulphate adsorption capacity of the podzolic Bs horizon in these Welsh soils must be in equilibrium with the concentrations in current precipitation inputs.

Cation leaching, and aluminium mobilisation, in the soils is driven by anions. In the upper soil, chloride, sulphate and organic anions probably all contribute to leaching; in the lower soil, organic anions will contribute little to leaching, the organic acids having been largely degraded or immobilised in the B horizon. The importance of chloride in leaching probably varies; in so-called salt events, when very high concentrations of sea salts are deposited in

Table 14.3. Annual mean solute concentrations in bulk precipitation, throughfall, soil waters and groundwaters from a *Nardus-Festuca* grassland on stagnopodzols over Ordovician mudstones at Plynlimon, mid Wales (μeg l^{-1} except pH, Si μmol l^{-1}). For soil waters, arithmetic means are given, for other waters, volume weighted means.

	pH	H	Na	K	Ca	Mg	Al	Si	NH_4	NO_3	SO_4	Cl	DOC
Bulk precipitation	4.63	23	83	2	9	20	6	8	15	12	51	105	1
Throughfall	4.62	24	94	38	30	35	6	8	10	4	53	124	10
O	4.11±.02	77±4	100±8	8±2.0	14±1.0	39±4	29±4	18±1	n.d	29±6	68±10	124±12	9.5±1.1
Soil E water	4.26±.03	55±4	187±15	4±0.4	20±0.8	44±2	55±3	42±2	n.d	39±5	87±5	209±18	6.0±0.7
Bs	4.36±.03	44±3	173±11	4±0.4	14±0.6	39±2	44±4	36±2	n.d	28±4	71±5	185±15	3.7±0.4
C	4.70±.04	20±3	194±13	6±0.5	17±0.4	47±2	66±5	34±1	n.d	44±6	77±5	202±5	2.3±0.4
Groundwater	6.2	1	152	5	144	126	6	15	n.d	3	144	155	

Arithmetic means for soil waters; otherwise volume weighted means.

Table 14.4. Mean annual concentrations of polymeric, inorganic and organic aluminium concentrations in solutions from a stagnopodzol under *Nardus-Festuca* grassland at Plynlimon ($\mu mol\ l^{-1}$).

Horizon	Polymeric Al	Labile monomeric monomeric Al	Non-labile monomeric Al
O	0.31 ± 0.07	1.96 ± 0.23	3.60 ± 0.38
E	1.11 ± 0.21	12.47 ± 1.23	3.19 ± 0.25
Bs	1.09 ± 0.21	9.92 ± 0.73	2.27 ± 0.16
C	1.54 ± 0.21	21.50 ± 1.60	0.81 ± 0.11

precipitation, it becomes very important with sodium and magnesium retained in the 0 horizon and H^+ generated by ion exchanges.[12] In between 'salt events', the role of sulphate will be more important than chloride.

An important change in the composition of waters as they move through soils is not shown in the results presented in Table 14.3. The content of dissolved carbon dioxide in the waters increases markedly in the soils. This is because the partial pressure of CO_2 in the soil is higher than in the atmosphere, as a result of root and microbial respiration. The content of CO_2 in solution can be an important control on solution acidity but the most important influence is when soil waters enter streams and lose the CO_2 by degassing.[2, 16]

2. *Temporal variations in solute concentrations*

Most solutes in the soil waters show considerable variation in concentration with time. Two types of variation are present; broad cyclical patterns and short term, event related variation. Most solutes show both types of variation with the short term superimposed on the cyclical patterns. Any given solute shows the same pattern of variation in all soil horizons but the amplitude of the variations declines with increasing depth.

Concentrations of sodium, magnesium and chloride are highest in autumn and winter (e.g. Fig. 14.4). This pattern of variation parallels that found in precipitation; the variation in the soil solution concentration is driven by inputs. In contrast, variations in solution concentrations of dissolved organic carbon (DOC), nitrate and potassium seem to be driven by temperature interacting with biological processes. DOC concentrations are highest in summer (Fig. 14.4) probably reflecting increased biological activity and production of decomposition products. Potassium and nitrate concentrations peak in winter and early spring; available potassium and nitrate are probably utilised by plants in summer. Sulphate levels in solution mainly reflect specific events, with high input concentrations in throughfall, but high sulphate levels also follow dry periods. The latter may reflect wash-off of dry deposits accumulated in the dry period plus production of sulphate in organic soil horizons by oxidation. The pH of soil solutions shows no clear cyclical pattern but low values usually coincide with high total ion concentrations, such as large sea salt inputs or flushing of dry deposits accumulated during dry periods.

Impact of land use on soil water chemistry

a. Afforestation

Waters from stagnopodzols below coniferous plantations at Plynlimon and Llyn Brianne have significantly higher concentrations of sodium, sulphate, chloride, aluminium and silica than those from beneath moorland vegetation (Fig, 14.5). Aluminium concentrations in the forest soil waters are 3–4 times greater than in the grassland soils; aluminium concentrations of almost 300 $\mu eq\ l^{-1}$ (= 3.3 mg l^{-1}) have been recorded in forested stagnopodzols at Llyn Brianne. Sodium remains the dominant cation in the waters from the forested soils, accounting for about 45% of the total cationic charge, but aluminium now accounts for about 28% of the cationic charge compared with 15% in the moorland podzol. The soil waters from the forested soils are also more acidic. Chloride is still the dominant anion in the forest soil waters but sulphate now accounts for more than 35% of the total anion charge, compared with about 25% in the moorland soils.

These differences in chemistry are thought to reflect the high concentrations of sodium, chloride and sulphate in the spruce throughfall and stemflow, compared with the throughfall from the moorland vegetation. This, in turn, partly reflects the greater atmospheric deposition on to the coniferous canopies. There may also be increased mobilisation of sulphate from the organic horizons of the forest soils as a result of drying. The increased concentrations of mobile anions and increased acidity lead to the mobilisation of additional aluminium and silica by ion exchange and weathering.[9] Similar differences in soil solution chemistry between forest and moorland soils have been shown to occur in other soil types in the Llyn Brianne area.

b. Pasture improvement

Pasture improvement on stagnopodzols also produces marked changes in soil water chemistry. The improvement generally involves additions of lime and fertilisers; the precise changes in soil water chemistry reflect the amount and type of lime and fertiliser added and the time since treatment.[8]

The main impact, however, is an increase in solute concentrations of calcium and magnesium, and reductions in the concentrations of H^+ and aluminium (Fig. 14.5). Bicarbonate is also present in solution following liming and in some cases becomes the dominant anion. The waters from all the soil horizons are usually less acid and contain more calcium, magnesium and bicarbonate than precipitation.

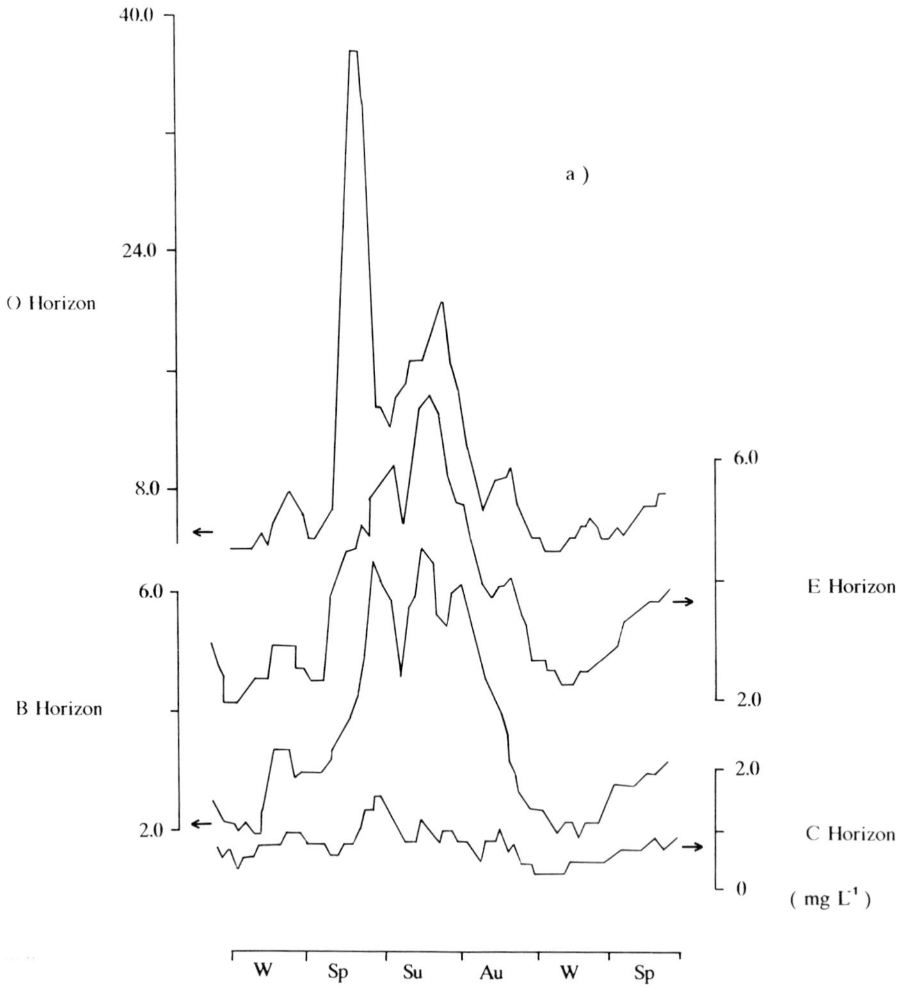

Fig. 14.4. Temporal variations in concentrations of (a) DOC and (b) chloride in soil solutions from different horizons of a stagnopodzol at Plynlimon.

Solute chemistry and soil type

Soil water chemistry also varies between soil type in the uplands of Wales. For example, drainage from interfluve peats at Plynlimon is more acid and contains less aluminium than even the stagnopodzol 0 horizon,[17] and a major part of the acidity is due to organic acids. The soil waters from brown podzolic soils near Llyn Brianne are less acid and contain less aluminium, but higher concentrations of calcium and magnesium, than those from the stagnopodzols: this reflects the slightly higher base status of the brown podzolic soils,

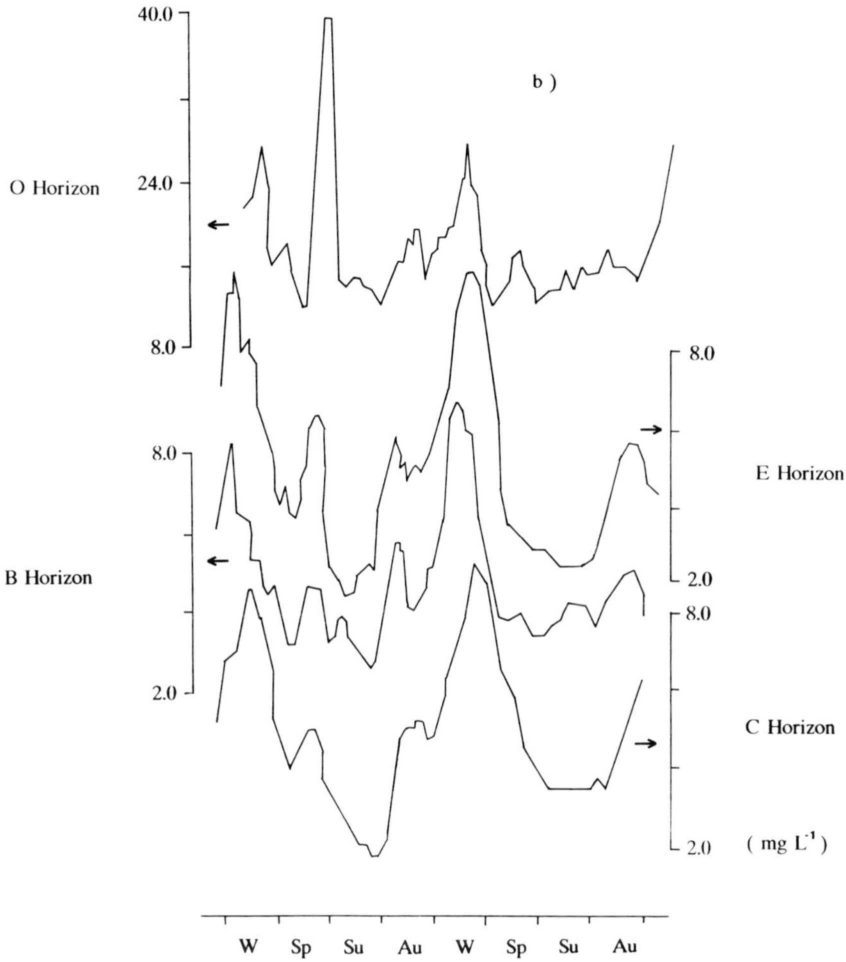

some buffering of acidity taking place by ion exchange. The brown podzolic soils do not have a surface peaty horizon comparable to that of the stagnopodzols.

The waters from all the widespread soil types in the uplands of Wales are, however, similar in acidity to, or more acid than, incoming precipitation. Where acid, organic rich surface horizons have developed, these are a source of protons while the mineral horizons are a source of aluminium. The major buffering mechanism in the acid mineral soils is linked to mobilisation of aluminium.

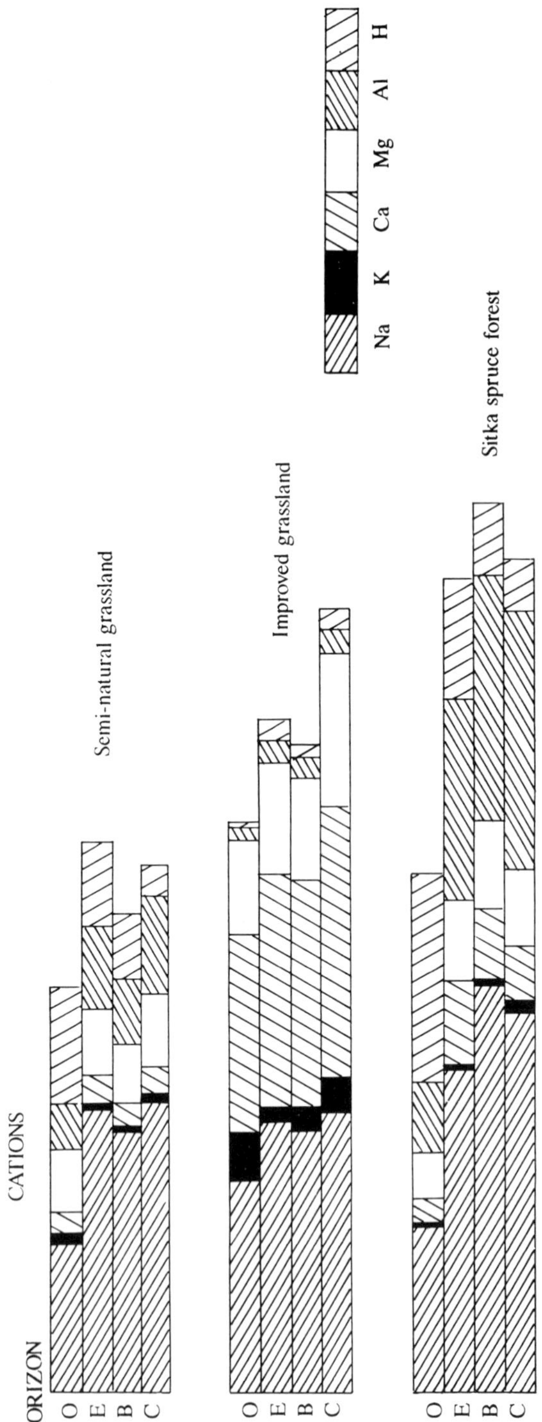
HORIZON
CATIONS
O
E
B
C
Semi-natural grassland
Improved grassland
Sitka spruce forest
Na
K
Ca
Mg
Al
H

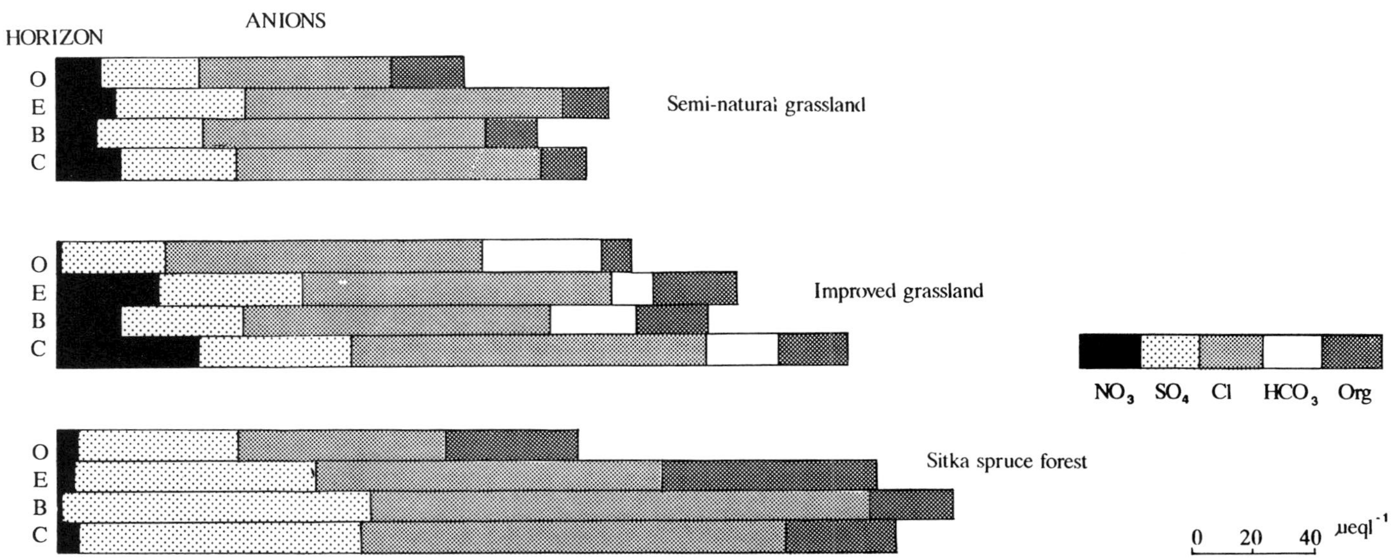

Fig. 14.5. Mean solute concentrations in soil solutions from stagnopodzols in semi-natural grassland, improved grassland and Sitka spruce forest.

Solute-bedrock interactions: groundwater chemistry

The main factors influencing groundwater acidity are the acidity of input waters, usually derived from overlying soils or drift, and the content of carbonate and weatherable silicates in the bedrock. The bedrock underlying the study sites in upland Wales is dominated by Ordovician and Silurian slates and mudstones. These do not normally contain carbonates and have few readily weatherable silicate minerals. The chemistry of input waters is, therefore, modified relatively little by reactions with these rocks. Comparison of the chemistry of waters from the stagnopodzol C horizons with that from very low flows in streams, here used as an analogue for groundwater, does, however, show a slight reduction in acidity and in aluminium concentrations. Some catchments in the Plynlimon area are, however, underlain by mudstones which contain small calcite veins associated with secondary mineralisation. In these catchments, there is a major difference between the chemistry of soil waters and the analogue groundwater chemistry (Table 14.3).[8, 18] The pH was 4.70 in the C horizon waters but 6.2 in the 'groundwater'; calcium and magnesium concentrations were 17 and 42 $\mu eq\ l^{-1}$ respectively in the C horizon waters but 144 and 126 $\mu eq\ l^{-1}$ in the 'groundwater'. Solution of carbonates in the veins has released Ca^{2+} and consumed H^{+} ions; aluminium has been precipitated out as pH has increased.

Conclusions

Significant changes in solute chemistry, including acidity, take place as waters move through terrestrial ecosystems from precipitation to stream. The acidity of precipitation may be increased or decreased as it moves through vegetation canopies but the concentration of mobile acid anions is almost always increased. This may be more important than throughfall acidity in influencing reactions in the acid upland soils of Wales, and in acidification of surface waters.[14] In upland Wales the replacement of moorland by forest plantations results in an increase in the concentration and flux of mobile anions into soils.

The chemistry of soil waters varies both between soils and between horizons of a given soil. Soil waters from the surface horizons of acid soils are more acid than the throughfall input; mobilisation of organic acids is an important factor in peats and in soils with organic surface horizons. Solution acidity decreases in the lower soil; this is linked to degradation of organic acids, weathering of aluminium and silicate minerals, and ion exchange. These latter processes mobilise aluminium and silica into solution. The surface layers of peats and organic surfaced soils are therefore a source of very acid waters (pH about 4.0) with relatively low concentrations of aluminium, calcium and magnesium. The lower, mineral horizons are sources of acid water (pH about 4.5), with high concentrations of aluminium and silica but low levels of calcium and magnesium.

The chemistry of the soil drainage waters is modified by land use. Waters from soils below coniferous plantations have higher solute concentrations than those from similar soils beneath moorland; the largest increases are shown by aluminium, sodium and mobile acid anions. Liming, in connection with land improvement, reduces soil water acidity and the soils become a source of water with relatively high concentrations of calcium, magnesium and bicarbonate (see Chap. 19).

Little buffering usually takes place as waters move through the slates, mudstones, grits and acid igneous rocks of upland Wales. Groundwaters have slightly higher pH and calcium contents than drainage waters from subsoil horizons. In a few areas, however, small amounts of carbonate minerals in veins within the bedrock, or of feldspars, have an important effect on groundwater chemistry and can produce significant buffering and precipitation of aluminium.

The streams are a mix of waters derived from a variety of sources within a catchment, the proportions varying with flow (Chap. 13). The chemistry, including the acidity, of the streamwater will change in response to the changes in the proportions of the inputs. The chemistry of the streamwater is not, however, a simple composite of the mixed waters. During the mixing, chemical reactions may take place. Any rise in pH will itself be determined by the proportions of the various inputs. In areas with feldspars or calcite veins in the bedrock, bicarbonate in the groundwater input will buffer the acidity in the soil-derived waters at all but the highest flow levels.[18] Elsewhere, the acid soil waters are poorly buffered during mixing in the stream channel. Loss of dissolved CO_2 from input soil waters (degassing) can produce a dramatic rise in pH in some situations, specifically if the pH of the input soil is between 4.5 and 5.0.[2, 16] The pH of the acid soils of much of upland Wales are below 4.5 so that this mechanism is not important. It may, however, become important where liming has raised the pH of the naturally acid soils and hence of the soil drainage waters.

References

1. Calder, I.R. and Newson, M.D. (1979). Land use and water resources in Britain – a strategic look. *Water Resour., Bull.*, 15: 1628.
2. Cresser, M.S. and Edwards, A.C. (1987). Acidification of freshwaters. Cambridge: Cambridge Unversity Press.
3. Cronan, C.S., Reiners, W A., Reynolds, R.C. and Lang, G.E. (1978). Forest floor leaching: contributions from mineral, organic and carbonic acids in New Hampshire subalpine forests. *Science*, 200: 309–311.
4. Cronan, C.S. and Schofield, C.L. (1979). Aluminium leaching response to acid precipitation; effects on high elevation watersheds in the Northeast. *Science*, 204: 304–305.
5. Dollard, G.J., Unsworth, M.H. and Harvey, M.J. (1983). Pollutant transfer in upland regions by occult precipitation. *Nature*, 302: 241–243.
6. Driscoll, C.T. (1980). Chemical characterization of some dilute acidified lakes and streams in the Adirondack region of New York State. *PhD thesis*. Ithaca, New York: Cornell University.

7. Fowler, D. (9184). Transfer to terrestrial surfaces. *Phil. Trans. Roy. Soc. Lond. B*, 305: 281.
8. Hornung, M., Adamson, J.K., Reynolds, B. and Stevens, P.A. (1986). Influence of mineral weathering and catchment hydrology on drainage water chemistry in three upland sites in England and Wales. *J. geol. Soc. Lond.*, 143: 627–634.
9. Hornung, M., Reynolds, B., Stevens, P.A. and Neal, C. (1987). Increased acidity and aluminium concentrations in streams following afforestation: causative mechanisms and processes. In: *Acidifiation and water pathways*. Oslo, Norway: Norwegian National Committee for Hydrology.
10. Johnson, D.W. and Cole, D.W. (1977). Sulphate mobility in an outwash soil in western Washington. *Water, Air and Soil Pollution*, 7: 489–595.
11. Johnson, D.W., Cole, D.W., Gessel, Singer, M.J. and Minden R.W. (1977). Carbonic acid leaching in tropical, temperate, sub-alpine and north forest soil. *Arct. Alp. Res.*, 9: 329–343.
12. Langan, S.J. (1987). Episodic acidification of streams at Loch Dee, SW Scotland. *Trans. Roy. Soc. Edinburgh: Earth Sciences*, 78: 393–398.
13. Miller, H.G. and Miller, J.D. (1980). Collection and retention of atmospheric pollutants by vegetation. In: *Ecological impact of acid precipitation*, (eds) D. Drablos and A. Tollen (Proc. Int. Conf., Sandefjord, Norway). SNSF: Oslo-As.
14. Miller, H.G. (1985). The possible role of forests in streamwater acidification. *Soil Use and Management*, 1: 28–29.
15. Parker, G.G. (1983). Throughfall and stemflow in the Forest Nutrient Cycle. *Advances in Ecological Research*, 13: 58–135.
16. Reuss, J.O. and Johnson of soils and waters. New York: Springer-Verlag.
17. Reynolds, B., Hornung, M. and Hughes, S. (1983). Some factors controlling variations in chemistry of an upland stream in mid-Wales. *Cambria* 10: 130–145.
18. Reynolds, B., Neal, C., Hornung, M. and Stevens, P.A. (1986). Baseflow buffering of streamwater acidity in five mid Wales catchments. *J. Hydrol.*, 87: 167–185.
19. Reynolds, B., Cape, N. and Patterson, I.S. (1989). Comparison of element fluxes in throughfall below larch and Sitka spruce at two contrasting UK sites. Forestry.
20. Reynolds, B., Neal, C., Hornung, M. and Stevens, P.A. (In press). The role of cation exchange in determining the impact of afforestation on the solution chemistry of stagnopodzols in mid Wales. *Water, Air and Soil Pollution.*
21. Stevens, P.A. (1987). Throughfall chemistry beneath Sitka spruce of four ages in Beddgelert Forest, north Wales, UK. *Plant & Soil*, 101: 291–294.
22. Stevens, P.A. and Hornung, M. (1989). Nitrate leaching from a felled Sitka spruce plantation in Beddgelert Forest, north Wales. *Soil Use & Management.*
23. Tukey, H.B. Jr. (1970). The leaching substances from plants. *Annual Rev. Pl. Physiol.*, 21: 305–324.
24. Unsworth, M.H. (1984). Evaporation from forests in cloud enhances the effects of acid deposition. *Nature* (Lond)., 312: 262–264.

15. The influence of land management on stream water chemistry

S.C. BIRD, S.J. BROWN and E. VAUGHAN

Introduction

Previous studies[1, 9, 10] have shown that upland areas of Britain with base-poor bedrock and soils are vulnerable to acid deposition and that different types of land use can lead to significant differences in the chemistry of streams draining similar lithologies. In particular, those streams draining areas of conifer forest are often more acidic and contain elevated levels of toxic dissolved aluminium compared with streams draining equivalent moorland areas.

In order to quantify short term, seasonal and longer term trends in stream chemistry with respect to land management practices, a number of comparative catchment studies have been established in Scotland and in Wales. This chapter describes the typical patterns of stream water chemistry observed for different types of land use in upland Wales, using data from the Llyn Brianne study.

The chemistry of moorland and forest streams

Water chemistry data for twelve of the Llyn Brianne study catchments are used to illustrate the differences between the streams in relation to land use, including five reference catchments and seven catchments where land management techniques have subsequently been applied (See Chap. 19). Consideration is given to descriptions of seasonal changes, using data from a routine (weekly 1985–1986; fortnightly 1987) spot sampling programme. Short-term changes resulting from storm episodes are assessed using continuous monitoring data and associated intensive sampling records.

To facilitate comparison between the study catchments, the streams are classified into groups according to the system devised by Stoner, Gee and Wade,[10] described in Chap. 1 and detailed in Table 15.1.

R.W. Edwards et al. (eds) Acid Waters in Wales. 241–253.

Table 15.1. Summary of stream chemistry (mean and range) for the Llyn Brianne study catchments (before treatments).

Determinand (all mgl^{-1} except pH)	STREAMS (AND GROUP) 1 (IIIB)	2 (IIIB)	3 (IIB)	4 (IIB)	5 (I)	7 (IIB)	9 (IIIA)	10 (IIIA)	11 (IIIA)	12 (IIIA)	13 (IIIA)	14 (IIIA)
pH	4.93 4.1–6.2	4.57 4.3–4.8	5.33 4.4–6.5	5.03 4.3–5.5	6.82 6.0–7.5	5.33 4.4–6.5	5.19 4.10–6.10	5.37 4.5–6.4	5.27 4.7–6.2	5.76 4.8–6.7	5.97 4.8–7.0	6.2 5.8–6.7
Total hardness (as $CaCo_3$)	6.0 2.6–13.1	5.0 3.1–8.3	7.4 3.1–11.0	7.4 4.4–9.5	14.5 4.5–45.2	7.2 3.3–31.2	4.8 3.1–8.2	4.7 2.3–10.4	5.1 2.5–11.0	5.5 2.8–9.3	6.8 3.8–11.4	7.7 4.0–14.3
Chloride	7.80 4.0–11.0	7.59 5.0–21.0	7.49 5.0–10.0	6.64 4.0–8.0	4.83 3.0–7.0	6.33 3.0–10.0	4.63 3.0–8.0	4.94 2.0–7.0	5.11 3.0–8.0	5.02 2.0–9.0	5.62 3.0–9.0	6.40 4.0–9.0
Nitrate	0.15 0.1–1.1	0.24 0.1–1.0	0.30 0.1–0.7	0.59 0.3–4.0	0.17 0.1–0.4	0.27 0.1–3.4	0.34 0.1–1.6	0.17 0.1–0.4	0.24 0.1–1.1	0.25 0.1–0.7	0.14 0.1–0.9	0.27 0.1–1.7
Sulphate*	6.18 3.38–9.38	6.00 4.19–8.39	6.43 3.80–8.43	6.32 1.92–8.58	3.94 1.10–7.52	5.05 3.41–7.80	3.91 3.04–4.85	3.30 2.30–5.31	4.16 2.90–5.64	3.75 2.51–5.57	4.00 2.05–11.66	5.57 4.08–7.40
Calcium*	1.25 0.48–2.34	1.04 0.50–1.50	1.90 0.30–3.05	1.88 0.28–2.5	2.81 0.98–5.55	1.31 0.78–3.20	0.82 0.59–1.53	0.85 0.33–1.37	0.87 0.52–1.47	1.16 0.69–1.89	1.30 0.68–2.18	1.80 1.47–2.15
Aluminium*	0.33 0.05–0.76	0.53 0.27–0.79	0.26 0.01–0.68	0.18 0.02–0.42	0.05 0.01–0.17	0.21 0.02–0.52	0.12 0.05–0.25	0.11 0.04–0.17	0.15 0.03–0.38	0.10 0.02–0.24	0.06 0.03–0.09	0.03 0.01–0.08
Magnesium*	0.70 0.36–1.22	0.59 0.43–0.78	0.71 0.49–1.30	0.75 0.58–0.96	1.61 0.53–3.14	0.75 0.48–1.50	0.60 0.41–0.84	0.63 0.39–0.90	0.61 0.40–0.87	0.69 0.42–1.07	0.95 0.51–1.66	0.84 0.64–1.01
Sodium*	4.26 2.52–7.07	3.97 2.86–5.05	4.12 2.44–5.05	3.63 0.10–4.79	2.88 0.83–4.10	3.43 2.08–5.24	2.94 1.79–3.89	2.82 1.90–3.87	3.02 1.88–4.01	3.03 1.62–4.08	3.37 1.97–4.88	3.96 2.98–4.56
Potassium*	0.20 0.03–1.32	0.12 0.02–0.36	0.19 0.09–0.66	0.22 0.06–1.00	0.42 0.10–1.75	0.21 0.04–1.18	0.18 0.07–0.55	0.23 0.02–0.95	0.21 0.02–0.67	0.34 0.05–5.60	0.35 0.02–2.96	0.30 0.10–0.94

* Dissolved.

Moorland streams

Six moorland streams are considered and clear differences between the groups (Group I, Stream 5; Group IIIA, Streams 9, 10, 11, 12, 13), defined on the basis of mean hardness, can be demonstrated. The Group I unimproved, unacidic control is characterised by well-buffered waters of comparatively high hardness and hence circumneutral pH, together with low dissolved aluminium levels (Table 15.1). Sulphate, nitrate and sodium levels are also low. In contrast, the Group IIIA unimproved, acidified moorland streams have poorly buffered waters of low hardness. As a consequence of this, pH levels are lower and dissolved aluminium levels are elevated. However, sulphate, nitrate and sodium levels are broadly similar to those of the Group I stream (Table 15.1).

Despite the generally well buffered conditions in Streams in Group I, seasonality patterns (illustrated by deviation of quarterly means from the annual mean over a three year period) are still discernible, particularly for pH, dissolved calcium and nitrate (Fig. 15.1). Reduced pH and calcium levels characterise the winter period, together with increased nitrate as typically observed for upland moorlands elsewhere.[7] A statistical comparison of winter (October to March) and summer (April to September) stream chemistry using Student's 't' test undertaken on both transformed ($\log_{10}$) and untransformed data also demonstrates significant seasonal differences in certain key parameters for Group IIIA (Stream 5) (Table 15.2).

Elevated hydrogen ion, nitrate, sulphate and sodium concentrations are confirmed in winter whereas during summer, higher potassium and marginally higher aluminium concentrations occur. However, the latter are not likely to be of ecological significance at prevailing levels. Seasonal patterns in chemistry also occur in the Group IIIA moorland streams as is shown for Stream 12, with respect to aluminium, calcium and nitrate (Fig. 15.1). The existence of statistically significant seasonal differences for other determinands have also been identified, although these differences are not common for all the streams classified within the Group. For example, significantly higher hydrogen ion concentrations occur in Stream 9 in winter than in summer although this trend has not been identified for the other Group IIIA streams (Table 15.2). Similarly, higher levels of sodium occur in some streams (9 and 13) in winter than in summer although, in the case of aluminium for Stream 10, higher levels occur in summer than in winter.

Clearly, a simple winter/summer six-monthly split may not be very sensitive in defining all seasonal differences, particularly in unusually wet seasons, as was experienced in summer 1985. However, the data demonstrate that, even within groups of streams with similar hardness, significant differences in stream chemistry may occur.

Associated, in part, with the existence of seasonal patterns in stream water chemistry, significant chemical inter-relations may be demonstrated for the moorland streams. In the case of the Group I (Stream 5), pH is significantly

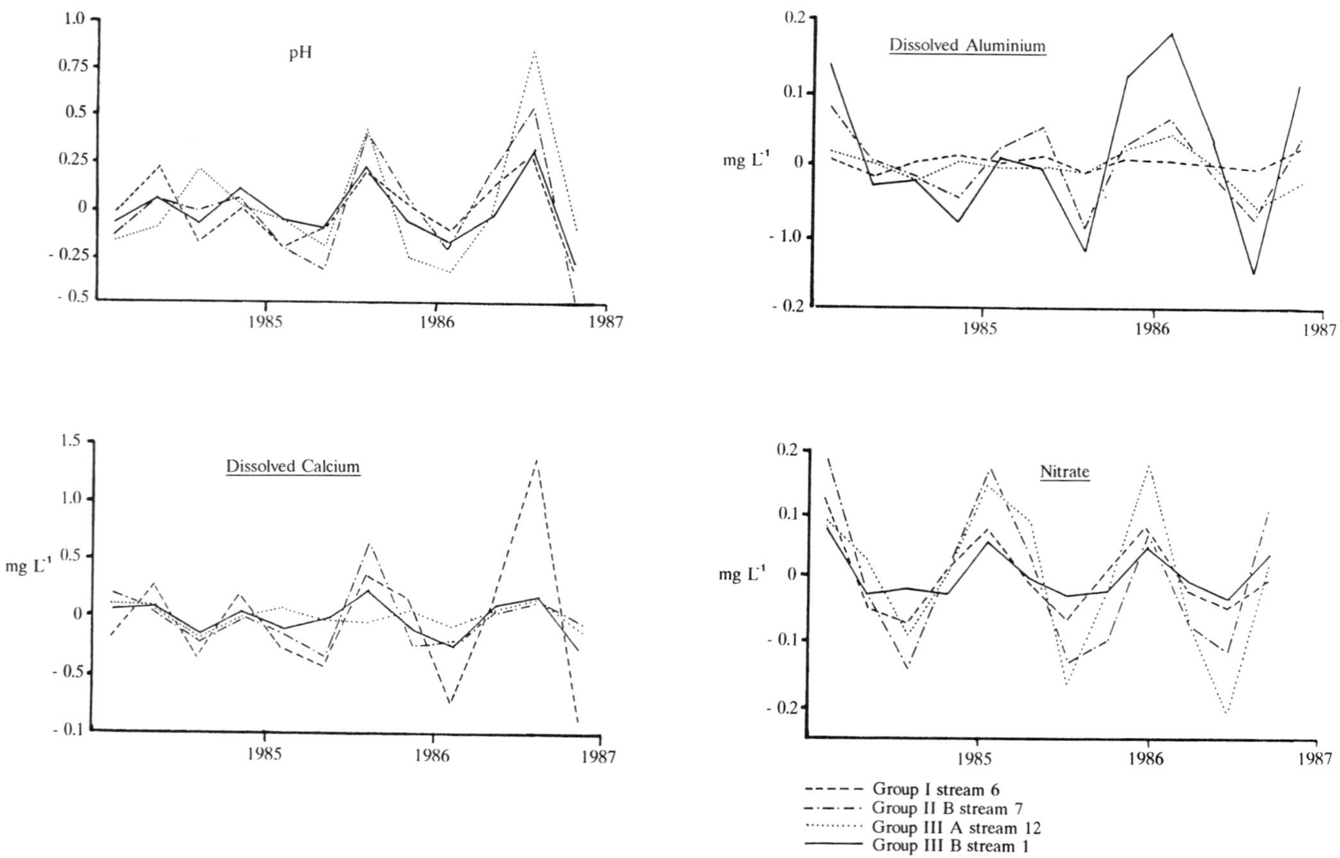

Fig. 15.1. Seasonality (variation between quarterly and annual means) of pH, dissolved aluminium, dissolved calcium and nitrate for group I, IIB, IIIA and IIIB streams, 1985–7.

Table 15.2. Statistical comparison of winter and summer stream chemistry using student's 't' tests.

Stream (and group)		Determinand H^+	Ca^{2+}	Al^{3+}	Na^+	K^+	SO_4^-	NO_3^-	Cl^-
1	N	W>S**				S>W***	S>W**	W>S***	
(IIIB)	L					S>W**	S>W***	W>S**	
2	N						S>W**		S>W***
(IIIB)	L						S>W***		S>W***
3	N					S>W***	S>W*		
(IIB)	L		W>S*			S>W**	S>W***		
4	N	W>S***	W>S**		W>S***			W>S***	
(IIB)	L		W>S***		W>S***		S>W**	W>S*	W>S*
5	N	W>S*		S>W**	W>S*	S>W***		W>S*	
(I)	L			S>W**	W>S***	S>W***	W>S*		
7	N	W>S**	W>S**	W>S**				W>S***	
(IIB)	L								
9	N	W>S***			W>S**				
(IIIA)	L				W>S**				
10	N					S>W***			
(IIIA)	L			W>S*		W>S***			W>S*
11	N					W>S*	W>S*	W>S**	
(IIIA)	L						W>S*		
12	N					W>S***			
(IIIA)	L								W>S*
13	N				W>S*	W>S***		W>S***	
(IIIA)	L				W>S**				
14	N	W>S***		S>W**		S>W**		W>S***	
(IIIA)	L	W>S***			W>S*	S>W*			

W – winter;
S – summer;
* – $p < 0.05$;
** – $p < 0.01$;
*** – $p < 0.001$;
N – untransformed
L – log

correlated with calcium, aluminium, nitrate and magnesium ($P < 0.05$ in each case; Fig. 15.2). However, no significant correlations appear to exist between stream flow and stream chemistry on the basis of the spot sample data. Furthermore, Stream 5 does not experience the acidic episodes typically observed in streams with lower buffering capacity (see Chap. 13), even during large storm events.

In contrast, statistically significant ($p < 0.05$) negative correlations between pH and logFlow are exhibited by Group IIIA Streams 10, 11 and 13. Moreover, these correlations generally give rise to useful regression equations, explaining up to 76% of the variance (Table 15.3). Significant negative correlations are also apparent between pH and aluminium and between pH and nitrate in all five Group IIIA streams (Fig. 15.2), pH explaining up to 69% of the variance in aluminium levels in Stream 12 (Table 15.3). Aluminium levels are also positively correlated with flow, yielding regression equations explaining up to 72% of the variance in aluminium concentrations (Table 15.3). In addition, Stream 13 also exhibits a significant correlation between flow and sulphate concentrations yielding a regression which explains some 30% of the variance in sulphate levels.

Conifer forest streams

Conifer forest streams within Groups IIB (Streams 3, 4 and 7) and IIIB (Streams 1 and 2) drain, with the exception of Stream 7, areas of 'closed-canopy' conifer plantations.

The most notable feature of the streams draining conifer forests is their greatly increased acidity compared with streams draining moorlands within similar hardness categories. The Group IIB streams, for example, exhibit similarly acidic conditions to the Group IIIA moorland streams, despite having higher buffering capacities reflecting their higher calcium and magnesium levels. Moreover, significantly higher levels of dissolved aluminium and sulphate occur in the Group IIB streams compared with the Group IIIA moorland streams (Table 15.1).

The Group IIIB streams (1 and 2) are very poorly buffered, as in the case of the Group IIIA moorland streams. However, in contrast with the Group IIIA streams, they are consistently acidic throughout the year. Moreover, the Group IIIB streams have considerably higher aluminium concentrations. Both sodium and sulphate concentrations are also generally higher than in any of the other groups, whereas nitrate levels are similar to the Group I and Group IIIA streams but much lower than in the Group IIB streams (Table 15.1).

As with the moorland streams, the conifer forest streams display a number of seasonal variations. For example, Stream 7 (Group IIB) displays significant seasonality in pH and calcium (lower in winter) and aluminium and nitrate (higher in winter) (Fig. 15.1, Table 15.2). Statistical analysis, comparing winter and summer periods, suggests the existence of significantly increased

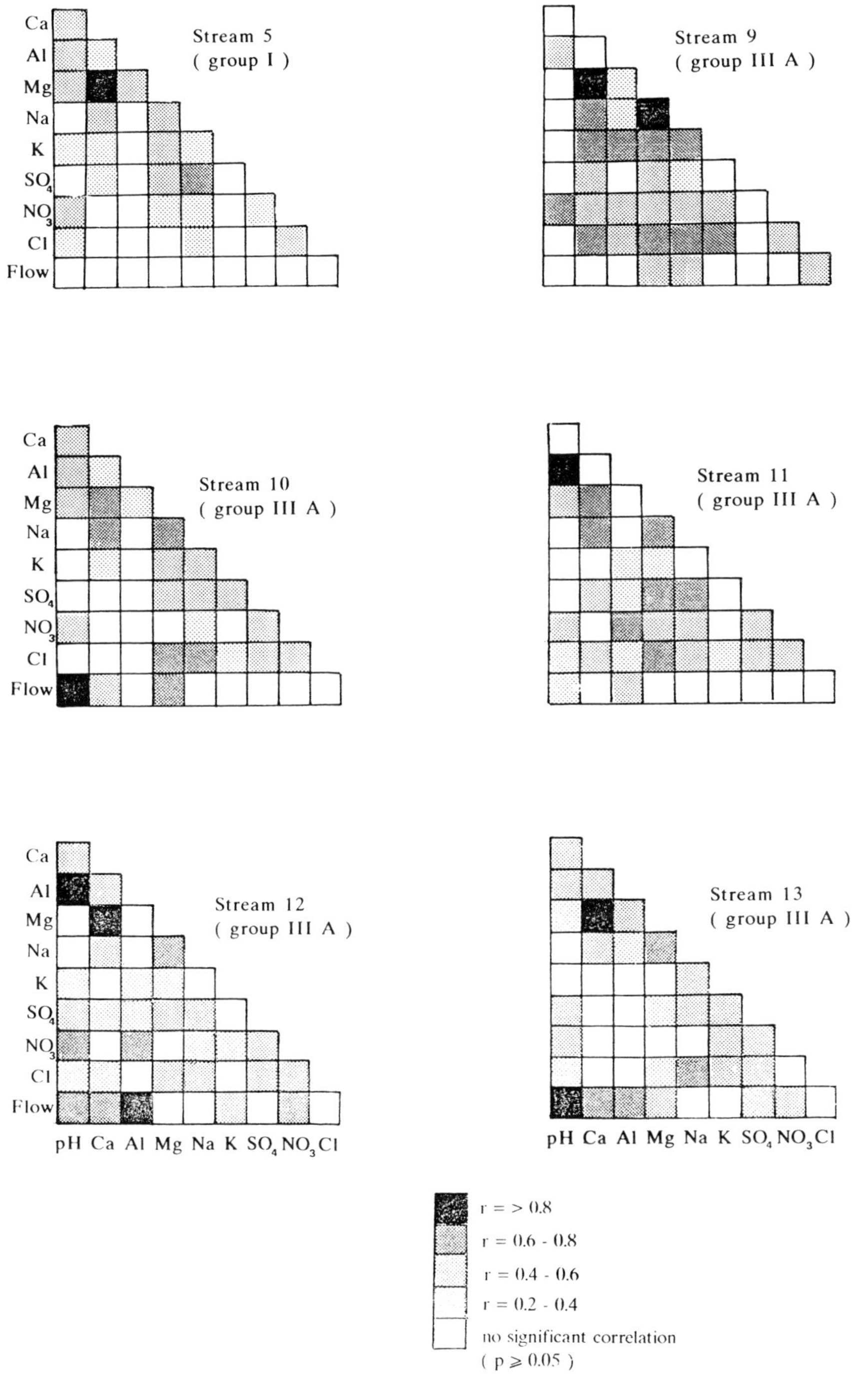

Fig. 15.2. Correlation matrices for selected determinands: Moorland streams.

Table 15.3. Statistically significant regression equations for selected determinands for the Llyn Brianne study streams.

Regressions Stream	Log Al (y) (mgl^{-1}) v pH (x)	pH (y) v log flow (x) ($1s^{-1}$)	log Al (y) (mgl^{-1}) v log flow (x) ($1s^{-1}$)	log SO_4 (y) (mgl^{-1}) v log flow (x) ($1s^{-1}$)
1	log Al = −0.663 pH + 2.691 (77%)	pH = −0.737 log flow + 6.26 (64.5%)	log Al = 0.555 log flow −1.598 (60.4%)	
2	log Al = −0.503 pH + 2.01 (48.7%)	pH = −0408 log flow + 5.237 (95.2%)		
3	log Al = −0.634 pH + 2.605 (72.4%)	pH = −0.612 log flow + 6.02 (43.4%)	log Al = 0.738 log flow −1.616 (83.3%)	
4	log Al = −0.73 pH + 2.85 (38.8%)		log Al = 0.357 log flow −0.999 (50.6%)	log SO_4 = −0.76 log flow + 0.858 (39.7%)
5	log Al = −0.284 pH + 0.791 (29.5%)			
9	log Al = −0.225 pH + 0.188 (25.9%)			
10	log Al = −0.095 pH − 0.477 (20.9%)	pH = −0.97 log flow + 6.632 (69.8%)		
11	log Al = −0.684 pH + 2.683 (61.7%)			
12	log Al = −0.332 pH + 0.838 (68.9%)	pH = −0.673 log flow + 6.459 (62.5%)	log Al = 0.262 log flow −1.344 (72.2%)	
13	log Al = −0.10 pH −0.65 (22.3%)	pH = −1.08 log flow + 7.79 (75.8%)	log Al = 0.189 log flow −1.577 (46.1%)	log SO_4 = 0.173 log flow + 0.319 (30.1%)

All regressions significant at $p < 0.001$.
Values for Corrected R^2 in brackets.
Log e.

sulphate levels in summer (Streams 3 and 4) and increased calcium levels in winter (Streams 3, 4 and 7) (Table 15.2). Seasonality patterns are also evident in the Group IIIB streams with reduced sulphate and pH, and increased aluminium and nitrate levels, occurring in winter (Table 15.2).

The seasonality plots for Stream 7 (Group IIB) in an 'open canopy' catchment give further support to the maturation effect of conifers, the pH seasonality of this stream being closer to, but less than, that of a Group IIIA moorland stream. Aluminium seasonality is greater than in the IIIA moorland stream but less than in the IIIB conifer forest stream (Fig. 15.1).

Correlation matrices for the Group IIB streams (Fig. 15.3) confirm the existence, in common with Group IIIA streams, of consistent negative correlations between pH and flow and between pH and aluminium. As in Group IIIA, stream flow and aluminium are also consistently positively correlated.

The above correlations give rise to a number of regression equations (Table 15.3) which relate, for example, aluminium levels with both pH and flow conditions. In the case of Stream 3, this relationship explains up to 83% of the variance in aluminium on the basis of stream flow variations alone.

Correlation matrices for the Group IIB streams (Fig. 15.3) confirm significant correlations between pH and flow, pH and aluminium, and pH and calcium. Regression equations can be derived which relate aluminium to pH and flow (Table 15.3), as well as pH based on flow. For example, 95% of the variance in pH for Stream 2 may be explained by flow variations.

On the basis of hardness data for the period of 1985–87 (Table 15.1), streams 3, 4 and 7 should be classified as Group IIIB (mean hardness less than 8 mgl^{-1}), rather than Group IIB as originally defined (Chap. 1). However, trends in mean hardness during the period 1981–87 demonstrate that the hardness values during the 1985–87 period have been depressed by wet summer periods, in particular during 1985 and 1986 when the typical summer increases in hardness levels compared to the winter periods did not occur. Longer term studies will establish whether canopy closure, with subsequent increases in interception of atmospheric pollutant levels, leads to a reduction in hardness due to any increased removal rate of base cations from the soils.

Acid oak woodland stream

There is only one stream in this category (Stream 14 Group IIIA) and this displays less acid conditions than all but Stream 5 (Group I). It is sufficiently well buffered for aluminium concentrations to remain at levels even lower than those of Stream 5 (Table 15.1). Sulphate levels in Stream 14 are significantly higher than in the other Group IIIA catchments, although slightly lower than those found in the Group IIB and IIIB streams (Table 15.1). Significant seasonality patterns (Table 15.2) are restricted to slightly higher hydrogen ion levels and slightly lower potassium concentrations in winter.

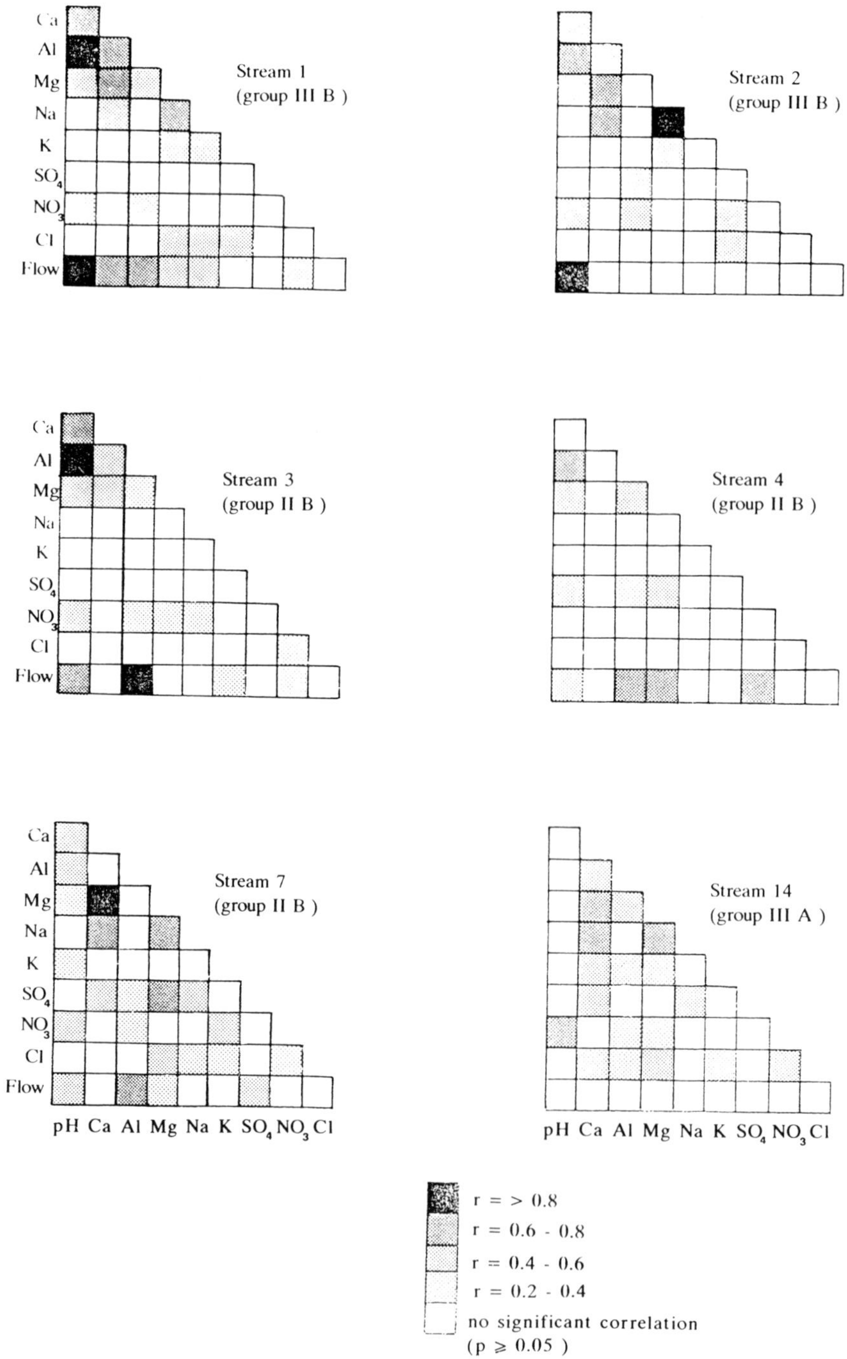

Fig. 15.3. Correlation matrices for selected determinands: Conifer Afforested and Deciduous Woodland Streams.

No correlations exist between flow, pH and aluminium for this stream (Fig. 15.3), unlike the other Group IIIA and Group IIIB streams. Hence, stream water chemistry in this catchment, whilst broadly similar to the other IIIA streams, exhibits few of the seasonal patterns or important chemical interrelationships found in these other streams.

Stream chemistry during storm events

The most acidic conditions often occur during the winter months of higher rainfall and streamflow, with the Group IIB and IIIB streams remaining acidic for longer periods than Group IIIA streams. Such patterns are often more pronounced in small streams.[6, 11] Short term acidic episodes related to storm events may often lead to rapid changes in stream chemistry with low pH and elevated aluminium levels.[4, 10] Such episodes may be particularly acidic in the early autumn following dry summer periods[8] or during snowmelt[2, 6] and are often accompanied by flushes of sulphate, particularly in Group IIIB streams.[3]

These changes are dependent on a number of factors including antecedent flow and chemistry, storm intensity and duration, deposition quality and quantity and the hydrological pathways through the vegetation and soil cover (see Chap.13). This is exemplified by contrasting a winter 'rain on snow' episode with a summer episode resulting in a comparable stream hydrograph for both Group IIIA moorland (5) and Group IIIB conifer forest (1) (Fig. 15.4).

Whereas pre-episode conditions were broadly similar in both streams, the responses of each were quite different for the two sets of conditions. The rain-on-snow episodes occurred following a period of one month when easterly airstreams dominated, leading to sub-zero temperatures, frozen soils and snow cover. From analysis of snow samples (pH as low as 3.5) it is clear that substantial deposition of atmospheric pollutants occurred during this period, resulting in considerably elevated levels of sulphate, nitrate, ammonia and chloride in the snow compared with typical rainfall quality. Snowmelt was induced by heavy rainfall leading to dramatic impacts on stream chemistry, with important differences between the streams resulting from the different land use within the catchment areas (Fig. 15.4). In both moorland and forest streams pH declined, accompanied by increases in dissolved aluminium concentrations as buffering capacity provided by calcium was exceeded by anions such as sulphate. Whilst the timing of flow changes was similar in both streams, the flush of solutes to the moorland stream, illustrated by conductivity changes, was more rapid than in the forest streams, reflecting differences in hydrological pathways. Also, whereas both streams have similar buffering capacity, the changes in forest stream chemistry were much greater (pH reduction of 2 units and dissolved aluminium levels exceeding 1 mgl^{-1}), reflecting the greater flux of anions. The key factor leading

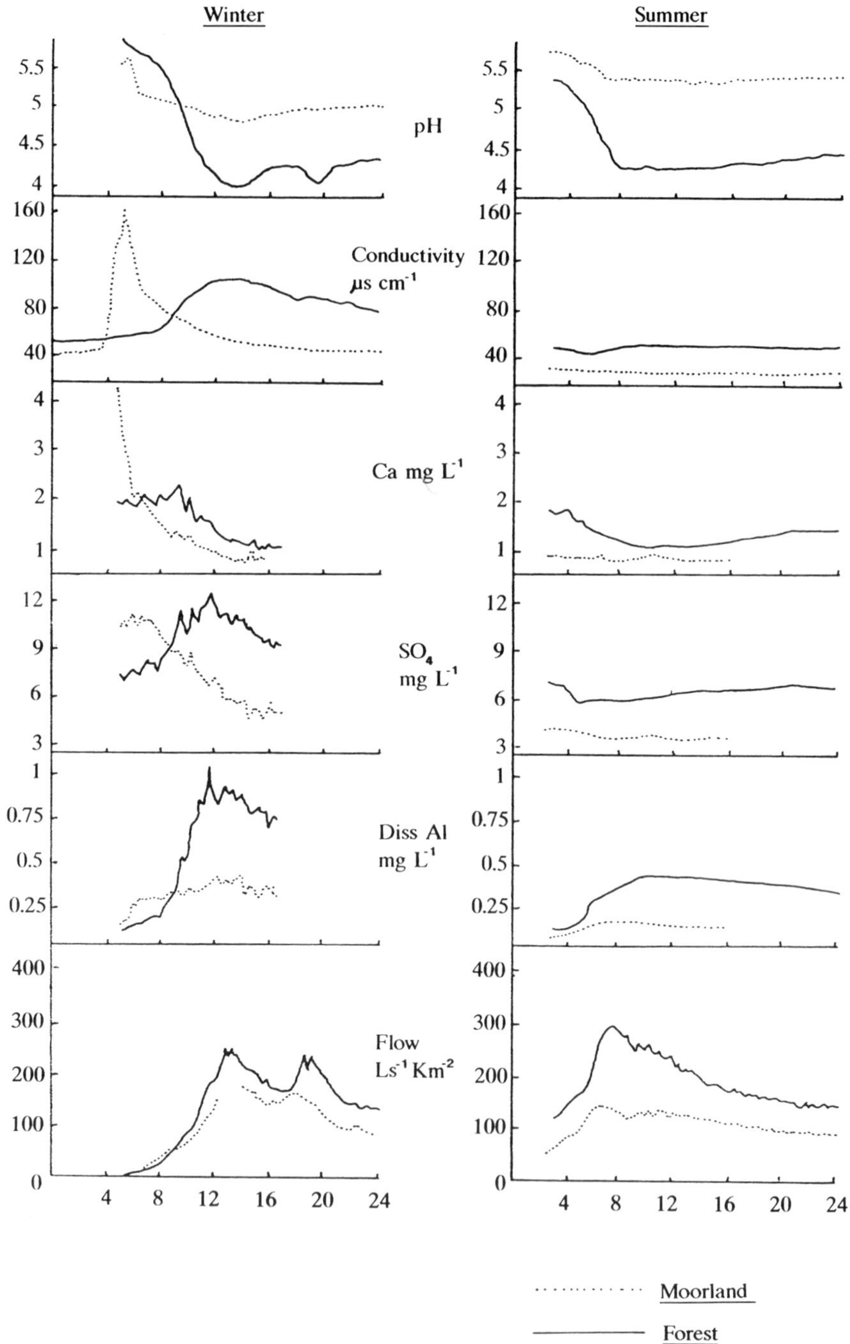

Fig. 15.4. Response of moorland (Stream 11) and forest (Stream 1) streams to winter (rain on snow) and summer storm episodes.

to such dramatic changes in both streams was the predominance of surface run-off and reduced buffering from frozen soils. The summer episode, resulting from a high rainfall event that produced comparable stream flows to the winter episode, was reflected in different responses between the two streams. The forest stream demonstrated a greater reduction in pH and higher increase in dissolved aluminium than the moorland stream. However, in this instance the concentrations of sulphate were much lower in both streams compared with the winter episode and less aluminium was thereby mobilised to maintain the ionic balance. The lower sulphate levels reflect the different preceding deposition patterns and higher antecedent flows. Also the available buffering capacity from soil cover was higher during this summer period despite the likelihood of dominant near-surface hydrological pathways (see Chap. 13). The importance of such episodes in determining the ecological status of streams has been demonstrated[5, 10] (see also Chap. 7).

References

1. Harriman, R. and Morrison, B.R.S. (1982). Ecology of streams draining forested and non-forested catchments in an area of Central Scotland subject to acid precipitation, *Hydrobiologia*, 88: 251–263.
2. Johanessen, M. (1980). Aluminium, a buffer in acidic waters? In, Drablos, D. and Tollen, A. (eds): *Ecological impact of acid precipitation*, pp. 224–225, SNSF Project.
3. Likens, G.E., Bormann, F.H., Pierce, R.S., Eaton, J.S. and Johnson, N.M. (1977). *Biogeochemistry of a forested ecosystem*, Springer Verlag.
4. Lynch, J.A., Hanna, C.M. and Corbett, E.S. (1986). Predicting pH, alkalinity and total acidity in streamwater during eposidic events. *Wat. Resar. Res.*, 22: 6: 905–912.
5. Ormerod, S.J., Boole, P., McCahon, C.P., Weatherley, N.S., Pascoe, D. and Edwards, R.W. (1987). Short-term experimental acidification of a Welsh stream: comparing the biological effects of hydrogen ions and aluminium. *Freshwater Biology* 17: 341–35.
6. Overrein, L.N., Seip, A.M. and Tollon, A. (1981). *Acid precipitation effects on forest and fish*, final report of the SNSF project 1972–1980.
7. Roberts, G., Hudson, J. and Blackie, J. (1983). *Nutrient Cycling in the Wye and Severn at Plynlimon*. Institute of Hydrology Report No. 86.
8. Stoner, J. (1987). Llyn Brianne Acid Water Project. First Technical Summary Report, April, 1987, WWA, 98 pp.
9. Stoner, J. and Gee A. (1985). The effects of forestry on water quality and fish in Welsh rivers and lakes, *J. Inst. Water Eng. Sci.*, 39: 27–454.
10. Stoner, J.H., Gee, A.S. and Wade K.R. (1984). The effects of acidification on the ecology of streams in the Upper Tywi catchment in West Wales. *Environ. Poll.* (A) 35: 125–157.
11. Sutcliffe, D.W. (1984). *Acid precipitation and its effects on aquatic systems in the English Lake District (Cumbria)*. 51st Annual report on the F.B.A. Report of the Director. pp. 31–62.

16. Hydrochemical modelling of acidification in Wales

P.G. WHITEHEAD, T.J. MUSGROVE and B.J. COSBY

Introduction

Acidification may be regarded as essentially a problem over two very different time scales. Short-term fluctuations in acidification caused by the flushing of near surface waters, or snowmelt, are generally driven by the hydrological processes operating in the catchment. The timescale of these events is in the order of hours, or at most, days, and the level of acidity will be largely controlled by the ability of the catchment to buffer incoming acidity within the catchment hydrological response time. On the other hand, there are long-term changes in catchment acidity that occur over decades (see Chap.3). These trends arise from the slow chemical processes operating on the catchment soils which reduce the catchment buffering capacity.

From a modelling point of view it is important to recognise the two differing time scales so that appropriate models can be developed and applied.

Recently there has been considerable use of mathematical models to describe the dominant interactions and processes operating at differing time scales and to simulate catchment behaviour. Steady state models have been used prescriptively to investigate the long-term consequences of changes in the industrial emissions of SO_x.[5, 6, 7, 8, 14]

Correspondingly, dynamic models have been successfully applied to several catchments. Christopherson, Siep and Wright[3] and Christopherson, Rustad and Seip[4] have developed a simple conceptual model that reproduces major trends in chemical and hydrological behaviour in Norwegian catchments. This model has been successfully extended[21] and applied to the Harp Lake catchment in Canada. The model has also been applied to two forested catchments in Sweden.[12]

In this chapter, the problems of applying models to short-term hydrochemical responses are briefly discussed and some applications described. Of particular interest, however, is the application of the model MAGIC (Model of Acidification of Groundwaters In Catchments) for predicting long-term changes in acidification, and this has been applied to specific catchments and in a regional analysis across Wales. The effects of afforestation and deposition reductions are investigated using the model.

R.W. Edwards et al. (eds) Acid Waters in Wales. 255–277.

Event type models of acidification

By event type models, we mean those models that aim to reproduce the short-term (hourly or daily) behaviour of catchments. They can range from relatively simple time series models[22] through lumped or semi-process based models such as the BIRKENES model[3, 4] to detailed, fully mechanistic and distributed models. Unfortunately, catchments are notoriously heterogeneous with complex flow paths, micropore and macropore flow with soil types varying both spatially and vertically. Even a fully distributed model cannot cope with the true complexity of catchment behaviour and therefore the approach adopted here has been to apply the semi-process based models initially before moving to a semi-distributed modelling approach, such as TOPMODEL.[2]

Time series approaches

The time series analysis techniques provide a rapid assessment of catchment hydrology and chemistry data. For example, the techniques have been applied to data from the Plynlimon catchment to produce a much simplified model of catchment behaviour.[23] Similarly they have been applied to data from Birkenes in Norway and Loch Dee in Scotland indicating that the technique has wide applicability. Typically, the model fitted is an autoregressive moving average type of the form.

$$x_t = -a_1 x_{t-1} + b_o u_t$$

where x_t is the hydrogen ion concentration (μeq l^{-1}) and u_t is the flow (m^3 sec^{-1}) in the stream at time t. The parameters a_1 and b_0 are estimated from the data. Results for the Llyn Brianne catchments are presented in Fig. 16.1), and this shows the simulated hydrogen ion concentration compared with the observed levels. Further detailed analysis and interpretation of the time series techniques applied to Llyn Brianne are given elsewhere[15] (see also Chap. 13).

Birkenes model

The BIRKENES model (see Fig. 16.2) has also been applied to Welsh catchments.[18] The model consists of a simple two-reservoir hydrological model operating on a daily timestep upon which has been superimposed the key chemical processes that control acidification. Figure 16.2 shows the principal hydrological and chemical process operating. The model outputs are daily concentrations of hydrogen ion, aluminium, sulphate, calcium + magnesium, and bicarbonate in the stream along with predicted flow.

Hydrologically the model is relatively robust providing a simple but effective description of the catchment hydrology. However, difficulty was found in reproducing the extreme chemical damping that appears to occur in the Plynlimon catchments.[18] This aspect of the model is currently being investigated

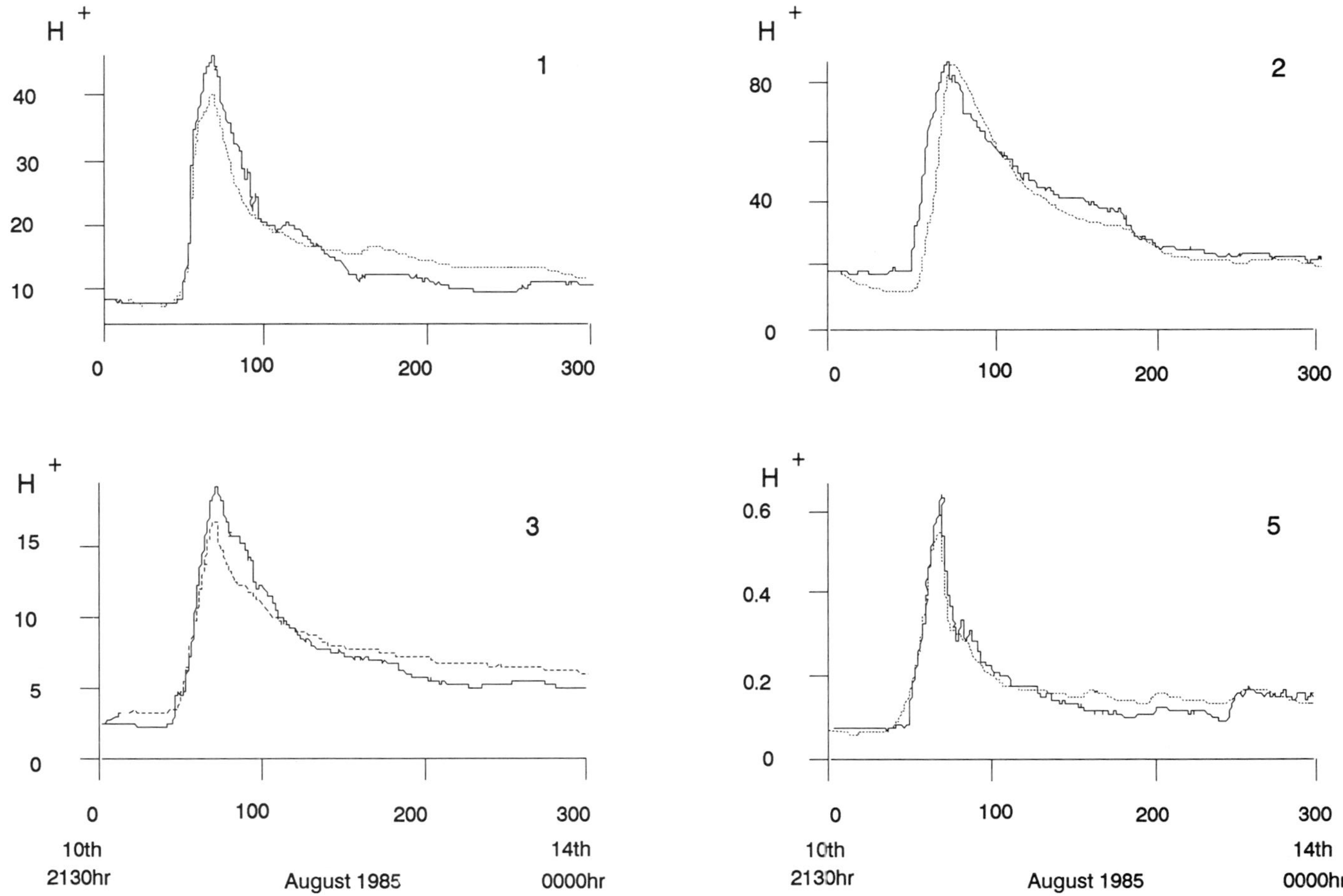

Fig. 16.1. Simulated (. . .) and Observed (____) H^+ for four Llyn Brianne streams during a storm event.

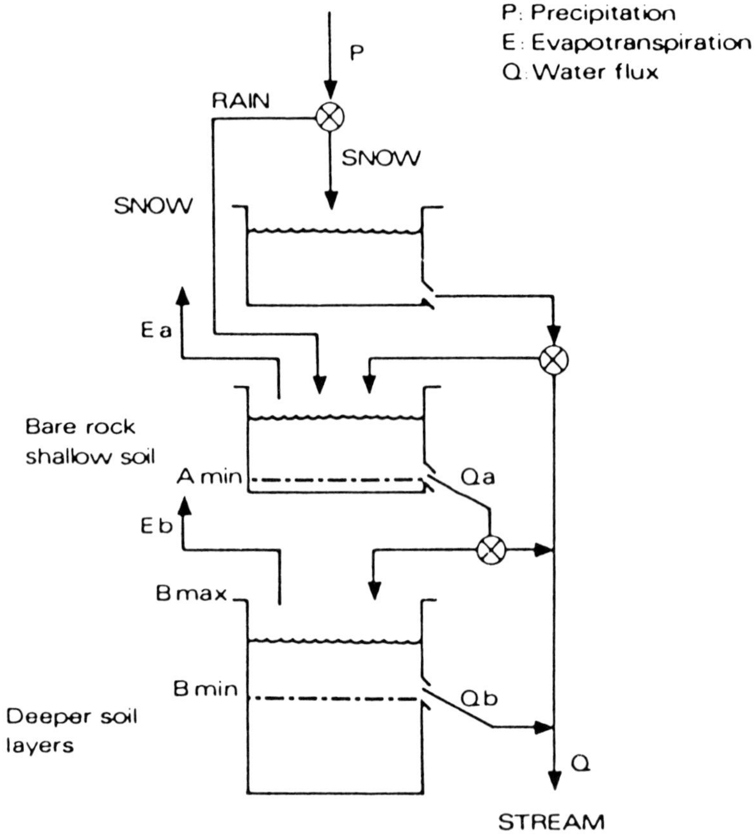

Processes operating

Shallow soil reservoir		Deeper soil reservoir
H_2O	precipitation, evapotranspiration, infiltration to lower reservoir, discharge to stream	infiltration, evapotranspiration, discharge to stream, piston flow
SO_4^{2-}	wet + dry deposition, adsorption/desorption, mineralisation	adsorption/desorption, reduction
$Ca^{2+} + Mg^{2+}$	ion exchange	release by weathering, adsorption/desorption
H^+	ion exchange and equilibrium with gibbsite	consumption by weathering, adsorption/desorption, equilibrium with gibbsite
Al^{3+}	equilibrium with gibbsite	equilibrium with gibbsite, adsorption/desorption
HCO_3^-		equilibrium with a seasonal varying CO_2 pressure

Fig. 16.2. The BIRKENES model: hydrological flow paths and chemical processes.

using Oxygen 18 as a conservative tracer on the Llyn Brianne catchments. An interesting feature of the BIRKENES model is the separation of hydrology into two components; surface runoff and baseflow. The chemistry of these two waters can be significantly different and the subsequent mixing in the stream can produce a wide variety of behaviour. The model has also been used to assess the sensitivity of stream acidity to hydrological parameters and changes in baseflow.

Flow movement between the soil and groundwater compartments is restricted by a percolation equation as follows:

$$A_{SIG} = K - (B - B_{min})/B_{max} \quad \text{for } B > B_{min}$$
$$A_{SIG} = 1.33P - 0.33K(B/B_{min})^3 \text{ for } B < B_{min}$$

where B refers to the groundwater compartment water level and B_{min} and B_{max} refer to minimum and maximum water levels respectively (see Fig. 16.2). The parameter, K, can be considered as a percolation parameter so that increasing P increases the fraction of flow, A_{SIG}, routed to the lower reservoir. This leads to an increase in the baseflow contribution to the stream. The changes in stream water concentrations for H^+ and Al^{3+} to changes in the baseflow contribution is highly non linear. This is illustrated in Fig. 16.3 which shows H^+ and Al^{3+} means over a range of baseflow conditions; all concentration values fall as the baseflow increases, the rate of decrease varying from one variable to another. In general, increase in baseflow results in significant reductions in H^+ and Al^{3+} concentrations. Other studies[20] show a similar non-linear behaviour when upper and lower soil horizon waters are mixed. Further studies[23] confirm the sensitivity of the simulated chemistry to baseflow, changes. These results have important implications for land use change since ploughing prior to afforestation on upland pasture improvement often improves drainage. The effect of improved drainage is to reduce the baseflow contribution and Fig. 16.3 suggests that such changes will increase aluminium and hydrogen ion concentrations. Thus careful consideration should be given to methods of ploughing and planting to minimise hydrological and hence chemical changes.

Long-term acidification models

In order to estimate surface water chemistry in response to historical and future deposition patterns, mathematical models must (1) be based on physical, chemical and biological processes that control catchment response, (2) treat interacting processes simultaneously and (3) be capable of representing long-term responses. MAGIC (Model of Acidification of Groundwater In Catchments) provides a tool by which soil processes can be simultaneously and quantitatively linked to examine the impact of acid deposition on surface water chemistry over time scales of several years to several decades. The model was originally developed and tested for catchments in Shenandoah

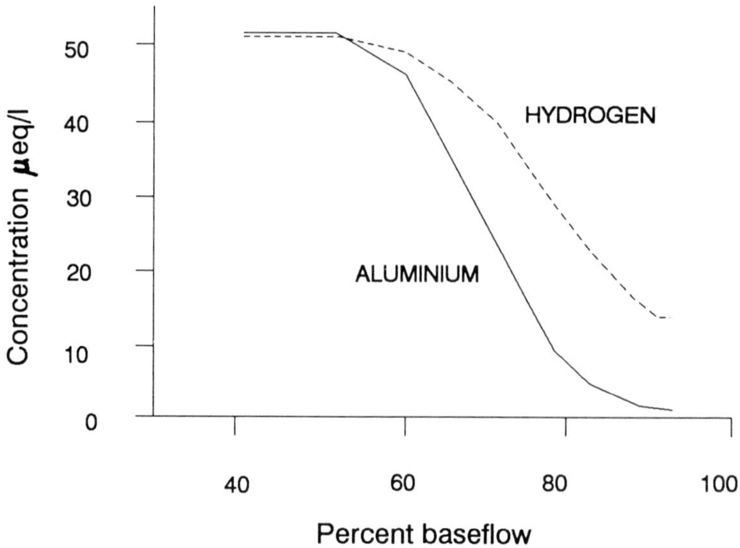

Fig. 16.3. H^+ and Al^{3+} concentrations in a stream under a range of baseflow conditions.

National Park, Virginia (USA) and has recently been adapted for the Loch Dee and Loch Grannoch catchments in Scotland, the Plynlimon[24] and Llyn Brianne[25] catchments in Wales, and Lake Hovvatn and the RAIN project catchments at Sogndal and Risdalsheia in Norway. The model has also been applied in regional studies of Norway, Wales and Scotland.

The processes on which the model is based are:

- anion retention by catchment soils (eg. sulphate adsorption);
- adsorption and exchange of base cations and aluminium by soils;
- alkalinity generation by dissociation of carbonic acid (at high CO_2 partial pressures in the soil) with subsequent exchange of hydrogen ions for base cations;
- weathering of minerals in the soil to provide a source of base cations;
- control of Al^{3+} concentrations by an assumed equilibrium with a solid phase of $A1\ (OH)_3$.

MAGIC simulates these processes using:

- A set of equilibrium equations which quantitatively describe the equilibrium soil processes and the chemical changes that occur as soil water enters the stream channel;
- A set of mass balance equations which quantitatively describe the catchment input-output relationships for base cations and strong acid anions in precipitation and streamwater;
- A set of definitions which relate the variables in the equilibrium equations to the variables in the mass balance equations.

The principal data requirements for MAGIC are a knowledge of rainfall and runoff quantity and quality, the sequence of deposition over time (taken from emission records or from atmospheric transport modelling studies), information on soil characteristics (such as bulk density, depth, cation exchange capacity, sulphate adsorption rates and base saturation), temperature and CO_2 levels. Parameters in the models such as selectivity coefficients, weathering rates, nitrate and ammonia uptake rates are determined by calibrating the model against the catchment data. Land use is an important factor in acidification because of the scavenging effect of dry particles and acid mist by vegetation. For example, forested catchments enhance the acid loading by up to 80% compared with moorland catchments. Such effects are incorporated into MAGIC and their importance is illustrated in the next section in an application to the Llyn Brianne catchments, details of the equations and the model structure being given elsewhere.[5, 6, 7, 9]

Application to the Llyn Brianne catchments

MAGIC has been applied to three distinctly different catchments at Llyn Brianne. Catchment 1 (see Fig. 16.4) is the largest catchment being modelled (2.53 km^2) and is afforested. Catchment 11 is one of the smallest (0.34 km^2) and is an acidified moorland catchment. Catchment 5, an unacidified moorland catchment, although fairly small (0.68 km^2), exhibits the highest drainage density (2.74 km km^{-2}) and channel slope (194 m km^{-1}), and hence exhibits a distinctly more rapid hydrological response.

All three catchments are underlain by lower Silurian shales, mudstones, greywackes and grits, with the shales and mudstones being dominant (Fig. 16.4). The drift materials present are only locally derived and lie mainly on the interfluves and upper slopes in thin layers (<1m), although some of the lower slopes and valley bottoms have thicker masses of up to 5 m in depth, particularly in Catchment 5 where the drift appears especially base rich.

Details of rainfall and runoff chemistry, hydrology and soil types are given elsewhere (Chap. 13).

Simulation results for catchment 11

MAGIC has been applied to Catchment 11 assuming a sulphate deposition pattern as shown in Fig. 16.5; a significant increase in sulphate loading has occurred since 1900 with a peak in 1970, since when levels have fallen.

An optimisation routine was applied initially to provide best estimates of the key parameters in the model. These include E_{mx} the maximum sulphate adsorption rate, nitrate and ammonia uptake rates, weathering rates, selectivity coefficients and the partial pressure of CO_2. From the optimisation runs the parameters shown in Table 16.1 were obtained.

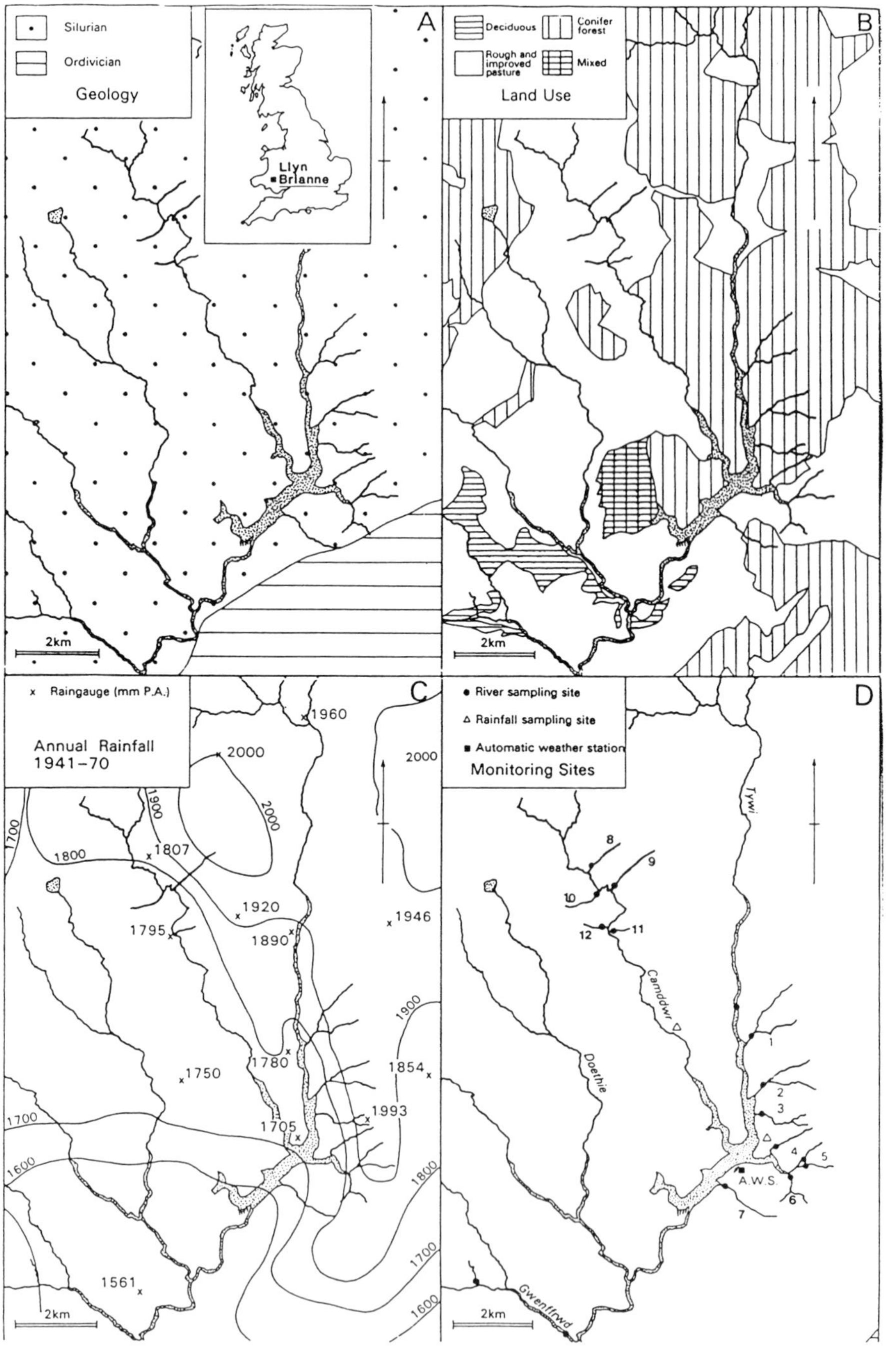

Fig. 16.4. Llyn Brianne geology, land use, rainfall and monitoring sites.

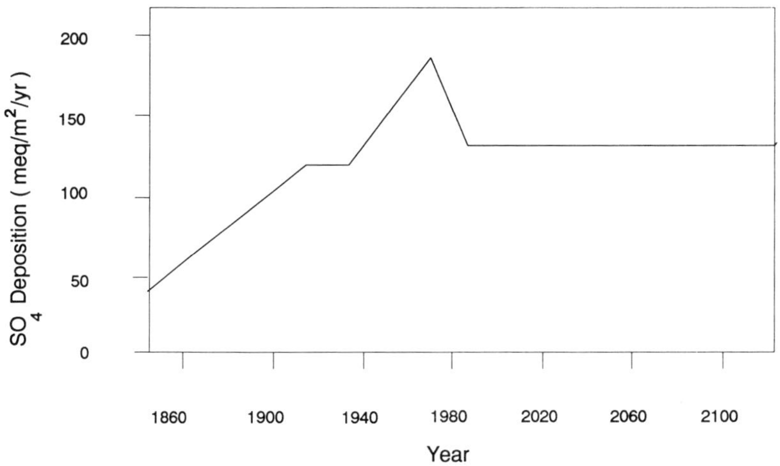

Fig. 16.5. Sulphate deposition pattern assumed for MAGIC.

E_{mx} is particularly low, suggesting that the soils at Llyn Brianne have a relatively low capacity to adsorb sulphate, compared with other catchments in the UK. Nitrate and ammonia uptake rates are high and reflect nutrient uptake by the vegetation. Weathering rates are low and this, coupled with the low base saturation levels, indicates the limited ability of the soils to buffer incoming acidity. The dry/occult deposition factor reflects the relatively low scavenging rate of moorland compared with forest catchments. In the moorland situation only 20% additional sulphate to that deposited by wet deposition enters the catchment via deposition of dry particles, aerosols and droplets of mist, fog and cloud (occult deposition). As previously discussed, in the forest catchment this 'additional' deposition can increase to 80% or more.

Table 16.2 shows that the model-simulated chemistry matches closely the observed values for Catchment 11, Fig. 16.6 shows the historical reconstruction of pH, base saturation and aluminium trends for the catchment. The significant decrease in the period 1940–1960 is very similar to the trend in South West Scotland.[1, 9] Recent research indicates that similar trends are observed in lakes in Wales (Chap. 3). Figure 16.6 suggests that base saturation levels and aluminium concentrations have changed significantly over the same period, with base saturation falling to very low levels and aluminium concentrations increasing to 19 $\mu eq l^{-1}$. The future predictions, assuming constant future deposition of sulphate at 1984 levels, as illustrated in Fig. 16.5, suggest a further slight deterioration in catchment pH, although, as shown in Table 16.2, aluminium levels will continue to rise significantly (see Fig. 16.6). These future changes are accompanied by the continuing reduction in base saturation levels shown in Fig. 16.6.

Table 16.1. Optimal parameters for MAGIC applied to catchment 11.

Emx	=	0.01 meq kg^{-1}
Nitrate uptake rate	=	68.9 meq m^{-2} $year^{-1}$
Ammonia uptake rate	=	99.1 meq m^{-2} $year^{-1}$
Weathering rates Ca	=	25.0 meq m^{-2} $year^{-1}$
Mg	=	15.0 meq m^{-2} $year^{-1}$
Na	=	10.0 meq m^{-2} $year^{-1}$
K	=	1.0 meq m^{-2} $year^{-1}$
Selectivity coefficients		
Log10 K Al/Ca	=	1.94
Log10 K Al/Mg	=	1.67
Log10 K Al/Na	=	–2.10
Log10 K Al/K	=	–5.33
pCO_2 in the soil	=	0.02 atm
Dry/occult deposition factor	=	1.2

Table 16.2. Observed and simulated runoff chemistry for catchment 11 (all units μeql^{-1} except pH, Al assumed to be trivalent).

	Observed	Simulated		Simulated assuming forest growth from 1958	
	1984–5	1984–5	2124	1984–5	2124
Ca	44.0	43.7	36.8	54.1	44.8
Mg	56.0	55.3	43.0	71.6	56.0
Na	149.0	149.4	141.2	202.4	194.8
K	6.6	8.0	7.7	10.4	9.7
NH_4	1.5	1.6	1.6	2.3	2.3
SO_4	102.0	98.8	98.4	147.1	146.2
Cl	168.0	168.3	168.3	235.7	235.7
NO_3	15.0	15.3	15.4	21.5	21.5
Alkalinity	–	19.0	9.0	–64.8	–98.2
Al	18.0	19.2	41.9	53.7	85.0
pH	5.2	4.8	4.7	4.7	4.6
Soil base saturation %	5–10	9.6%	8.0%	9.0%	7.0%

Effects of afforestation on catchment 11

The effects of afforestation can be illustrated using the model. Table 16.2 and Fig. 16.6 show the chemistry of the runoff and base saturation over time assuming a forest is grown on Catchment 11 from 1958. This is achieved in the model by increasing the dry/occult deposition factor and by increasing evapotranspiration. These enhance the sulphate input to the system by 80% compared with 20% for the moorland situation. Sea salts are also increased by 40% during the transition from moorland to forest. The transition is presumed to occur linearly over a fifteen year period from 1958 to 1973 when canopy closure is assumed to occur. The effects of uptake of cations by trees

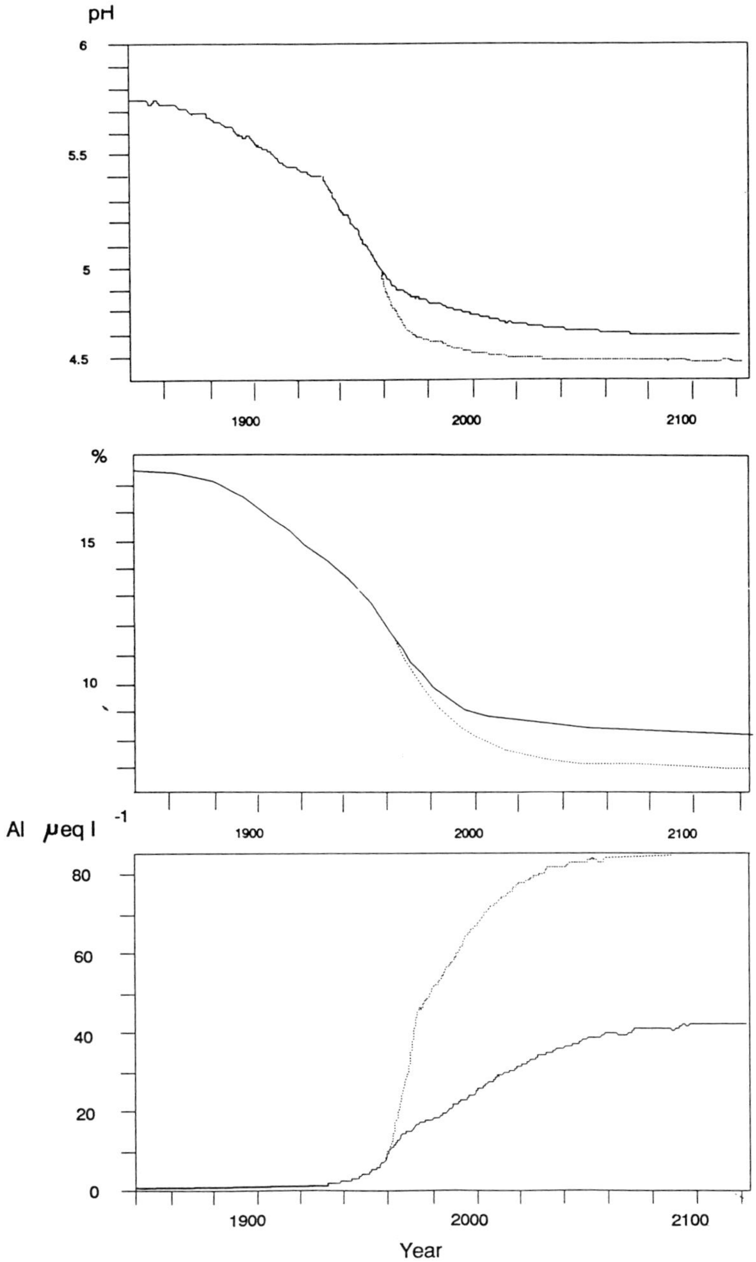

Fig. 16.6. Simulated pH(a), base saturation (b) and aluminium concentrations (c) in Catchment 11 from 1844 to 2124 for moorland and forest conditions. Forest effects from 1958 are shown with a dotted line.

Table 16.3. Observed and simulated chemistry for catchment 1 (for units, see Table 16.2).

	Observed	Stimulated	
	1984	1984	2124
Ca	59.0	62.9	46.5
Mg	60.0	60.9	41.4
Na	204.0	205.9	191.0
K	5.0	5.2	0.0
NH_4	0.4	0.0	0.0
SO_4	154.0	152.4	150.7
Cl	247.0	249.9	249.9
NO_3	11.0	14.0	14.0
Alkalinity	–	–83.3	–139.6
Al	42.0	70.8	124.7
pH	4.8	4.6	4.5
Soil base saturation %	4–8	6.2	3.8

during the early stages of growth are not included in this simulation; also excluded are the effects of hydrological changes caused by increased drainage immediately prior to afforestation. The latter can have significant effects on stream quality as the proportion of surface runoff to baseflow is increased by the additional drainage.[23] However, for the purposes of the current modelling exercise, these two effects are ignored. Despite these omissions, the simulated chemistry shown in Table 16.2 compares well with observed chemistry of the forest Catchment 1 shown in Table 16.3. The chemistry of the forest catchment differs markedly from that of the moorland catchment and the fact that the model can reproduce the principal changes in the anions and cations is very encouraging. It suggests that the MAGIC model does indeed capture the main components controlling stream acidification. The simulated long-term responses of the moorland catchment with and without the forest are illustrated in Fig. 16.6. The effect of the forest is to reduce pH and base saturation slightly but to increase aluminium levels rapidly. It has been shown that fish are particularly sensitive to aluminium levels[19] and the results of the simulation suggest that aluminium levels will rise even further in the long term.

Effects of reduced future deposition on catchment 11

The effects of reducing deposition in the future are illustrated in Fig. 16.7, which shows the pH in Catchment 11 assuming a 50% reduction in deposition from 1984 levels phased over a 20 year period. The effect on pH is not particularly striking; a slight recovery in pH or the continuation of current pH levels is shown. This poor recovery is not surprising given that base saturation and weathering rates are low. A similar result for pH is obtained in the afforested catchment simulation. However, as shown in Fig. 16.8, aluminium levels are significantly changed, with a 50% reduction in deposition for the

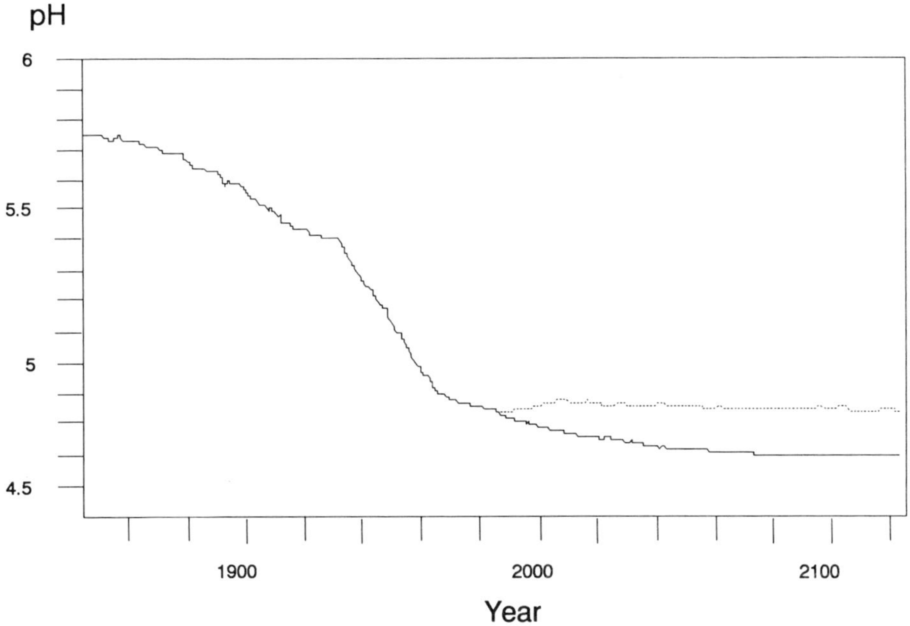

Fig. 16.7. Simulated pH in Catchment 11 from 1844 to 2124 for moorland assuming constant deposition from 1984 (continuous line) and deposition reduced by 50% since 1984 (dotted line).

afforested catchment. This is of more significance from a biological point of view (see Chap. 17).

Simulation of catchments 1 and 5

The afforested Catchment 1 and the 'unacidified' moorland Catchment 5 have been simulated using the same procedure as for Catchment 11. In the case of Catchment 1 the forest effect is simulated from 1958 with canopy closure occurring after 15 years. The dry/occult deposition factor increases linearly from 1.2 in 1958 to 1.7 for sulphate over this forest growth period. Sea salts are presumed to increase from 1.0 to 1.6 over the same period. In the case of Catchment 5, the moorland catchment is highly buffered and this effect is simulated by increasing the weathering rates of both Ca and Mg to 170 μeq m^{-2} yr^{-1}.

With these three major changes, the simulated current and future chemistry obtained from the model is illustrated in Tables 16.3 and 16.4 for Catchments 1 and 5 respectively. The simulated 1984 chemistry for these catchments compares well with the observed chemistry (Tables 16.3 and 16.4). The contrast between the catchments is enormous, with Catchment 1 exhibiting very acid conditions, low alkalinities and high aluminium levels,

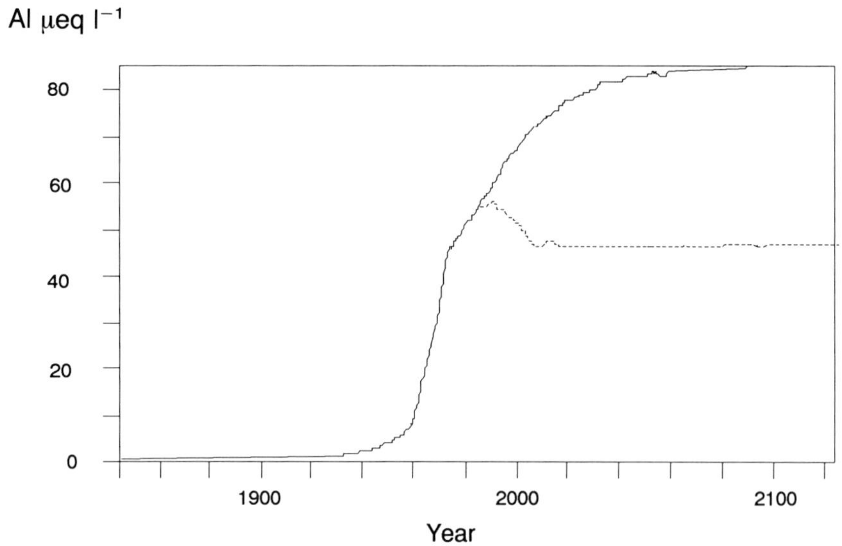

Fig. 16.8. Simulated aluminium in Catchment 11 from 1844 to 2124 assuming forest growth from 1958 and with constant deposition from 1984 (continuous line) and deposition reduced by 50% since 1984 (dotted line).

Table 16.4. Observed and simulated chemistry for catchment 5 (for units see Table 16.2).

	Observed	Simulated	
	1984	1984	2124
Ca	146.0	146.9	144.2
Mg	146.0	145.3	141.1
Na	138.0	138.9	133.8
K	11.0	10.4	11.5
NH_4	1.6	2.1	2.1
SO_4	103.0	109.2	109.3
Cl	155.0	156.2	156.2
NO_3	10.0	12.8	12.8
Alkalinity	–	165.1	153.9
Al	7.0	1.6	1.5
pH	6.9	6.4	6.4
Soil base saturation %	–	13.4	12.4

and with Catchment 5 showing highly buffered waters rich in cations, high alkalinities and very little aluminium. The long-term simulations for Catchment 5 show that even growing a forest on the catchment would only reduce pH from 6.4 to 6.3, as illustrated in Fig. 16.9. Thus the internal sources of alkalinity in the catchment, probably generated by calcite intrusions, will buffer incoming acidity in the long term.

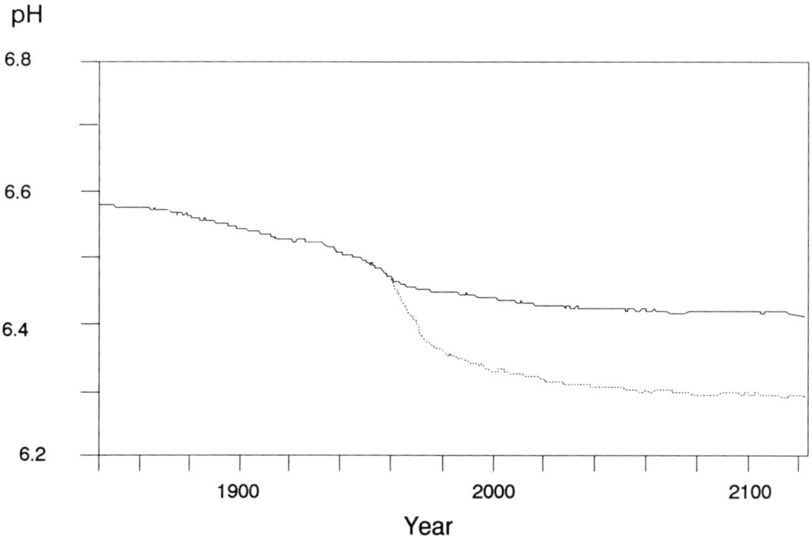

Fig. 16.9. Simulated pH in Catchment 5 from 1844 to 2124 for moorland and forest conditions. Forest effects from 1958 are shown with dotted line.

Applying magic in a regional study

Between October 1983 and December 1984, WWA undertook a regional survey of streams and rivers within an area of Wales believed to be sensitive to acid deposition. One hundred and twenty streams were sampled on a weekly basis with sixteen of them being sampled at more than one point. During the same period rainfall quality was monitored at forty-four sites throughout Wales (Chap. 4). The mean yearly rainfall concentrations are summarised in Table 16.5. A breakdown of the results by region shows the most acidic rainfalls falling in the uplands of mid and North Wales.[11] Eastern Wales receives the highest concentrations of the acid pollutants, such as sulphate and nitrate. It also receives high levels of ammonia and calcium, which partially buffer the rainfall.

The study area is underlain by rocks of the Cambrian, Ordovician and Silurian Series. In the upland areas the soils are thin and base-poor. The combination of slowly weathering rocks and thin base-poor soils gives the region its vulnerability to acid deposition. Prolonged acid deposition may induce a leaching of the base cations from the soil, reducing the buffering ability and allowing the surface waters to become acidic. The study area lies mainly in the region of high rainfall acidity and is consequently of prime interest in the investigation of acidification in Wales.

The mean stream chemistry results are presented in Table 16.5. pH varies from 4.6 to 7.2 across the region whilst the mean pH is 6.1. The variability in

Table 16.5. Regional rainfall and stream chemistry in Wales 1983/84 (for units see Table 16.2).

	Rainfall		
	Mean	Minimum	Maximum
pH	4.7	4.4	5.0
Ca	47.7	9.9	102.2
Mg	42.6	17.4	177.8
Na	137.0	62.5	510.9
K	8.6	4.7	21.8
SO_4	118.7	52.0	217.1
Cl	176.1	96.8	525.0
HCO_3	51.3	22.9	114.9
Al	8.8	1.1	57.7
	Stream water		
	Mean	Min	Max
pH	6.1	4.6	7.2
Ca	130.9	40.4	862.3
Mg	111.5	44.4	1272.7
Na	220.1	149.1	640.0
K	13.4	5.4	89.0
SO_4	153.7	49.1	507.4
Cl	267.4	167.0	918.0
HCO_3	75.5	0.7	604.0
Al	10.0	1.1	68.8
Alk	73.3	–15.7	603.8

pH suggests that the region contains a mixture of sites in terms of acid sensitivity.

Regional modelling procedure

In the regional approach, the MAGIC model is repeatedly run using different sets of parameter values chosen randomly from given distributions (see Table 16.6). The ensemble of model runs is then evaluated and compared with the observed distributions of water quality across the region obtained from the analysis of the WWA survey data. This technique, termed Monte Carlo Analysis, allows the region as a whole to be simulated. It has been used in an analysis of 700 Norwegian lakes[13] and in the Galloway region of Scotland.[16] In order to calibrate the regional model, the input parameter distributions are adjusted until the observed stream quality distributions are matched. The model can then be used to evaluate changes in distribution over time.

Simulation results and regional chemistry changes over time

Details of the simulation results are given elsewhere.[17] Altogether 3,000 Monte Carlo runs were obtained, and of these, 532 were successful in

Table 16.6. Ranges for the parameters in the simulations for Wales.

Parameter	Monte Carlo ranges Units	Minimum	Maximum
Qp	cm	120.00	230.00
Cl	$\mu eq\ l^{-1}$	167.00	400.00
SO_4	$\mu eq\ l^{-1}$	40.00	190.00
NO_3	$\mu eq\ l^{-1}$	2.40	30.20
CEC	$meq\ kg^{-1}$	10.00	300.00
WE CA	$meq\ m^{-1}\ y^{-1}$	1.00	250.00
WE MG	$meq\ m^{-1}\ y^{-1}$	1.00	75.00
WE NA	$meq\ m^{-1}\ y^{-1}$	1.00	50.00
WE K	$meq\ m^{-1}\ y^{-1}$	1.00	20.80
SALCA	$\log_{10}$	–1.00	5.00
SALMG	$\log_{10}$	–1.00	5.00
SALNA	$\log_{10}$	–3.00	3.00
SALK	$\log_{10}$	–6.00	0.00
KAL	$\log_{10}$	–8.00	11.00

* Where WE is weathering rate and S is selectivity coefficient.

Table 16.7. Behaviour criteria plus a summary of simulated water chemistry from the 532 acceptable Monte Carlo runs (all units are μ eq.1^{-1}).

	Behaviour Criterion		Accepted results			
	Minimum	Maximum	Ave	S.D.	Min	Max
Ca	40.5	300.0	109.0	39.5	41.7	268.5
Mg	40.5	175.0	87.3	24.4	41.1	170.7
K	6.5	35.0	14.9	5.4	6.5	34.8
Na	150.0	400.0	244.2	46.9	157.1	393.4
SO_4	85.0	250.0	152.3	33.3	85.4	249.9
Cl	167.0	400.0	256.2	50.6	168.9	3883.0
H^+	0.1	25.0	2.2	2.2	0.3	10.3
Alkalinity	–12.5	250.0	41.9	35.7	–12.5	175.1

producing variables that matched the observed stream chemistry across the region. The range of stream chemistry, specified as behaviour criteria, is presented in Table 16.7 together with a summary of the water chemistry of the acceptable simulations. The simulated chemistry matched the observed chemistry range in all cases. Although the mean simulated values for calcium, magnesium and alkalinity appear low when compared with the observed means, they are comparable to the median of the observed values, indicating that the simulation is satifactory apart from a long tail end weighting. Fig. 16.10 shows the simulated fits to the distributions of calcium, magnesium, alkalinity, and sulphate. In general, the simulated distributions match closely the observed distributions and the regional model can be used to investigate how water quality over the region has changed from 1840. As shown in Fig. 16.11, the 1840 distributions for sulphate and alkalinity are very different

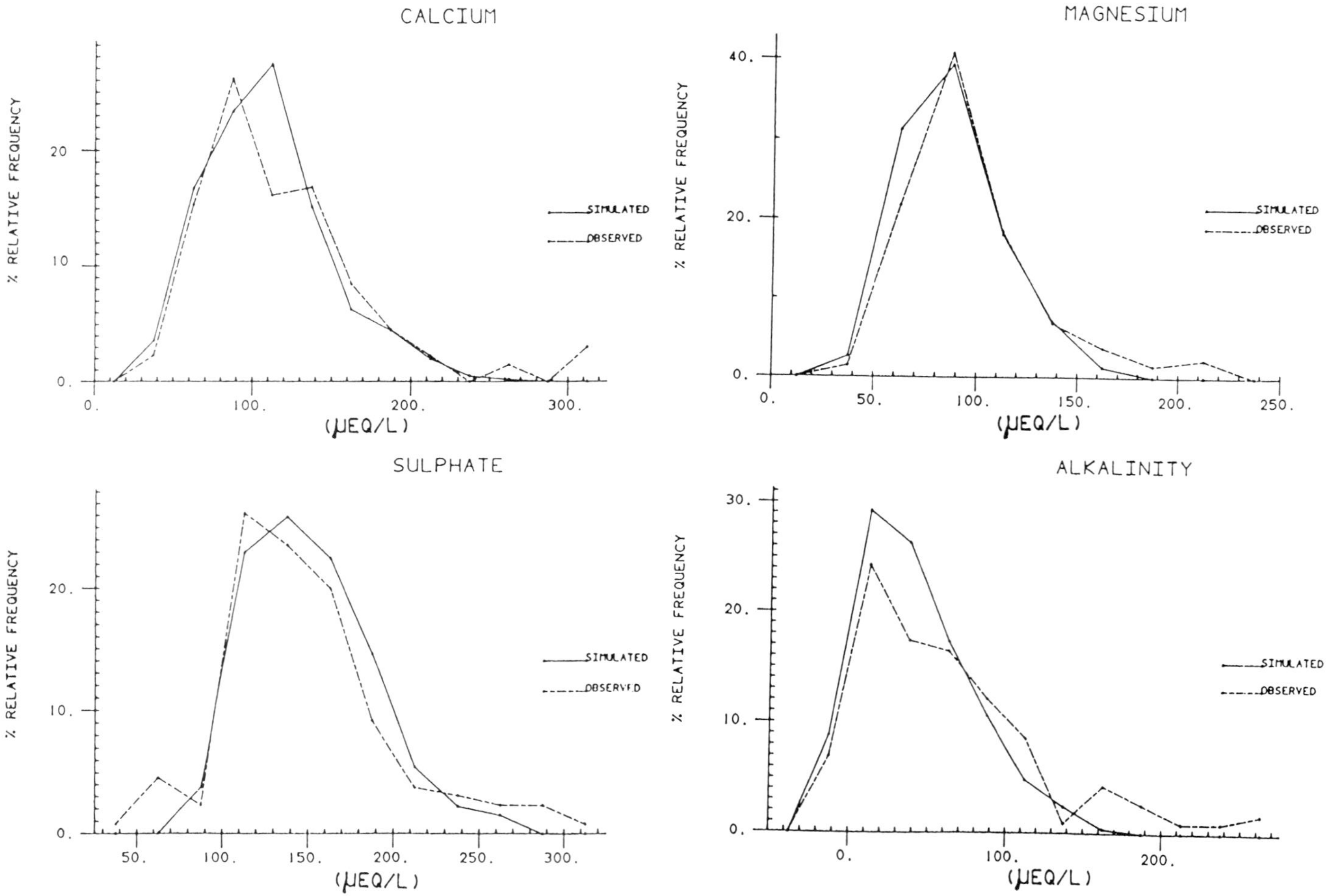

Fig. 16.10. Simulated and observed distributions of calcium, magnesium, sulphate and alkalinity in the regional study.

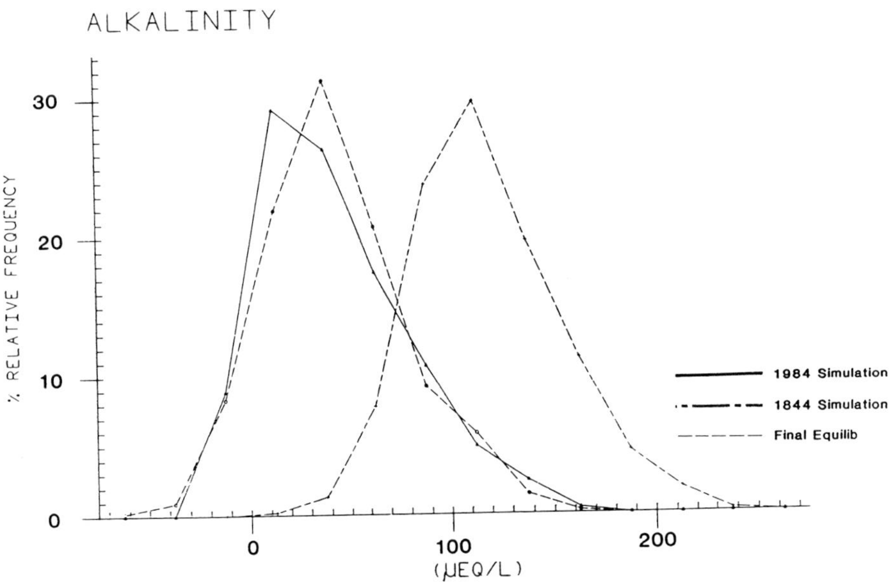

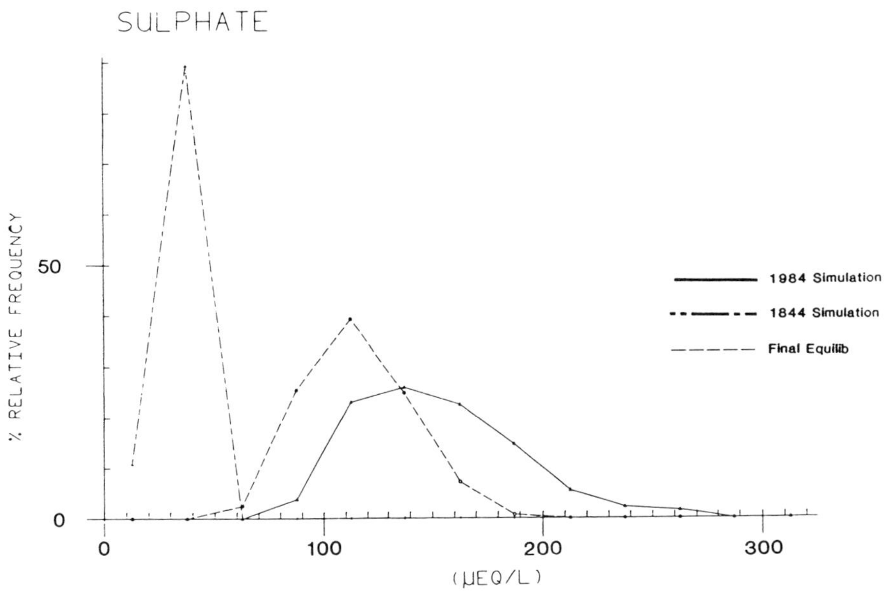

Fig. 16.11. Background (1844), Present (1984) and Future (2124) distributions for sulphate and alkalinity.

from present day distributions. Sulphate levels are much lower and alkalinity is significantly higher.

Effects of reduced deposition on regional chemistry

The temporal changes in the entire regional distributions were also investigated under an assumed driving deposition sequence of 30% reduction in excess sulphate deposition linearly between the present day and 2,000 A.D. Fig. 16. 11 shows the effects of changing excess sulphate concentrations in rainfall on the stream sulphate and alkalinity distributions. The initial, present day and final equilibrium distributions are shown. A slight recovery in alkalinity occurs for the medium/high alkalinity sites, but for the lowest alkalinity catchments ($< -10\ \mu eql^{-1}$) there is a decline of up to 48 μeql^{-1}. These represent the most acid sensitive sites whose base cation replenishment from soil weathering is too slow to cope with the remaining 70% sulphate deposition.

Impact of reduced deposition on the most sensitive sites

The Reuss-Johnson concept of soil acidification proposed that the most acidic stream chemistry will result in sites whose alkalinity is less than zero μeql^{-1}. Their definition of alkalinity is the one used by the MAGIC model and is given by:

$$\text{Alkalinity} = HCO_3^- - H^+ - Al^+$$

The portion of the simulated region that has alkalinity < 0 was subjected to various hypothetical future deposition patterns in an attempt to look at the likely response of the worst affected catchments. Of the 532 simulations, 53 lay within this category. Three future deposition scenarios were used, these being linear reduction in non-marine origin sulphate deposition over a 20 year period to 30%, 60% and 90% of the 1984 value, thenceforth remaining constant. Table 16.8 describes the resultant equilibria for each deposition scenario. The major feature of the results is the trend in increased pH and alkalinity and decreased sulphate in the streams as the deposition is decreased. Reductions in deposition of 30% and 90% lead to mean stream sulphate concentrations of 115 and 43.7 μeql^{-1} respectively. This reduction of sulphate concentrations raises alkalinity, from a mean of 9.4 to 75.2 μeql^{-1}. Coupled with the change in alkalinity, the pH rises from a mean of 5.5. to 6.2. A fuller description of this model application can be found in Musgrove et al.[17]

Conclusions

The current research has illustrated the ability of the MAGIC model to reproduce catchment chemistry in both moorland and forest streams in Wales.

Table 16.8. Stream concentrations in equilibrium with reduced depositons for 10% most sensitive sites (all units $\mu eq l^{-1}$ except pH).

(a) *Effect of 90% reduction in deposition*

Variable	Mean	Standard Deviation	Minimum	Maximum
pH	6.2	0.2	5.6	6.5
Na	230.4	43.2	146.8	322.0
Ca	68.7	27.8	25.8	149.0
Mg	62.3	14.0	36.6	106.3
SO_4	43.7	6.4	34.5	58.5
Cl	253.8	47.2	168.9	347.1
Alk	75.2	32.0	14.5	150.7

(b) *Effect of a 60% reduction in deposition*

Variable	Mean	Standard Deviation	Minimum	Maximum
pH	5.9	0.3	5.1	6.3
Na	231.5	42.8	154.5	322.0
Ca	26.8	26.8	28.0	149.0
Mg	64.8	13.7	40.3	107.1
SO_4	79.8	12.9	58.4	106.9
Cl	253.8	47.2	168.9	347.1
Alk	43.9	25.8	–6.9	96.6

(c) *Effect of a 30% reduction in deposition*

Variable	Mean	Standard Deviation	Minimum	Maximum
pH	5.5	0.4	4.7	6.0
Na	231.9	42.4	159.2	322.0
Ca	71.8	26.4	28.8	149.0
Mg	65.8	13.6	41.4	107.2
SO_4	115.8	21.3	80.1	161.3
Cl	253.8	47.2	168.9	347.1
Alk	9.4	23.7	–59.9	44.6

(d) *Current chemistry*

Variable	Mean	Standard Deviation	Minimum	Maximum
pH	5.1	0.4	4.9	5.3
Na	236.6	41.3	167.0	333.1
Ca	79.2	23.4	41.9	147.6
Mg	76.4	19.1	47.0	138.7
SO_4	153.2	30.2	98.4	216.8
Cl	253.8	47.2	168.9	347.1
Alk	–4.9	3.6	–11.9	0.4

The model reconstructs historical trends in acidification, compares well with the trends in acidification derived from palaeoecological analysis and provides some measure of confidence in using the model to predict future trends.

The poorly buffered upland catchments in Wales are significantly affected by acid deposition, and are likely to be affected for some time even if deposition is reduced. Any short-term improvement will probably be effected by liming and land management (see Chap. 19).

Afforestation in the poorly buffered Welsh Uplands causes a major increase in acidification following the increased scavenging of sea-salts and anthropogenic sources of acidity.

Small variations in catchment hydrology, soils and geochemistry can have significant effects on the long-term behaviour of stream chemistry. It is essential therefore to collect detailed data on hydrology, geology and soils prior to modelling studies.

The survey region within Wales has been simulated quite accurately by the MAGIC model. Close reproduction of all the major variables was achieved in terms of their distribution across the region.

The present analysis indicates that Wales is particularly vulnerable to further pollutant deposition.

The buffering capacity has declined in the region and the acidification of its streams is significant. The survey region as a whole is likely to show little change in the rate of acidification under an assumed deposition decrease of 30%. A deposition reduction of the order of 50% is required to reverse acidification in Wales. For the most acid-sensitive sites, a strong recovery in pH and alkalinity is likely to occur following major reductions in deposition of strong acid pollutants.

References

1. Batterbee, R.W., Flower, R.J., Stevenson, B.C. and Rippey, B. (1985). Lake Acidification in Galloway: A palaeoecological test of competing hypotheses. *Nature*, 314 (6009): 350–352.
2. Beven, K.J., Kirby, K.J., Schofield, N. and Tagg, A.F. (1984). Testing a Physically Based Flood Forecasting Model (TOPMODEL) for three UK Catchments, *J. Hydrology*, 69: 119–143.
3. Christophersen, N., Seip, H.M. and Wright, R.G. (1982). A model for stream-water chemistry at Birkenes, a small forested catchment in southernmost Norway. *Water Resour. Res.*, 18: 977–966.
4. Christophersen, N., Rustad, S. and Seip, H.M. (1984). Modelling stream-water chemistry with snowmelt. *Phil. Trans. r. Soc. B.*, 305: 427–439.
5. Cosby, B.J., Wright, R.F., Hornberger, G.M. and Galloway, J.N. (1984). Model of acidification of groundwater in catchments. *Project Completion Rep., EPA/NCSU Acid Preci-Program*, Project number E2–14.
6. Cosby, B.J., Wright, R.F., Hornberger, G.M., and Galloway, J.N. (1985). Modelling the effects of acid deposition: assessment of a lumped parameter model of soil water and streamwater chemistry. *Water Resour. Res.*, 21: 51–63.
7. Cosby, B.J., Wright, R.F., Hornberger, G.M., and Galloway, J.N. (1985). Modelling the

effects of acid deposition: estimation of long-term water quality responses in a small forested catchment. *Water Resour. Res.*, 21: 1591–1601.
8. Cosby, B.J., Hornberger, G.M., Galloway, J.N., and Wright, R.F. (1985). Fresh water acidification from atmospheric deposition of sulphuric acid: a quantative model. *Environ. Sci. Technol.*, 19: 1145–1149.
9. Cosby, B.J., Whitehead, P.G. and Neale, R. (1986). A preliminary model of long-term changes in stream acidity in south western Scotland. *J. Hydrol.*, 381–401.
10. Cosby, B.J., Hornberger, G.M., Wright, R.G. (1987). A regional model of surface water acidification in southern Norway: Calibration and validation using survey data. In: *Env. Impact Models to Assess Regional Acidification* (Ed. J Kamari) Klouwer, Amsterdam.
11. Donald, A.P. and Stoner, J.H. (1989). The quality of atmospheric deposition in Wales. *Arch. Environ. Contam. Toxicol.*, 18: 109–110.
12. Grip, H., Jansson, P.E., Honsonm, H. and Nilsson, S.I. (1986). Application of the Birkenes model to two forested catchments on the Swedish west coast. *Ecol. Bull.*, 37: 176–192.
13. Hornberger, G.M., Cosby, B.J., Wright, R.F. (1987). Analysis of historical surface water acidification in southern Norway using a regionalized conceptual model (MAGIC). In: *Systems Analysis in Water Quality Management*, M.B. Beck, ed., Pergamon Press, New York, 1987.
14. Kamari, J., Posch, M. and Kauppi, L. (1984). Development of a model analysing surface water acidification on a regional scale: application to individual basins in southern Finland. *Proceedings of the Nordic h.p. Workshop. (nhp report No. 10.)* Uppsala.
15. Littlewood, I. (1987). Streamflow – pH dynamics in small moorland and conifer afforested catchments in the Upper Twyi Valley, Wales. *Proceedings of British Hydrological Society Meeting*, Hull.
16. Musgrove, T.J., Whitehead, P.G. and Cosby, B.J. (1988). Regional modelling of acidity in the Galloway region in south east Scotland. In *Env. Impact Models to Assess Regional Acidification* (Ed. J. Kamari), Klouwer, Amsterdam. (In press)
17. Musgrove, T.J., Whitehead, P.G. and Cosby, B.J. (1990). Regional Model of Acidification for Wales. J. of Hydrology. In press.
18. Neal, C., Christophersen, N., Neale, R., Smith, C., Whitehead, P.G. and Reynolds, B. (1988). Chloride in precipitation and streamwater for the upland catchments of the River Severn in mid-Wales: Some consequences for hydrochemical models describing the effects of acid deposition. *J. of Hydrological Processes*, 2: 155–165.
19. Ormerod, S.J., Boole, P., McCahon, C.P., Weatherley, N.S., Pascoe, D. and Edwards, R.W. (1987). Short-term experimental acidification of a Welsh stream: comparing the biological effects of hydrogen ions and aluminium. *Freshwater Biology*, 17: 341–356.
20. Seip, H.M. and Rustad, S. (1983). Variations in surface water pH with changes in sulphur deposition. *Water, Air and Soil Poll.*, 21: 217–223.
21. Seip, H.M., Seip, R., Dillon, P.J. and de Grosbois, E. (1986). Model of sulphate concentration in a small stream in the Harp Lake catchment, Ontario. *Can. J. Fish. Aquat. Sci.*, 42: 927.
22. Whitehead, P.G., Neal, C., Seden-Perriton, S., Christophersen, N. and Langan, S., (1986). A time series approach to modelling stream acidity. *J. Hydrol., (Amst.)*, 85: 281–304.
23. Whitehead, P.G., Neal, C. and Neale, R. (1986). Modelling the Effects of Hydrological changes on Stream Acidity *J. Hydrol., (Amst.)*, 84: 353–364.
24. Whitehead, P.G., Reynolds, B., Hornung, M., Neal, C., Cosby, J. and Paricos, P. (1988). Modelling long-term Acidification in Wales at Plynlimon. *Hydrological Processes*, 2:.357–368.
25. Whitehead, P.G., Bird, S., Hornung, M., Cosby, J., Neal, C. and Paricos, P. (1988). Stream Acidification Trends in the Welsh Uplands: A modelling study of the Llyn Brianne catchments. *J. of Hydrol.*, 101: 191–212.

17. Modelling the ecological impact of changing acidity in Welsh streams

S.J. ORMEROD, N.S. WEATHERLEY and A.S. GEE

Introduction

Models have a potentially large and important role in the assessment and management of ecosystems. They act as a focus for conceptualising ecological phenomena, provide a basis for developing and testing hypotheses, facilitate the interpretation of past or present changes and the simulation of future conditions. Once models are mathematically formulated and validated, experimental manipulation and perturbation can be achieved in a time frame much shorter than is usually necessary for ecological research in real-world systems.

The advantages of model simulations are becoming particularly apparent in studies of surface water acidification, where changes occur often over decades or centuries, but where management decisions on potential acidifying influences (e.g. air pollution or change in land use) are required in the short term. Some of the hydrochemical models, initially developed to gain insight into the acidification process, are being used increasingly for this purpose.[21] However, the reconstruction or prediction of ecological change in acid waters has not been widely attempted,[10] despite being a highly desirable adjunct to chemical modelling. Biological resources are amongst those most at risk from surface-water acidification, and model simulations are required to provide information on losses or gains which might already have occurred, or which might occur under different future scenarios.

This chapter examines the development of ecological models for Welsh streams. It describes how they have been used in conjunction with the hydrochemical model MAGIC to generate temporal patterns in three catchments around Llyn Brianne, and in the Welsh 'acid sensitive' region as a whole. It also points out some weaknesses which must be overcome before this approach becomes widely accepted.

The development and concept of ecological models for Welsh streams

A model is a formulation, either verbal or mathematical, that mimics a real-world phenomenon and permits prediction.[13] Many such models in

R.W. Edwards et al. (eds) Acid Waters in Wales. 279–298.

ecology are derived from highly complex, sometimes process oriented, 'systems' analysis,[7, 29] but simple and empirical models are also frequently used. The latter group are usually derived from ANOVAR, regression or multivariate analysis, and they may be descriptive, predictive or both.

In simple terms, the approach to modelling the effects of acidification on Welsh streams has been to use the output from a process model (MAGIC) to drive linear empirical models which 'predict':

1. The toxic response of brown trout, derived from a linear regression of survival time on aluminium concentration.
2. Brown trout density, derived from a multiple regression on aluminium concentration, total hardness and stream size.
3. Changes in invertebrate fauna, derived from a multiple discriminant analysis of environmental variables which differentiated most strongly between invertebrate assemblages identified using a cluster analysis (TWINSPAN).

These procedures are presented schematically in Fig. 17.1, and described more fully by Ormerod et al.[19] and Whitehead and Ormerod.[33]

MAGIC – a brief outline

A Model of Acidification of Groundwater In Catchments (MAGIC) has been used widely in north America and Europe, both to assess temporal trends in surface-water acidity, and to aid understanding of the processes involved. Its conceptual basis is that atmospheric deposition, mineral weathering and exchange processes in the soil and interstitial solutions are responsible for the observed chemistry in surface waters. Output from the model includes pH, alkalinity, the concentrations of strong acid anions (SO_4^{2-}, Cl^-, NO_3^- and F^-), base cations, and aluminium. In simple terms (explained more fully in Chap. 16), bicarbonate buffering and aluminium mobilisation/retention are key processes in the soil, whilst CO_2 degassing and aluminium precipitation influence pH and alkalinity in the stream. Long-term trends are controlled partly by the exhaustion of base cations in the soil, in turn affected by leaching and weathering due to the atmospheric deposition of strong acids.

As a further, recent, extension to MAGIC, some of the acidifying influences of conifer forest have been simulated by increasing the magnitude of dry and occult deposition, and evapotranspiration[12] (see Chap. 16). These effects are particularly important in the context of this study (see below).

MAGIC has also been used to simulate regional changes in chemistry under different deposition scenarios in Norway and Scotland.[33]

The fish models

Most fishery models in previous studies of acidification have been used to simulate population changes expected under given chemical conditions, often

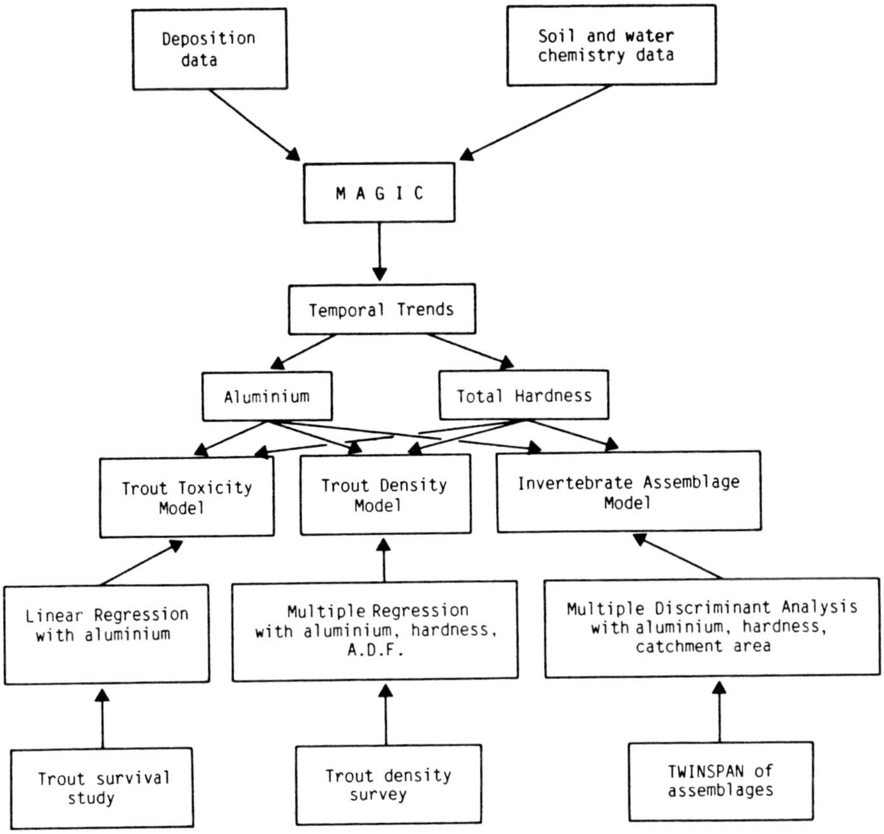

Fig. 17.1. A schematic representation of the modelling procedure applied to Welsh catchments.

extrapolating toxic responses in the laboratory to nature.[6, 27] For example, Sadler[22] used data from laboratory bioassays to construct Leslie matrices from which he assessed the capacity of Norwegian trout populations to withstand egg and fry mortalities expected at pH 4.5. More recent models of fish presence/absence or population dynamics used data collected in the field.[8, 20] However, few studies have involved simulating the likely effects of different scenarios of deposition on surface water chemistry and fish biology. For example, Minns, Kelso and Johnson[10] attempted to assess temporal trends in fish populations by simulating changing alkalinity and total dissolved solids in lakes of the Canadian Shield. The chemical data were then used to drive an empirical model of potential fish yield. Small and Sutton[23] simulated changes in the pH and alkalinity of lakes in the Adirondack region of New York, and thereby predicted changes in fish distribution under different acid loadings. The models described here are conceptually similar, but produce values for trout survival time and density, instead of yield or presence/absence.

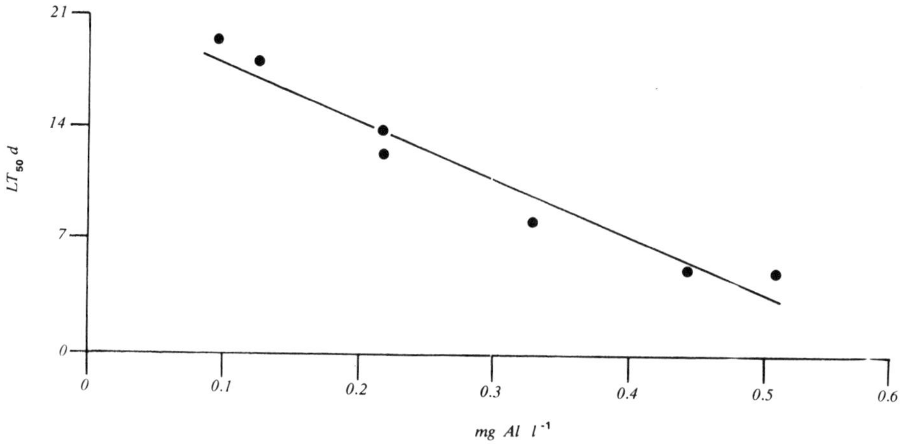

Fig. 17.2. The survival time of caged brown trout (0+) in relation to the concentration of dissolved aluminium in streams at Llyn Brianne.[25]

Derivation

Both models were derived empirically from spatial relationships between water chemistry and the biology of brown trout in the study area. For the toxicity model, O^+ fish from wild stock were held in cages in streams of different pH and aluminium concentration.[25] Aluminium and pH were highly correlated during the exposure period, and mean aluminium concentrations (filterable at <0.45μm) explained 96% of the variance in survival time (Fig. 17.2) giving the highly significant regression:

$$LT_{50} = 22.2 - 36.2\,[Al_{filt}]$$
(LT_{50} in days, aluminium in mg l^{-1})

Most aluminium in streams of pH < 5 in the study area occurs in the labile form, and hence $[Al_{filt}]$ is equivalent to this fraction (see Chap. 11).

For the trout density model, 88 streams were electro-fished between June and August, 1984, and trout densities were estimated using catch-removal methods.[32] Densities were related to water quality (from weekly spot samples) and other environmental data using stepwise multiple regression, and the following model explained 50% of the variance:

$$\text{Log } 2 = -1.24 - 1.08\,[Al_{filt}] + 1.33 \text{ Log Hardness} - 0.22 \text{ Log A.D.F.}$$
(Density in n. 100 m^{-2}, aluminium in mg l^{-1}, hardness in mg $CaCO_3\,l^{-1}$, A.D.F. = average daily flow in m^3s^{-1}; Log indicates $\log_{10}$).

Alternative regressions including zinc and pH (but not hardness) as predictors gave similar fits to the data, but were not used in this study because zinc concentrations are not elevated in the streams being modelled, and because pH and aluminium concentrations are highly correlated. Additionally, the incorporation of total hardness was appropriate in view of the importance of base cations in MAGIC, and their role in influencing fish responses to acid stress.[1]

The invertebrate model

Several recent studies in the U.K., including Wales, have revealed strong and consistent relationships between acid-base status and the faunal composition of upland streams (see Chap. 7). In particular, invertebrate data reduced by ordination or classification have repeatedly shown striking linear trends in relation to pH, total hardness, or aluminium concentration (Fig. 17.3). Such relationships are clearly conducive to empirical modelling, and suited to methods recently developed by Green and Vascotto[4] and Wright et al.[34]

In simple terms the technique involves:

1. Classifying streams into groups according to their invertebrate faunas.
2. Assessing the environmental varibles which differ most strongly between the groups.
3. Combining these variables into a series of linear functions which then 'discriminate' between the *a priori* grouping.

Prediction involves using new values of the discriminant functions to indicate which type of invertebrate fauna would be expected under the new environmental conditions.

Derivation

The derivation of the invertebrate model has been described more fully by Weatherley and Ormerod.[30] Eighteen streams in the study area were sampled for macroinvertebrates in April 1985, with data being pooled from riffle and marginal habitats. Streams were then classified using Two-way Indicator Species Analysis (TWINSPAN),[5] on the basis of species composition and relative abundance, resulting in the following three groups:

Group A – Acidic streams (mean pH 5.0, total hardness 5.9 mg $CaCo_3$ l^{-1}), often draining afforested areas. Dominated by stonefly nymphs with mayflies absent.

Group B – Moderately acidic streams (mean pH 5.4, total hardness 4.8 mg l^{-1}), usually draining moorland catchments, and with a more diverse fauna than A including stonefly predators (*Isoperla grammatica*, *Diura bicaudata*), some mayfly nymphs (e.g. *Leptophlebia marginata*) and dragonfly nymphs (*Cordulegaster boltonii*).

Group C – Circumneutral streams (mean pH 6.7, hardness 13.0 mg l^{-1}, 0.05 mg Al l^{-1}), draining moorland or deciduous woodland, and with the most diverse fauna, dominated by grazing mayfly nymphs (*Baetis* spp., *Rhithrogena* sp.).

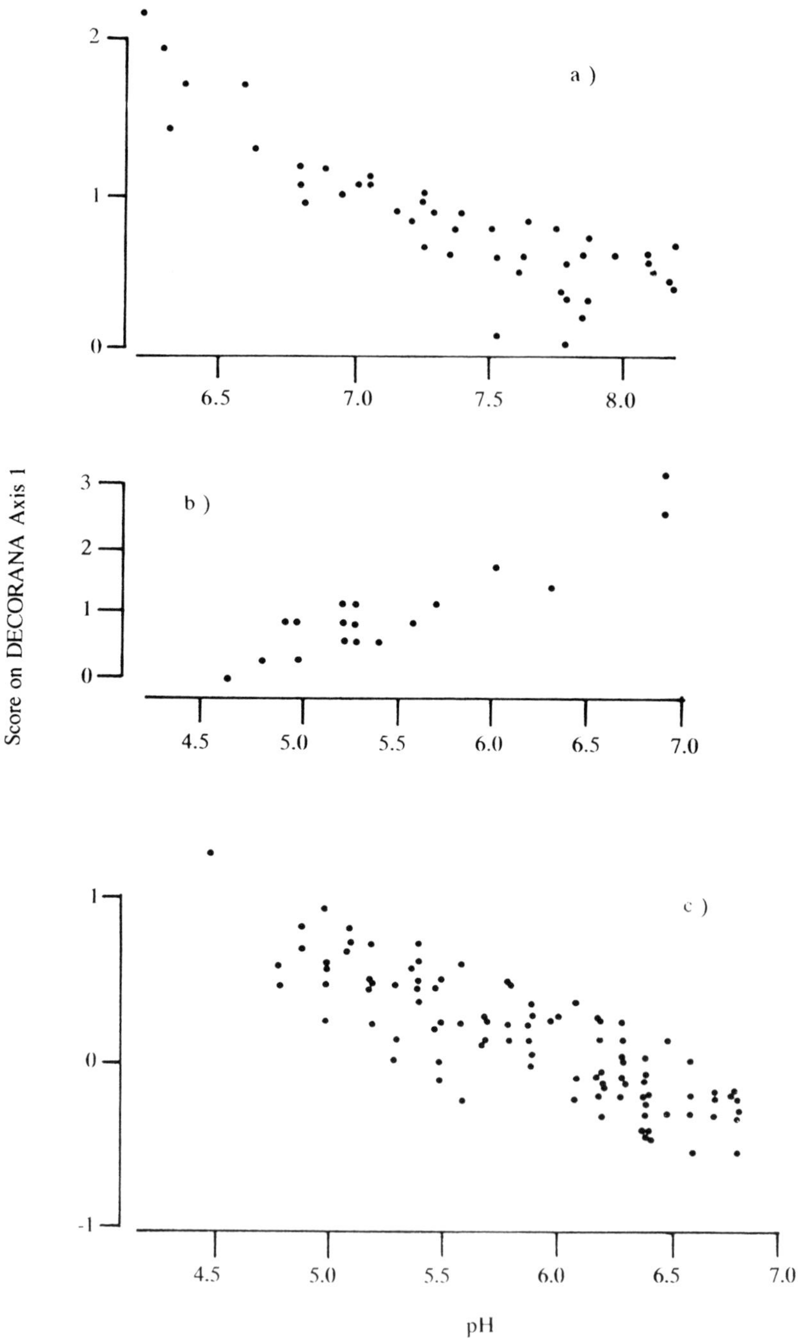

Fig. 17.3. The DECORANA ordination of macroinvertebrate assemblages from three Welsh studies in relation to acid-related factors: (a) the Wye catchment,[15] (b) the Upper Tywi catchment,[30] (c) Upland Wales.[28]

(Chemical characteristics were based on annual means from a weekly programme of spot sampling)

Environmental differences between those groups were best summarised by differences in aluminium concentration, total hardness and catchment area (as an expression of stream size). In multiple discriminant analysis, these three variables permitted 100% precision in placing each stream into its correct invertebrate groups using environmental data alone.[30] Output from MAGIC was used to locate each modelled stream on two discriminant functions, hence indicating the likely invertebrate assemblage (see Fig. 17.6), according to the following equations:

$$F1 = 6.58 \text{ Log[Al]} + 4.12 \text{ Log[hardness]} - 0.06 \text{ catchment area} + 2.72.$$
$$F2 = 9.93 \text{ Log[hardness]} + 1.23 \text{ Log[Al]} - 1.20 \text{ catchment area} - 4.35.$$
(concentrations in mg l^{-1}, area in hectares).

The discriminant functions also enable estimation of the probability of occurrence of individual species using the methods developed by Moss et al.[11] The probabilities depend on the distance of the new stream, or model conditions, from the centre of each invertebrate group in discriminant space. Similar procedures also enable a statistical assessment of the range over which the model involves extrapolation beyond its initial calibration.[19]

Applications to Welsh catchments

In this study, MAGIC was applied to three catchments around Llyn Brianne, and to the Welsh 'acid-sensitive' region. Both the specific sites and general area have been described elsewhere[25, 30, 33] (see also Chap. 16).

Of the three streams draining the Brianne catchments, two had soft-water (< 2 mg Ca l^{-1}, pH 4.8–5.2, 0.15–0.36 mg Al l^{-1} annual means) and catchments covered respectively by sitka spruce *Picea sitchensis* (Stream Type III B, catchment 1 on Fig. 1.2), and *Molinia/Festuca/Nardus* moorland (Type III A, catchment 9 on Fig. 1.2). A third stream had 4 mg Ca l^{-1}, pH 6.9 and <0.05 mg Al l^{-1} (Type I, catchment 5 on Fig. 1.2). Trout are currently absent from stream 1, whilst 9 and 5 had populations of 10 and 120–280 fish 100 m^{-2}, respectively, in 1984 and 1985 (Table 17.1). Invertebrate assemblages in these streams were represented by groups A, B and C.

Simulations and scenarios – Brianne applications

MAGIC was used to simulate trends in each catchment between 1844 and 2124. In hindcasts, between 1844 and 1984, deposition patterns were estimated on the basis of known trends in fossil-fuel consumption averaged over the U.K. (see Chap. 16). From 1984 onwards up to 2124, alternative future scenarios involved, firstly, continuing sulphate deposition at 1984 levels and,

Table 17.1. Some physico-chemical and biological characteristics of streams at Llyn Brianne used in the modelling exercise.

Catchment	Actual land use	pH	Aluminium (mg l^{-1})	Total hardness (mg l^{-1})	Catchment area (ha)	Slope (m km^{-1})	Salmonid density (n 100 m^{-2})	Invertebrate status
5	Moorland	6.9	0.07	12.9	82	122	120–280	C
9	Moorland	5.2	0.15	4.2	33	85	10	B
1	Conifer	4.8	0.36	5.3	264	96	0	A

secondly, a 50% reduction in sulphate deposition, beginning in 1984 but phased over a 20-year-period.

Additionally, in order to simulate the effect of land use changes on the catchments, scenarios from 1958 onwards involved either moorland or conifer forest. This date marked the time of major tree-planting activity in the Tywi catchment, and conifer presence in the model was simulated by linear increases in the dry/occult deposition of sulphate and sea salts over a 15-year-period of forest growth and canopy closure.

Thus, in summary, the scenarios involved on each catchment were:

1. Moorland with sulphate deposition constant at 1984 levels.
2. Moorland with sulphate deposition progressively reduced to 50% of 1984 levels.
3. Conifer forest from 1958 onwards, and sulphate deposition at 1984 levels.
4. Conifer forest from 1958 onwards, with sulphate deposition reduced to 50% of 1984 levels.

In effect, the alternative scenarios of land use involved simulating events which had actually occurred (moorland continuation at stream 9; forest presence at stream 1), besides permitting the return of stream 1 to moorland and alteration of stream 9 to forest.

Stream biology was simulated according to the fish and invertebrate models in the years 1844 (the origin), 1984 (following forest canopy closure and at a time when real data became available), 2004 (the end of the 50% sulphate reduction period and approximately the time at which the first forest crop would be harvested) and 2124. Catchment area for the invertebrate model was assumed to be constant over all the scenarios. However, A.D.F. (= average daily flow), in the fish density model, was manipulated according to land use by removing 10% when moorland was altered to conifer forest, and adding 10% when forest reverted to moorland. This change represented the average effect of conifer forestry on water yield.[2]

Simulations and scenarios – regional application

In the regional approach, the MAGIC model was run repeatedly using different sets of the key parameters chosen randomly from their known distributions. This technique is based on Monte Carlo analysis, and is described more fully in Chap. 16. The model was calibrated to give a close fit with chemistry observed across the Welsh region during a survey by WWA in 1984. Once this fit was achieved, temporal trends were simulated between 1844 and the present day, whilst future scenarios involved assessing the effect of a 50% reduction in deposition beginning in 1984 and phased over a 20 year period. An alternative scenario involved deposition continued at 1984 levels. The biological consequences of the patterns in this regional model are illustrated here using the fish data, simulated in 1844, 1984 and 2024 (shortly after conditions in the model stabilized).

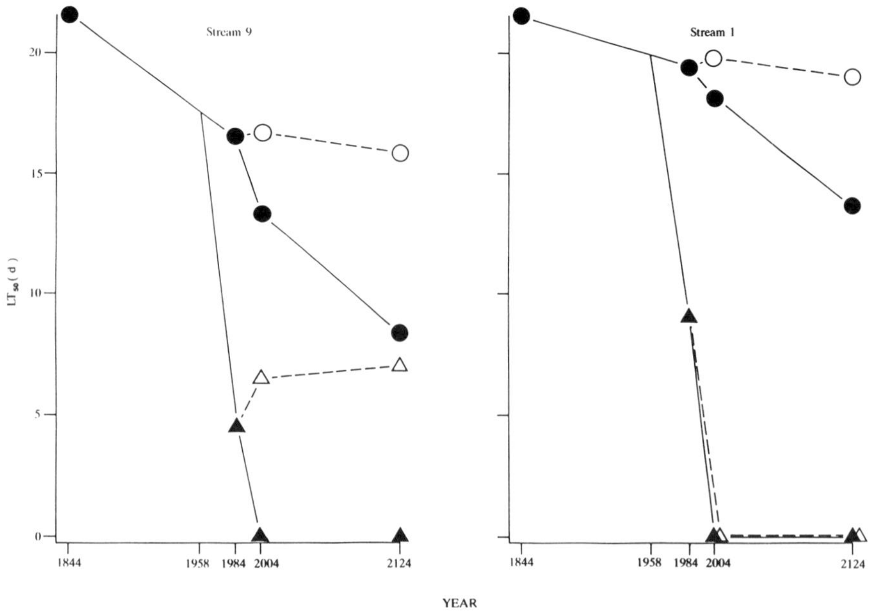

Fig. 17.4. Changes in the survival time of brown trout in two soft-water streams at Llyn Brianne according to MAGIC simulation (broken line, 50% reduction in deposition from 1984; solid line, continued deposition at 1984 levels; circles, moorland; triangles, forest from 1958 onwards.[19]

Results – Llyn Brianne

The fish models

Toxic response

MAGIC simulation indicated that chemical conditions in all the Brianne streams in 1844 would not have caused trout deaths (LT_{50}'s > 22 d; Fig. 17.4). This situation pertained in stream 5 through all tested scenarios of land use change and deposition, probably because of the elevated base status of this catchment (see Chap. 16). By contrast, simulation of both the soft water streams indicated progressive reduction in survival time between 1844 and 1984. The effect was particularly pronounced in the presence of forest, with LT_{50}'s by 1984 < 5 d, close to the actual value for stream 1 in 1982 of 9.2 d. Without forest, simulated LT_{50}'s for streams 9 and 1 in 1984 were 15.9 and 19.7 d, respectively, again close to the actual value for the moorland stream of 16.8 days.

Future simulations also indicated the importance of forest, with LT_{50}'s remaining below 5–10 days in both streams 9 and 1 under forest cover irrespective of a 50% deposition reduction from 1984 onwards (Fig. 17.4).

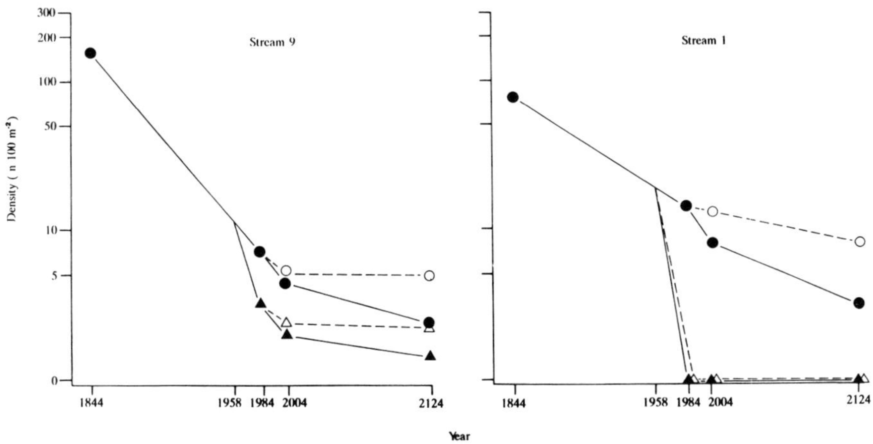

Fig. 17.5. Changes in the density of brown trout in two soft-water streams at Llyn Brianne according to MAGIC simulation.[19] Conventions as in Figure 17.4.

However, in the soft-water streams under moorland only, deposition reduction would, according to the model, prevent LT_{50}'s for trout falling below 15 d until 2124 A.D. at the earliest. At Llyn Brianne, this value of 15 days is one at which sites often become fishless.

Trout density

Trout density, according to the model, showed trends similar to those for survival time (Fig. 17.5). All the streams had 80–200 fish per 100 m^2 in 1844, values similar to those found at about 25% of the sites during the survey from which the model was derived. Densities in stream 5 remained high (193–300 per 100 m^2), similar to the actual values in 1984/85 of 120–280, throughout all the tested scenarios. By contrast, model densities in streams 9 and 1 declined considerably to 1984. Even under the moorland scenario both streams had only 8–15 fish per 100 m^2, close to the actual value for stream 9 in 1984 of 10 fish per 100 m^2. The decline was more pronounced under forest, with streams 9 and 1 having 3 and 0 fish per 100 m^2 by 1984, close to the actual value of 0 in stream 1.

In the simulation of future trends, only the moorland scenario with reduced deposition was sufficient to prevent further pronounced decline in trout density for either stream 9 or 1.

The invertebrate model

As with the fish models, neither historical change nor any future scenario was sufficient to alter the model fauna in stream 5 from group C, because of the base status of this catchment. More interestingly, simulation of historical

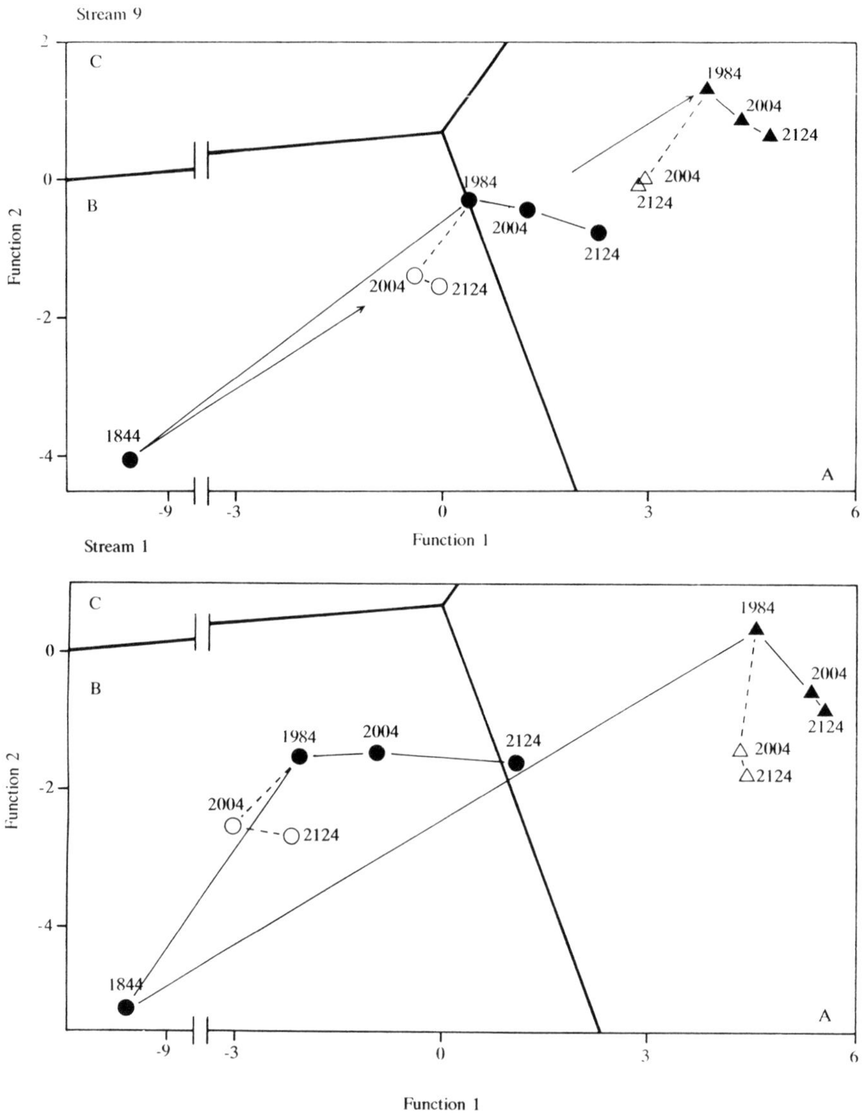

Fig. 17.6. Changes in the types of invertebrate assemblages in two soft-water streams at Llyn Brianne according to MAGIC simulation.[19] Conventions as in Fig. 17.4. The assemblages are described in the text.

patterns in streams 9 and 1 indicated that neither stream had invertebrate group C in 1844, reflecting low concentrations of calcium and magnesium at this time. Recreation of the fauna expected under these conditions involves extrapolation beyond the range of the model, since waters with the characteristics indicated by MAGIC (i.e. low calcium, but relatively high pH and low aluminium) are no longer present in Wales (Fig. 17.6). From 1940 onwards, the fauna of both streams 9 and 1 was type B, with this situation being maintained, under moorland scenarios at 1984 levels of deposition, until shortly after 1984 (stream 9) or 2124 (stream 1). The transition to the impoverished group A would be prevented in stream 9 and delayed until 2124 in stream 1 if continued moorland presence were accompanied by reduced deposition.

By contrast, simulated chemical changes due to the presence of forest advanced the impoverishment of the invertebrate fauna in both stream 1 and 9 to type A (Fig. 17.6). This model simulation is similar to the actual situation in stream 1, currently under forest, in 1984 and 1985. A 50% reduction in deposition, under the forest scenario, would not be sufficient to return the faunal type from A to B during any stage of future simulation.

Some species-specific patterns of change, expected from the model under the different scenarios, are illustrated in Fig. 17.7.

Results – regional application

According to the regional models, trout survival in all streams was high in 1844, none giving LT_{50}'s of less than 19 days (Fig. 17.8). Similarly, almost 80% of sites in the model had fish densities of >10 fish per 100 m^2. By 1984, however, survival values and densities had fallen, with approximately 20% of sites having LT_{50}'s <15 days and 60% having densities <10 per 100m^2. The survival and density patterns according to observed and simulated chemistry at this time were in close agreement.[33] The value of 60% for sites with <10 fish per 100 m^2 was, however, slightly in excess of the 40% of sites which actually had such low densities.

As with the catchment-specific applications, no marked increase in trout density occurred in the regional model, despite the 50% reduction in deposition. Some slight shifts were apparent across the range of sites with 1–10 fish per 100 m^2 (Fig. 17.8). Conditions for trout survival also improved only slightly by comparison with conditions in 1984. However, the alternative scenario, with continued deposition at 1984 levels, led to a marked reduction in LT_{50}'s across the region, with over 50% of sites having values of <15 days by shortly after the turn of the 21st century.

Discussion

As with all ecological models, the approach outlined here is imperfect and based on several assumptions. For this reason, the model's hindcasts and forecasts should be thought of as simulations and not necessarily representations of the real world.

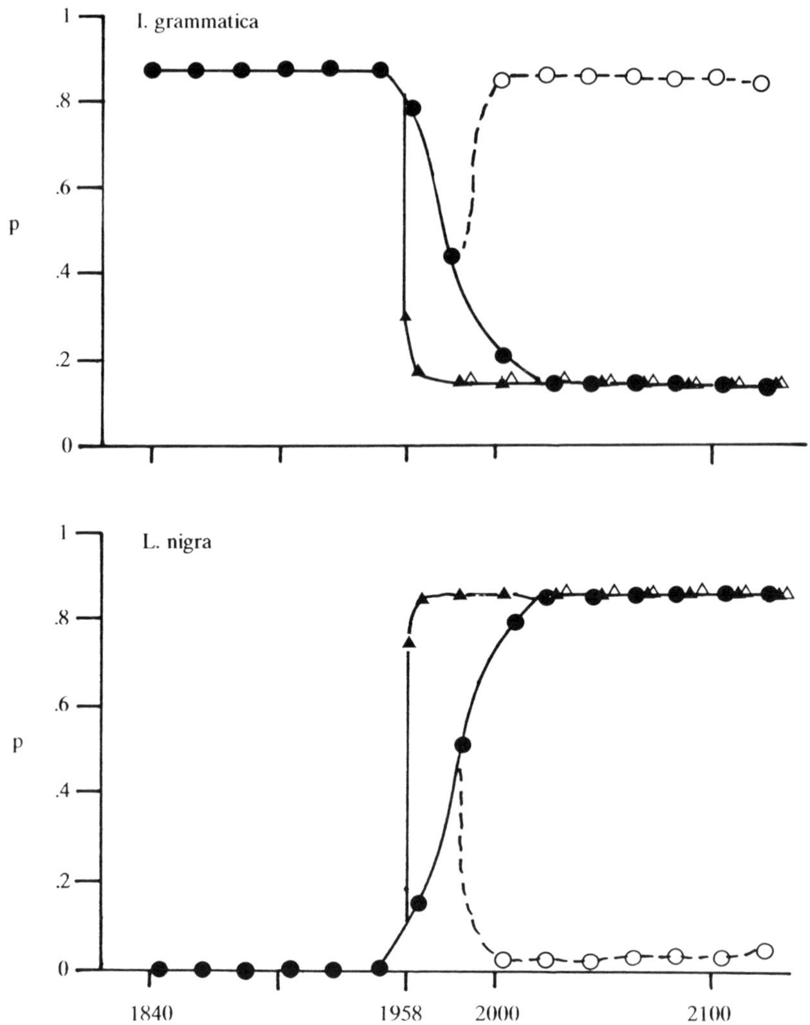

Fig. 17.7. Changes in the probability of occurrence of two stoneflies (*Leuctra nigra and Isoperla grammatica*) in stream 9 according to MAGIC simulation.[31] Conventions as in Fig. 17.4.

Problems with MAGIC

Even if the ecological models were shown to be correct, their accuracy is dependent on MAGIC, whose design and operation is subject to uncertainty.[21] For example, although acid deposition is of prime importance in MAGIC, real data on historic patterns of deposition are generally lacking, and absent for Wales. The temporal sequence of changing deposition is,

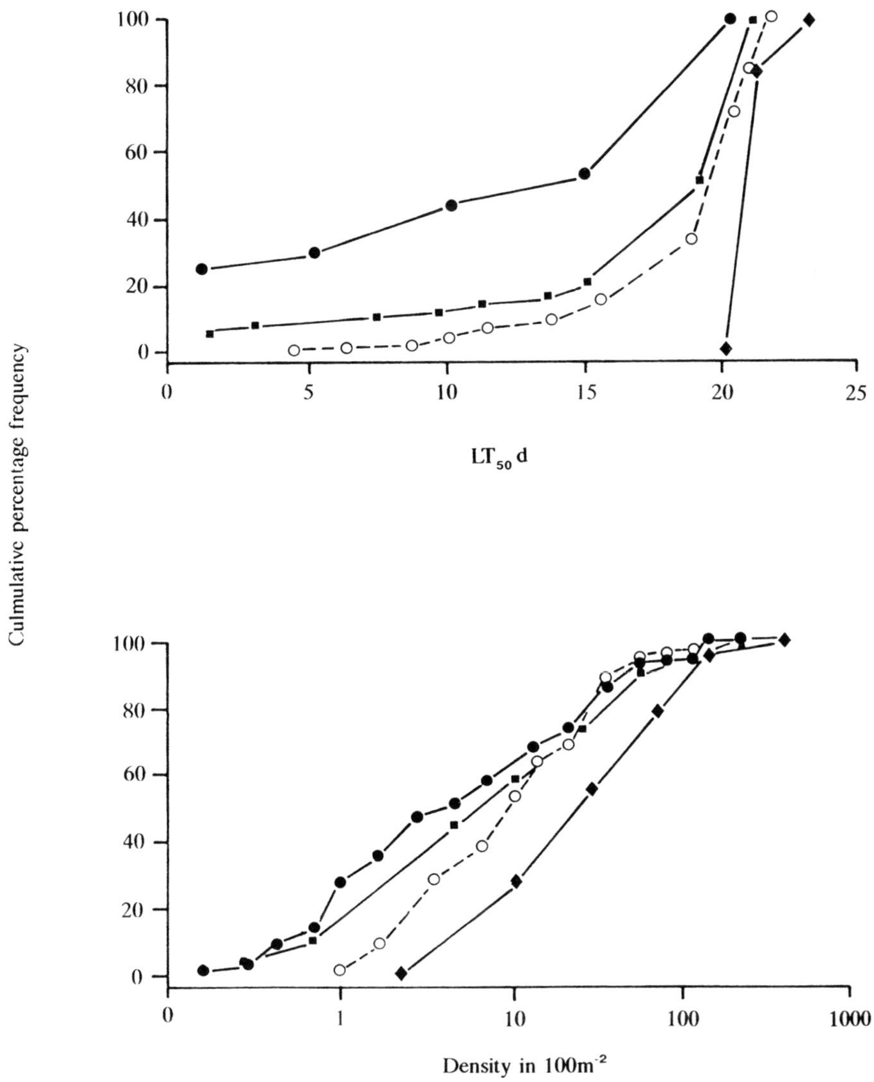

Fig. 17.8. Temporal changes in the frequency distribution for the survival and density of brown trout across Wales according to MAGIC (♦ 1844; ■ 1984; O 2024 with deposition reduced by 50%; ● 2024 with deposition constant at 1984 levels).

therefore, an estimate. Additionally, soil chemistry is lumped on a catchment basis, which might not truly mimic chemical behaviour in the different soil types and soil horizons which occur at the Welsh sites. The chemical processes themselves are subject to considerable debate. In MAGIC, the release of aluminium is controlled largely by reactions which assume the solubility of gibbsite ($Al(OH)_3$), which is now known to be a scarce mineral in many catchments to which MAGIC has been applied. The ecological models would be particularly sensitive to errors in aluminium chemistry, in view of its prominent place in the empirical relationships. However, this criticism of MAGIC may have been overstated.[21] Indeed, despite the many cautions governing MAGIC, it has recently been able to reproduce accurately the acidification of Scottish lochs as shown independently by changes in the diatom flora (T.J. Musgrove, pers. comm.). Similar trends to those given by MAGIC for Welsh streams were also reconstructed from diatom cores taken from adjacent lakes (Chap. 3), a feature of particular importance to this Welsh study.

The use of MAGIC in this and other studies which model acidification due to forestry carries the major assumption that the key acidifying process in these cases is due to increased evapotranspiration and dry/occult deposition.[12] However, other acidifying processes are possible in afforested catchments.[24] The reduced residence time of water because of plough furrows at afforested sites may be particularly important for MAGIC, altering the time available for weathering reactions, and changing the proportion of water derived from baseflow.[19, 21] Additionally, MAGIC appears to be particularly sensitive to changes in soil pCO_2, for which few field data are available from forested sites (P. Whitehead, pers. comm). The net effect of these other factors might indicate that the scenarios of change due to forest are underestimates of the real change.

Problems in ecological modelling

A major and implicit assumption in two of the ecological models is that the empirical relationships (essentially correlations) between stream chemistry and faunal distributions represent a causal influence. Through such causality, temporal changes in stream chemistry bring corresponding changes in biology. However, other abiotic and biotic effects could clearly be important influences on the fauna, both in the initial calibration of the model and in determining trends. For example, the effect of forestry on stream biology operates not only through chemistry, but probably also through changes in hydrology, sediment yields, energetic pathways and habitat structure.[16] Possible influences by these factors on stream invertebrates were masked in this analysis because aluminium concentration represented the major identifiable difference between afforested and moorland catchments.[30] There is also much evidence that salmonid abundance can be limited by habitat features[9] or density dependent regulation.[3] Both could restrict the upper limit of salmonid density in the study streams, keeping levels below those given by the model for 1844.

However, many data support an influence by acid related factors on both fish and invertebrates in Welsh upland streams. For example, ordination and classification of invertebrate assemblages has repeatedly revealed pronounced correlations with pH, or related factors (see Chap. 17). These correlations persist amongst the fauna from different stream habitats, which would be unlikely if chemistry did not exert some causal influence.[14, 30]

More importantly, experimental episodes of increased acidity and aluminium concentration have confirmed that some species are sensitive to such conditions (Chap. 7). In the case of salmonids, many data support direct toxicity by acid-related factors (Chap. 8). Low salmonid abundance in acid streams in Wales does not appear to reflect trophic effects,[26] nor limited habitat because densities are often lower than expected from habitat characteristics.[9]

Episodic changes in pH, aluminium and calcium concentrations are usually pronounced in acid streams during storm events or snow melt.[25] Field data and experimental investigations indicate that such events probably affect fish and some invertebrates (Chaps. 7 and 10). However, all the ecological models presented here were based on mean chemical conditions, partly because of difficulties in assessing from the data the relative biological influences of chronic and acute conditions. Additionally, MAGIC does not yet incorporate hydrological effects such as spates or their associated acid conditions, and hence long-term simulation involves only mean values.

In situations where episodes control the fauna, the predictive ability of the models described here could be limited, especially if such episodes occurred in streams of relatively high pH (~ 5.5–6.0). However, because mean and extreme values for pH and aluminium are closely correlated, the means probably indicate the likelihood of episodic change (Tyler and Ormerod, unpubl.). Moreover, there is some evidence that acidity affects some invertebrates only indirectly, through their food supply,[18] and this pathway is just as likely to involve chronic effects as episodic. Nevertheless, more investigations of biological processes are required in acid waters so that episodes can be built into ecological models and used with versions of MAGIC which forecast the chemistry of storm events.

Conclusions

Although governed by the above caveats, several features of this modelling exercise are of interest. Firstly, the patterns generated by the models indicate a considerable reduction in salmonid abundance in soft-water streams around Llyn Brianne, and throughout upland Wales, between the 19th century and present day. This feature is corroborated by the survival model, which suggests a reduction in the LT_{50} of brown trout over this period. The lack of invertebrate change between ~ 1940 and 1984, at least under the moorland scenarios, does not conflict with the fish pattern because salmonids show no trophic dependence on aquatic invertebrates in the study streams.[26]

Secondly, and importantly, the acidifying influences of conifer forest severely exacerbate aluminium concentrations in the soft-water streams at Llyn

Brianne, almost totally eliminating fish populations and causing change amongst invertebrates by 1984. Even though the density model indicated that some fish would be present in stream 9 under the forest scenario, the reduced LT_{50} indicated that their survival would be severely limited. Moreover, the influences of forest on fish and invertebrates would not be reversed, according to the model, in spite of deposition reduction by 50%. Thirdly, the models also suggest that, despite reduced deposition, pronounced increases in fish abundance will not occur in moorland streams at Llyn Brianne within the simulated time frame. However, the absence of forest with deposition reduced by 50% could at least support some salmonids in the Brianne streams studied. The extrapolation to other areas, in the regional analysis, also indicated that fish survival could be maintained following a 50% reduction in deposition. This feature is clearly important because of the nursery functions fulfilled by Welsh headwaters.

References

1. Brown, D.J.A. (1982). The effect of pH and calcium on fish and fisheries. *Water, Air and Soil Pollution*, 18: 343–351.
2. Calder, I. and Newson, M.D. (1979). Land use and upland water resources in Britain – a strategic look. *Water Resources Bulletin*, 15: 1628–1639.
3. Elliott, J.M. (1985). Population dynamics of migratory trout, *Salmo trutta*, in a lake district stream, 1966–83, and their implications for fisheries management. *Journal of Fish Biology, Supplement A*, 27: 33–43.
4. Green, R.H. and Vascotto, G.L. (1978). A method for the analysis of environmental factors controlling patterns of species composition in aquatic communities. *Water Research*, 12: 583–590.
5. Hill, M.O. (1979). TWINSPAN – A FORTRAN programme for arranging multivariate data in an ordered two-way table by classification of the individuals and attributes. Department of Ecology and Systematics, Cornell University, New York.
6. Howells, G.D., Brown, D.J.A. and Sadler, K. (1983). Effects of acidity, calcium and aluminium on fish survival – a review. *Journal of the Science of Food and Agriculture*, 34: 559–570.
7. Jeffers, J.N.R. (1978). An Introduction to Systems with Ecological Applications. Edward Arnold, London.
8. Lacroix, G.L. (1987). Model for loss of atlantic salmon from acidic brown waters of Canada. Acid Rain: Scientific and Chemical Advances (Eds. R. Perry, R.M. Harrison, J.N.B. Bell and J.N. Lester) pp. 516–522.
9. Milner, N.J., Hemsworth, R.J. and Jones, B.E. (1985). Habitat evaluation at a fisheries management tool. *Journal of Fish Biology, Supplement A*, 27: 85–108.
10. Minns, C.K., Kelso, J.R.M. and Johnson, M.G. (1986). Large-scale risk assessment of acid rain impacts on fisheries: models and lessons. *Canadian Journal of Fisheries and Aquatic Sciences*, 43: 900–921.
11. Moss, D., Furse, M.T., Wright, J.F. and Armitage, P.D. (1987). The prediction of the macroinvertebrate fauna of unpolluted running-water sites in Great Britain using environmental data. *Freshwater Biology*, 17: 41–52.
12. Neal, C., Whitehead, P., Neal, R. and Cosby, J. (1986). Modelling the effects of acidic deposition and conifer afforestation on stream acidity in the British uplands. *Journal of Hydrology*, 86: 15–26.

13. Odum, E.P. (1971). Fundamentals of Ecology. Saunders, Philadelphia.
14. Ormerod, S.J. (1987). The influence of habitat and seasonal sampling regimes on the ordination and classification of macroinvertebrate assemblages in the catchment of the River Wye, Wales. *Hydrobiologia*, 150: 143–151.
15. Ormerod, S.J. and Edwards, R.W. (1987). The ordination and classification of macroinvertebrate assemblages in the catchment of the River Wye in relation to environmental factors. *Freshwater biology*, 17: 533–546.
16. Ormerod, S.J., Mawle, G.W. and Edward, R.W. (1987). The influence of forest on aquatic fauna. Environmental aspects of forestry production in Wales (Ed. J.E.G. Good). ITE Symposium number 22, Institute of Terrestrial Ecology, Grange-over-Sands.
17. Ormerod, S.J., Weatherley, N.S. and Whitehead, P.G. (in press). Regional and temporal patterns of ecological change during the acidification and recovery of Welsh streams according to a hydrochemical model. Models to describe the geographic extent and time evolution of acidification and air pollution damage. Proceedings of a symposium held in Finland during July 1988.
18. Ormerod, S.J., Wade, K.R. and Gee, A.S. (1987). Macro-Floral assemblages in upland Welsh streams in relation to acidity and their importance to invertebrates. *Freshwater Biology*, 18: 545–557.
19. Ormerod, S.J., Weatherley, N.S., Varallo, P.V. and Whitehead, P.G. (1988). Preliminary empirical models of the historical and future impact of acidification on the ecology of Welsh streams. *Freshwater Biology*, 20: 127–140.
20. Reckhow, K.H., Black, R.W., Stockton, T.B., Vogt, J.D. and Wood, J.G. (1987). Empirical models of fish response to lake acidification. *Canadian Journal of Fisheries and Aquatic Sciences*, 45: 1432–1442.
21. Reuss, J.O., Christophersen, N. and Seip, H.M. (1986). A critique of models for freshwater and soil acidification. *Water, Air and Soil Pollution*, 30: 909–930.
22. Sadler, K. (1983). A model relating the results of low pH bioassay experiments to fishery status in Norwegian Lakes. *Freshwater Biology*, 13: 453–463.
23. Small, M.J. and Sutton, M.C. (1986). A direct distribution model for regional aquatic acidification. *Water Resources Research*, 22: 1749–1758.
24. Stoner, J.H. and Gee, A.S. (1985). Effects of forestry on water quality and fish in Welsh rivers and lakes. *Journal of the Institute of Water Engineers and Scientists*, 39: 27–45.
25. Stoner, J.H., Gee, A.S. and Wade, K.R. (1984). The effects of acidification on the ecology of streams in the upper Tywi catchment in West Wales. *Environmental Pollution (Series A)*. 36: 125–157.
26. Turnpenny, A.W.H., Sadler, K., Aston, R.J., Milner, A.G.P. and Lynam, S. (1987). The fish populations of some streams in Welsh and northern England in relation to acidity and associated factors. *Journal of Fish Biology*, 31: 415–434.
27. Van Winkle, W., Christensen, S.W. and Breck, J.E. (1986). Linking laboratory and field responses of fish populations to acidification. *Water, Air and Soil Pollution*, 30: 639–648.
28. Walters, C.J. (1971). Systems ecology: the systems approach and mathematical models in ecology. Fundamentals of Ecology (E.P. Odum). Saunders, Philadelphia.
29. Wade, K.R., Ormerod, S.J. and Gee, A.S. (1988). The ordination and classification of macroinvertebrate assemblages to predict stream acidity in upland Wales. *Hydrobiologia*, 171: 59–78.
30. Weatherley, N.S. and Ormerod, S.J. (1987). The impact of acidification on macroinvertebrate assemblages in Welsh streams: towards an empirical model. *Environmental Pollution*, 46: 223–240.
31. Weatherley, N.S. and Ormerod, S.J. (1989, in press). Modelling ecological impacts of the acidification of Welsh streams: temporal changes in the occurrence of macroflora and macroinvertebrates. *Hydrobiologia*.
32. Welsh Water (1986). Regional acid waters electrofishing survey interim report. Report No. SW/86/19. Welsh Water, Haverfordwest.

33. Whitehead, P.G. and Ormerod, S.J. (in press). Modelling the chemical and biological response of Welsh streams to changes in atmospheric deposition and land use. The influence of land use on the acidification of streams (Ed. H. Barth). Commission of the European Communities.
34. Wright, J.F., Moss, D., Armitage, P.D. and Furse, M.T. (1984). A preliminary classification of running water sites in Great Britain based on macroinvertebrate species, and the prediction of community type using environmental data. *Freshwater Biology*, 14: 221–256.

18. Llyn Brianne – Acid deposition modelling

S.E. METCALFE and R.G. DERWENT

Background to the model

A model of atmospheric pollution transport has been applied to examine the relationship between pollutant emissions and deposition with particular reference to the Llyn Brianne catchment area (see Fig. 1.1). The aim of the exercise was to identify the current major sources of pollutants (by direction and source height) and to investigate the possible effects of various emission reduction strategies.

The computer model applied in this study is one of the 'receptor oriented' type[5] in which the nature and quantity of pollutants of nitrogen and sulphur are followed until arrival at a predetermined receptor site (Llyn Brianne). The model follows the changing chemical composition of a box, or parcel, of air as it is carried along by the wind (a Lagrangian trajectory model). In this case, a base area for the 'box' of 100 × 100 km has been used and its height has been taken to be 800 m, equal to the assumed atmospheric boundary layer depth. Generally, the model is based on that developed by Eliassen[4] and is a slightly modified version of that described in Derwent, Dollard and Metcalfe.[3]

Each air parcel has a travel time of up to 96 hours, with the trajectories followed from the time they enter the land areas of the United Kingdom or the continent of Europe, over a distance of up to 2500 km through to Llyn Brianne. The trajectory directions were chosen at 15° intervals forming a wind rose, starting at −15° (or 345°). The results for the 15° intervals are then aggregated to produce eight sectors of 45° (Fig. 18.1). The model generates output for each of these 45 sectors in a cumulative fashion from sectors 1 to 8. The relative contribution of different source areas to long term deposition can, thus, be identified by weighting the sectoral contributions with the frequency of wind from that sector.

In applying the model, a number of the key underlying assumptions must be recognised. The trajectories of the air parcels are assumed to be straight. In reality, this is an extremely implausible assumption, as the calculation of back trajectories has shown.[7] The straight line assumption is, however, probably reasonable over the last few hundred kilometres of travel, particularly over the UK. A further key assumption is that of complete mixing. At

R.W. Edwards et al. (eds) Acid Waters in Wales. 299–309.

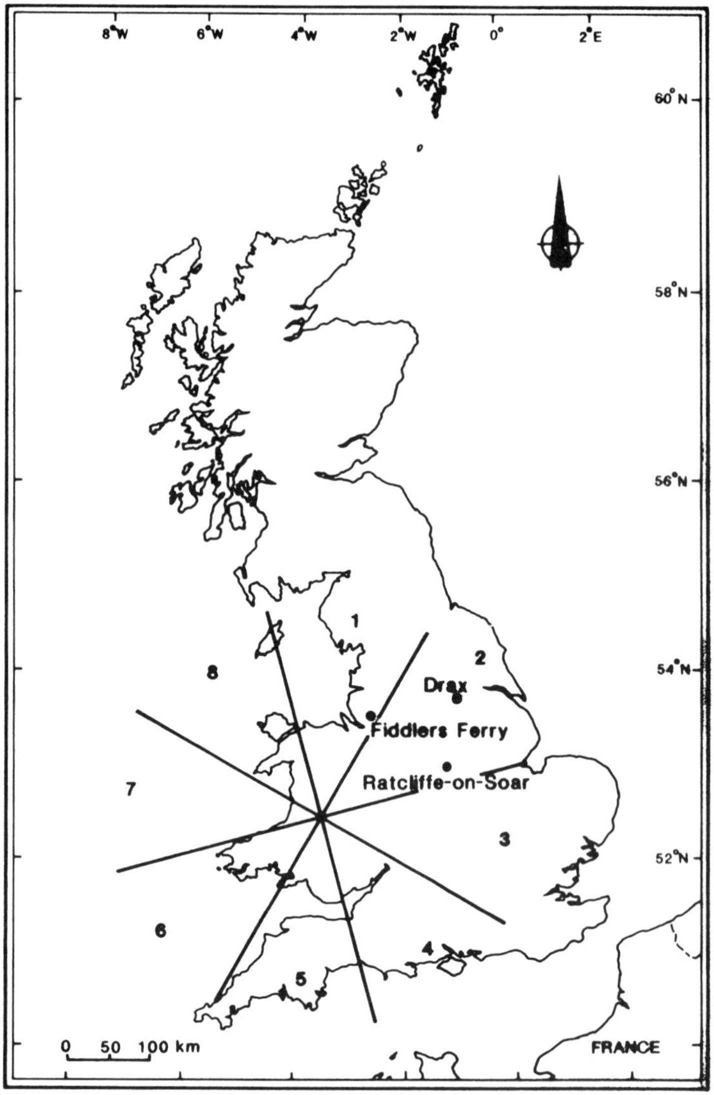

Fig. 18.1. Map showing the areas of the UK covered by sectors 1–8 employed in the model.

each time step within the model, the emissions received by an air parcel are assumed to be instantaneously mixed over the total volume of that air parcel. This is not a valid assumption over a spatial scale of 100 km, particularly during periods of high atmospheric stability such as can occur at night or when an inversion develops. Under such stable conditions, pollution can effectively be trapped at low levels in the atmosphere.

Further details of the model are provided elsewhere,[3] but some of its salient features are given below to facilitate the analysis of the model results. The model uses average values for wind speed (7.5 m s^{-1}) and an average wind rose for the UK[8] to provide wind direction for all points. The boundary layer depth is set at 800 m. Precipitation in the model is introduced as constant drizzle, producing 100 mm yr^{-1}. All factors such as wind speed and direction and precipitation would, in reality, vary continuously and co-vary to some extent.

Emissions

Emissions figures for the UK have been aggregated to produce an emissions grid of 100 × 100 km squares. Sulphur emissions have been divided into high level sources (mainly power stations) and low + medium level sources. Data for these sulphur emissions have been taken from the inventory compiled by Warren Spring Laboratory.[10] Total UK sulphur emissions in the model are 3.88 million tonnes of which 2.52 million tonnes comes from high level sources.

NO_x emissions from all emission heights have been taken together.[9] Total UK emissions are estimated to be 1.8 million tonnes and have again been distributed over the 100 × 100 km emissions grid. Ammonia emissions were taken from Buijsman, Maas and Asman[2] and based on the 150 × 150 km EMEP grid.

The spatial distribution of the emissions shows very strong concentrations of pollution sources in certain parts of the UK. Sources of high level sulphur are concentrated in the east Midlands (Trent valley), the north-east (Humberside), the north-west (Merseyside), the south-east and the central valley of Scotland. As might be expected, low and medium level sulphur emissions display a more homogeneous distribution pattern, but with the major urban/industrial conurbations still being marked by higher emissions levels (> 100,000 tones yr^{-1}). NO_x emissions also reflect major centres of population, being highest in the east Midlands, in the south-east and the north-east. By contrast, ammonia emissions are associated with important agricultural regions such as the west country, South Wales, the Midlands and East Anglia.

The chemical scheme of the model

The chemistry included in the model is illustrated in Fig. 18.2, which shows the life cycles for the various nitrogen and sulphur containing gases from emission to deposition. A simple linear chemistry scheme has been employed in which all the model species are set to zero at the boundaries of the grid. Ozone (O_3) is the only exception to this, which is set at 30 ppb.

The life cycle for sulphur compounds has been based upon that reported in Buckley-Golder and Derwent,[1] although the model has been expanded to include wet removal of SO_2. The equations used to describe the concentrations of SO_2 and sulphate aerosol, SO_4^{2-}, in the atmosphere are as those developed by Eliassen and Saltbones (1983). For SO_2:

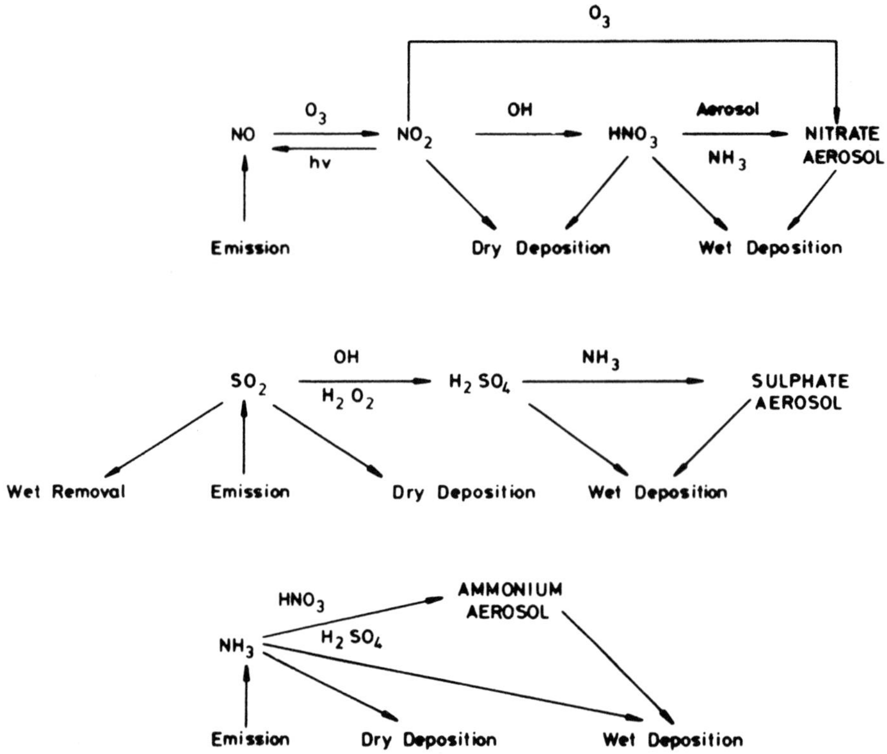

Fig. 18.2. The coupled life cycles of the sulphur- and nitrogen-containing trace gases.

$$\frac{dq}{dt}(SO_2) = (1 - \propto - \beta)\,\frac{Q}{h}\left(\frac{Vd}{h} + Kt\right) q$$

where q = the concentration of SO_2
Vd = dry deposition velocity of SO_2
h = mixing layer height
Kt = transformation rate of SO_2 to SO_4^{2-}
$\propto$ = local dry deposition of SO_2
β = sulphur emitted as SO_4^{2-}
Q = sulphur emissions/unit volume of air/unit time

For SO_4^{2-}:

$$\frac{dS}{dt}(SO_4^{2-}) = Ktq + \beta\,\frac{Q}{h} - \varkappa S$$

where S = the concentration of SO_4^{2-}
$\varkappa$ = dry deposition velocity of SO_4^{2-}

These equations can be used to describe the atmospheric concentrations and dry deposition of SO_2 and SO_4^{2-} at any point downwind of a source. To represent wet deposition, a scavenging ratio approach was used to remove aerosol sulphur and nitrogen species into precipitation and thence into wet deposition.

The model representation of nitrogen has been described by Derwent et al.[3] Coupled differential equations describing the production and loss of the various nitrogen compounds in the moving box of air are solved to yield atmospheric concentrations (NO, NO_2, HNO_3, NH_3 and nitrate aerosol) and wet and dry deposition rates in the same manner as for sulphur compounds. The general form of the differential equations was as follows:

$$\frac{dCi}{dt} = \frac{Ej}{Aj} + Pi - Qi.Ci - \frac{Vi}{h}.Ci - Li.Ci$$

where Ej = annual pollutant emission rate in mass units on the chosen grid j

Aj = conversion factor between mass units yr^{-1} and ppb s^{-1} for the chosen grid

Pi = total production rate of the species by chemical conversion processes

Qi.Ci = total removal rate by chemical conversion processes

Vi = dry deposition velocity of the species at a standard height (usually 1 m above the ground)

Li.Ci = total removal by wet processes

The model output, by 45° sector, indicates atmospheric concentrations (in ppb) for SO_2 and SO_4, NO, NO_x, NO_2, HNO_3, nitrate aerosol, NH_3 and NH_4 aerosol. Concentrations in precipitation (in mg l^{-1}) are shown for NO_3–N, SO_4–S and NH_4–N and wet and dry deposition in kg $ha^{-1}yr^{-1}$ is calculated for N, S and NH–N (ammonia and ammonium).

Application of the model

As indicated above, the aims of applying the model to the Llyn Brianne catchment area were 1) to identify the source area and, if possible, type of pollutants reaching Llyn Brianne at present and 2) to examine the effects of possible emissions control strategies on the amount of acidity deposited in the catchment. The modelled deposition of sulphur and nitrogen could then be used in models of catchment behaviour (such as MAGIC) to assess the possible effects of reductions in atmospheric pollutant loads on soil and surface water chemistry.

Past trends in SO_2 and NO_x emissions have been extensively reviewed elsewhere[10] and they give some guide as to how emissions may change over the next 10–15 years, to the year 2000. SO_2 emissions are likely to remain much the same as at present since they are dominated by emissions from large coal-fired power stations and on most projections, electricity demand is

Table 18.1. Comparison of measured and modelled precipitation concentrations for Llyn Brianne.

Year	NO_3^- μeq^{-1}	NH_4^+ μeq^{-1}	SO_4^{2-} μeq^{-1}	Precipitation mm	Sources
1981	17.0	31.0	54.0	2050	11
1982				1943	
1984	44.4	30.8	71.04	1699	12
1984	36.5	38.8	77.0	1699	12
1986	11.7	12.2	24.4*	1491	*non-sea salt
Model	19.3	31.4	41.2	1250	

anticipated to grow over this period. NO_x emissions are dominated by two contributions: from these same power stations and from motor vehicles. The former contribution is expected to remain much the same and any trend in total NO_x emissions is likely to be dictated by the motor vehicle contribution. As a contribution to national emissions of NO_x, that from vehicles has been growing because of increased traffic volumes and increased vehicle exhaust emissions factors. Emission legislation already in hand will reverse this trend in exhaust emission factors so that even with the projected growth in traffic, national total NO_x emissions will start to decrease during the 1990s.

Whilst it is feasible to consider how SO_2 and NO_x emissions might change in broad terms to the year 2000, two aspects of the problem cause particular difficulty. The first is to gauge the extent and timing of emission control measures on large sources, particularly the coal-fired power stations and motor vehicles. The second is to assess the spatial pattern of SO_2 and NO_x emissions in the year 2000. This requires detailed knowledge of the plans for retirement, refurbishment and the building of new large coal-fired plants, to which we have no access. The only approach that we could adopt is to perform some illustrative calculations which have no significance in terms of policies, programmes, requirements and legislation. They have been adopted solely to illustrate how models can be used to assign sources to deposition in Llyn Brianne to focus attention on the remaining model uncertainties.

The discussion of the modelling results which follows will focus upon the changing concentrations of the pollutants in rain and the wet and dry deposition loads for nitrogen, sulphur and ammonium.

Llyn Brianne has recently become one of the sites in the UK secondary acid deposition monitoring network. Prior to this, however, rainfall quantities and quality have been monitored as part of the project described elsewhere in this volume. Measured rainfall concentrations (in $\mu eq\ 1^{-1}$) are compared with the model results using full UK emissions in Table 18.1. Although the model output compared rather unfavourably with the 1986 data from the catchment, the results clearly lie within the range indicated by values from other years. On this basis it appears that the model produces a credible representation of the rainfall concentrations at Llyn Brianne, even with the simplifying assumptions made in its development and the differences in scale between the model and the catchment.

Given that the modelling results appear acceptable, the model can be employed to identify the major pollution sources affecting Llyn Brianne. In terms of sulphur, sectors 3 and 2, crossing the area of the Midlands, are the major sources of pollution, each sector contributing in excess of 23% of the SO_4^{2-} in rain and hence wet sulphur deposition. Dry sulphur deposition shows a similar pattern. As might be expected, sectors 7 and 8 are the cleanest. By cutting high level UK sulphur emissions out of the model, it is possible to gauge the importance of this source to the acid deposition at Llyn Brianne. Model results indicate that > 33% of the sulphur deposited (wet and dry) in the catchment area comes from these high level sources (i.e. mainly power stations). For SO_4^{2-} in rain and wet and dry sulphur deposition, the biggest reduction occurs in sector 2 (values 30% of full emissions) followed by sector 1. Sector 2 covers the major C.E.G.B. generating facilities in its Midlands Region (e.g. Ratcliffe-on-Soar) and North-Eastern Region (e.g. Drax).

NO_3^- in rain, wet N and dry N deposition show a very marked concentration of pollutant (> 30%) in sector 3 (Midlands and south-east). Sectors 2 and 4 also make important contributions (c. 20%). Ammonium in rainfall indicates a more even distribution of source areas with individual sectors contributing between 8 and 16% and most in sectors 3–6. Dry NH deposition has a rather different pattern from the others, with the highest proportions coming from sectors 6–8. It seems likely that this represents local ammonia sources.

It is important when designing emissions control strategies for the UK, to ascertain the role of European and other non-UK emissions. In order to investigate the significance of these 'external' sources for Llyn Brianne, the model was run with no UK emissions of sulphur or NO_x. For sulphur, the results indicated that non-UK sources provided less than half (c. 46%) of the S-pollutants reaching Llyn Brianne. Sector 3 is the major route for these 'external' pollutants, particularly for SO_4^{2-} in rain (54% loading on sector 3). This analysis highlighted the importance of UK sources in generating pollution for sector 2, which as we have seen above mainly comes from high level sources.

The contribution of non-UK emissions to deposited nitrogen compounds is considerably greater than for sulphur. For NO_3 in rain, non-UK sources provide 59% of the total, and for wet N deposition, the figure is 76% of the total. For NH_4 in rain and wet NH deposition the role of UK sources appears to be smaller still. In terms of dry deposition of N, there seems to be a more even balance between UK and and non-UK sources.

The results obtained from these model runs indicate the pollutants and their deposition routes which show the greatest sensitivity to domestic emissions controls. The benefits which would be derived by those countries receiving long-distance pollution from the UK (primarily of high level S) will

Table 18.2. Results of model runs using different emissions values.

Emissions	Concentration in mg l^{-1}			Deposition in kg ha^{-1}					
	NO_3^-	NH_4^+	SO_4^{-2}	Wet N	Dry N	Wet NH	Dry NH	Wet S	Dry S
Full emissions	0.27	0.44	0.65	3.32	2.11	5.47	4.32	8.19	10.06
Drax + Fiddlers Ferry reduced by 150 k tonnes	0.27	0.44	0.62	3.33	2.10	5.47	4.31	7.81	9.48
UK high S and NOx reduced by 30%	0.24	0.44	0.55	2.97	1.68	5.45	4.34	7.38	8.87
UK high S reduced by 50%	0.27	0.44	0.55	3.34	2.07	5.45	4.34	6.83	8.08
UK high S reduced by 90%	0.27	0.44	0.46	3.35	2.03	5.42	4.36	5.75	6.51

not be considered here. Although all forms of dry deposition, which usually occur close to sources, would be reduced by emission controls, reduction in wet deposition appears to be more complicated. The model output indicates the great importance of high level UK sources of S to the composition of rainfall reaching the Llyn Brianne catchment. As a consequence of this, it seems that the most direct approach would be to concentrate on reducing emissions from these high level sources.

Plans have been announced to reduce UK SO_2 emissions by 14% by 1997 using flue gas desulphurization. Drax B and A are to be fitted first, followed by the power station at Fiddlers Ferry. In terms of NO_x emissions, the combustion units of the UK's twelve largest power stations are due to be modified with a view to reducing emissions by 40% by the end of the century. These proposals have been used as the basis for a series of model runs. Acid deposition at Llyn Brianne has been modelled for reduced emissions at the grid squares including Fiddlers Ferry and Drax, for a reduction in high level S and NO_x of 30%, for high level S cut by 50% and by 90%. The last of these reflects a recent suggestion by the Nordic Council of Ministers that the countries of Europe should aim for a reduction of this magnitude . Table 18.2 shows the rainfall concentrations and wet and dry deposition values resulting from these model runs.

The reduction of emissions from the areas of Fiddlers Ferry and Drax by 150,000 tonnes each does lead to a decline in S levels at Llyn Brianne according to the model. The reduction for SO_4^{2-} in rain and for wet sulphur deposition is about 5%, and for dry sulphur deposition about 6%. As might be anticipated, this reduction is reflected in a decrease in pollution levels in sector 2, as well as in sector 1. The levels of NO_3^- in rain and NH_4 in rain show no change from full emissions. Wet N deposition increases very slightly and dry N and wet and dry NH deposition decrease very slightly.

An emissions reduction of 30% across the entire UK grid of high S and NO_x (reflecting about a 40% decrease in high level NO_x emissions) produces lower levels of all pollutants (except dry NH) for the study catchment. Sulphur levels (in rain, wet and dry deposition) are reduced to about 90% of full emissions values, with the decrease again centred on sector 2 (c. 79% of full). Sectors 1, 3 and 4, however, also show quite marked reductions as does sector 7 for SO_4^{2-} in rain. NO_3^- in rain and wet N deposition are reduced by about 6%. For NO_3^- in rain, sector 1 is the zone of greatest reduction, closely followed by sector 2. For wet N deposition sector 6 also shows a marked decline. Dry N deposition displays an even greater reduction to 80% of full, with sectors 1, 2 and 6 down to $< 70\%$ of full emissions level. Levels of ammonium concentration in rain and wet NH deposition levels were reduced only slightly, probably reflecting the importance of non-UK sources and the lack of direct emissions reductions assumed for this pollutant. Dry NH was again very slightly greater than the value from the full emissions run, probably reflecting the reduced production of ammonium sulphate in the model as a result of the drop in sulphur availability.

When UK high level S is cut by 50%, levels of S for Llyn Brianne are modelled to fall to $< 83\%$ of full emissions values. The reduction in sector 2 becomes more pronounced, falling to 65% of full emissions level. Values less than 90% of those from full emissions are also recorded for sectors 1, 4 and 6. An enhanced reduction of dry sulphur deposition, indicated by all the earlier runs of emissions control strategies, becomes more pronounced. The picture for the nitrogen compounds is not clear cut, with very slight increases in NO_3^- in rain, wet N and dry NH deposition and very slight decreases in NH_4^+ in rain, dry N and wet NH deposition.

The enhanced reduction of S levels in sector 2 is shown even more clearly when overall high level S emissions are reduced by 90%. Although S concentration and deposition levels decline to about 70% of full emissions levels, sector 2 shows only 36–37% of full emissions values. This clearly reflects the increasing impact of the reductions in S emissions on the area of major power station location. By contrast, sector 3 (the main source of non-UK sulphur) bears an increasing proportion of the sulphur load, over 30% of the total in both this and the previous model run for 50% reduction. Deposition of dry sulphur declines to $< 65\%$ of full. This confirms work on a larger scale model[1] which showed that the percentage reduction in dry deposition was equal to or larger than the percentage reduction in emissions. The trends for nitrogen compounds are the same as with the earlier run, although slightly accentuated.

The modelling exercise using different emission control strategies indicates that reduction in UK power station emissions can lead to a noticeable decline in pollutant levels within the country generally and specifically at Llyn

Brianne. The non-linear nature of emissions-deposition relationships is well illustrated. The response to emissions reductions is, however, most marked in the heavily polluted sector.

Appendix to chap. 18

Modelling the effects of the large combustion installation directive

The recent signing of the Large Combustion Installation Directive has major implications for emissions of S and NO_x in the UK in to the next century. The directive applies to any plant > 50 MW in size and therefore includes all coal-fired power stations, as well as many smaller installations. In terms of existing combustion plant (up to 1st July 1987) meeting the directive requires an overall reduction in S emissions of 60% by 2003 and a reduction in NO_x emissions of 30% by 1998.

It is clearly valuable to be able to assess the effects of these reductions in emissions on deposition loads of S and N in a given area, particularly in those regions known to be ecologically sensitive to acid deposition. Previous discussion of modelling with reference to the Llyn Brianne catchment was couched in rather general terms, but the UPDATE model has been further revised to assess the changes in total deposition of S and N resulting from the implementation of the terms of the directive. For S emissions two approaches were tried: Case 1 – a 60% reduction spread over all UK sulphur sources (high, medium and low); Case 2 – a 60% reduction achieved using high-level sources alone (basically, coal-fired power stations). Analysis of reductions in NO_x was a little more complicated, as the present model does not differentiate between source types for this pollutant and much comes from non-point, low-level sources (motor vehicles). It was estimated that if emissions from relatively large sources (> 50 MW) were to drop by 30%, then an overall reduction in NO_x emissions of 15% might be expected.

The initial modelling exercise produced a total S deposition (wet + dry) for Llyn Brianne of 18.25 kg ha^{-1} yr^{-1}. Under the two case studies, this figure was reduced to 11.63 and 12.14 kg S ha^{-1} yr^{-1} respectively. Total N deposition was reduced from 5.43 kg N ha^{-1} yr^{-1} to 5.03 kg N ha^{-1} yr^{-1}. For both pollutants, the percentage reduction in deposition achieved was about half that of the percentage reduction in emissions. The importance of sector 2 as a source of UK sulphur pollution for Llyn Brianne was again highlighted, with deposition in this sector reduced by 60% or more in both case studies. Of the two strategies for reducing S emissions by an overall figure of 60%, Case 1 was marginally the more effective (reduction in deposition 36.3%). The practical difficulties of implementing such universal emission control would, however, be very great. The reduction in N deposition (about 7%) was small, but expected in view of the nature of NO_x sources. Other techniques of pollution control are clearly needed in this case.

References

1. Buckley-Golder, D.H. and Derwent, R.G. (1986). Acidity in the Environment Part IV. ETSU-R-29, Harwell.
2. Buijsman, E., Maas, J.F.M. and Asman, W.A.H. (1985). Ammonia emissions in Europe. University of Utrecht, Utrecht. IMOU Report R-85-2.
3. Derwent, R.G., Dollard, G.J. and Metcalfe, S.E. (1987). On the nitrogen budget for the United Kingdom and north west Europe. AERE R 12616, Harwell.
4. Eliassen, A. (1978). The OECD study of long-range transport of air pollutants. Long-range transport modelling. *Atmospheric Environment*, 12: 479–487.
5. Eliassen, A. (1984). Aspects of Lagrangian Air Pollution Modelling. Air Pollution Modelling and Its Application III. Plenum Publishing Corporation, New York.
6. Eliassen, A. and Saltbones, J. (1983). Modelling of long-range transport of sulphur over Europe: a two year model run and some model experiments. *Atmospheric Environment*, 17: 1457–1473.
7. Fowler, D. and Cape, J.N. (1984). On the episodic nature of wet deposited sulphate and acidity. *Atmospheric Environment*, 18: 1859–1866.
8. Jones, J.A. (1981). The estimation of long range dispersion and deposition of continuous releases of radionuclides to atmosphere. National Radiological Protection Board Report NRPB-R-123.
9. PORG (1987). United Kingdom Photochemical Oxidants Review Group. Ozone in the United Kingdom – Interim Report. Department of the Environment and Department of Transport Publications Sales Unit, Ruislip.
10. RGAR (1987). Acid Deposition in the United Kingdom 1981–1985. A Second Report of the UK Review Group on Acid Rain. WSL, Stevenage.
11. Stoner, J.H., Gee, A.S. and Wade, K.R. (1984). The effects of acidification on the ecology of streams in the upper Tywi catchment in West Wales. *Environmental Pollution* (*series A*), 35: 125–157.
12. Whitehead, P., Neale, R., Williams, R., Neal, C., Brown, S., Bird, S. and Littlewood, I. (1986). The Lynn Brianne Acid Waters Study: Data Analysis and Modelling – First Progress Report.

drainage water. The most significant vegetation change likely to take place in upland Wales is from moorland to forest, or vice versa. Atmospheric deposition and evapotranspiration are greater for forests than moorland or grassland canopies. As a result, the fluxes and concentrations of solutes into soils can increase significantly following afforestation (see Chap. 14); the increased flux of mobile anions is thought to be the major factor causing the increase in surface water acidity following afforestation.

It has been suggested that the amount of afforestation should be limited in sensitive catchments because of the impact on surface water acidity.[15] Evidence from studies of soil solution chemistry suggest[5] that the magnitude of the increase in acidity and/or aluminium concentrations may vary between tree species and with soil type. For example, conifers have a greater impact than broadleaves on the same soil type and larch has a greater impact than spruce; the impact of conifers on stagnopodzols is greater than on brown podzolic soils.

On sensitive sites, removal of existing plantations might be seen as a means of ameliorating acidity. Data from Sweden[18] and northern England[1] show a rapid reduction in sodium and chloride concentrations and fluxes in streamwater following felling of experimental catchments. Both concentrations and fluxes of sulphate declined, the response being delayed in the study in northern England until the second full year after felling. These findings suggest that atmospheric inputs decrease once the forest canopy is removed. However, results from north America[20] and the Beddgelert catchments[17] show an increase in acidity and aluminium concentrations following felling. This is linked to increased nitrate release following the cessation of uptake, an effect which persists until groundflora becomes developed, after two to three years. Therefore, large scale felling, specifically to ameliorate acidity of streamwaters would not, necessarily produce benefits in the short term. Consideration should, instead, be given to whether or not given afforested catchments should be replanted following normal harvesting and if so, what proportion of the catchment and upon which soils and with which species.

Modifications to soil chemistry

1. Addition of neutralising materials

Farmers routinely apply lime to soils to overcome natural soil acidity and to create the required soil conditions for the growth of specific crops. In the uplands of Wales, lime has been used for hundreds of years to improve the quality of pastures.

Most upland pasture improvement this century in Wales has involved cultivation, additions of lime and fertilisers and re-seeding with more productive and nutritious forage species. The soil pH must be raised to at least 5.5 for successful growth of the introduced species. This generally requires the addition of between 5 and 10 t ha^{-1} ground limestone. In the 1930's, 40's and 50's lime and basic slag were generally used, the basic slag acting as a source of phosphate but also providing additional neutralising capacity. Since the mid 1960's basic slag has not been available in Britain. Following initial improvement, lime must be added at regular intervals to offset the natural acidifying processes, and any pollutant inputs.

Agricultural liming can have the further effect of improving surface water quality. The studies at the Plynlimon experimental catchments in mid Wales have shown that streams draining improved pasture had higher pH, Ca and Mg levels, and lower Al concentrations than streams draining unimproved moorland.[14] The impact of agricultural liming can also be seen at a regional scale. In Chap. 5 attempts were made to predict the occurrence of acidic surface waters in Wales. On the basis of soils and geology, acid waters would occur on much of the higher ground in south west Wales. Acid waters do not, however, occur over a large proportion of this area because of a long history of soil and pasture improvement.

The amount of lime added to upland soils/pastures in Wales has declined in the last two decades, since Government subsidies on lime were removed (See Chap. 6A).[25] Newbould[23] suggests that the removal of subsidies has led particularly to a decline in the use of lime needed to maintain soil pH levels. This decline in use many have had significant impacts on drainage water acidity in some areas.

Although much is known about addition of lime to improve crop growth, only recently have soils been limed specifically to ameliorate drainage water acidity; established agricultural liming practices may not be the most suitable for this purpose. Four main factors need to be considered when planning catchment liming to influence water quality.

a. type of neutralising material;
b. rate of application;
c. method of application;
d. placement of material.

a. Type of neutralising material

A wide range of materials has been suggested for use in catchment liming: ground limestone ($CaCO_3$ or $CaCO_3.MgCO_3$), chalk, slaked lime ($Ca(OH)_2$), quicklime/sodalime (CaO), calcium and magnesium silicates, sodium bicarbonate ($NaHCO_3$), soda ash (Na_2CO_3), fly ash and industrial slags. To date, almost all catchment liming has used ground limestone, usually calcium carbonate. Bell and Hockin[6] suggest that soda lime (CaO) should not be used in liming organic soils because its 'extreme alkalinity' causes leaching of humic acids and other degradation products. The same authors also suggest that use of slaked lime is inadvisable as a large volume of the material of low bulk density has to be applied, and this is 'unlikely to be visually acceptable'. Basic silicates, ashes and slags have not been widely used because of their relatively low solubility. The rate of solution depends on chemical composition but is also inversely related to grain size.[6, 24, 30] Warfvinge and Sverdrup[33] also suggest that the duration of the effect of soil liming on stream pH is prolonged if very fine material is used, although they stress that the benefits of using a finer grade must be weighed against the higher cost. Agricultural liming in the UK generally uses a material which is '4mm–dust' but in liming to ameliorate drainage water acidity a finer material is used, e.g. '50% <10 micron' mean diameter.

b. Rate of application

In agricultural liming, the application rates are calculated to ensure that soil pH is raised to a minimum of 5.5. The quantity required is determined by the chemical composition of the added material, the cation exchange capacity and base saturation of the soil being treated. Catchment liming studies in Sweden have shown a relationship between rate of application and the duration of the effect on stream pH.[33] They also showed a 'threshold effect', applications below a given level not providing a high enough calcium saturation in the soil significantly to influence stream pH and a 'saturation effect' which suggested that application of very large amounts of lime only gave a marginal increase in the duration of effect. It was concluded that sufficient lime should be added to exceed the exchangeable acidity in the upper 10cm of the limed soil. Ideally, therefore, data are needed on the chemistry of the soils being limed in order to calculate application rates.

c. Method of application

Ground spreading of lime, using wheeled or tracked vehicles is the simplest and cheapest method of application but is difficult in forests, on steep slopes in moorland and in remote areas. In these situations, aerial application has been used in the UK, Scandinavia and North America. Aerial application of the finely ground materials typically used in catchment liming is difficult in all but the slightest of winds. In forests, aerially applied fine material also remains on the foliage and may cause foliar damage. Pelletised finely ground calcium carbonate has been used to overcome these problems in Scotland and Sweden. The pellets are manufactured from ground limestone, starch and a clay filler; they successfully penetrate forest canopies and can be used in moderate winds. An alternative approach, used in a study in the Loch Fleet catchment in south west Scotland, pumped a slurry, made with finely ground calcium carbonate, under the trees of a plantation.[7] The use of slurries prevents aggregation of very fine powders, aggregation increasing the effective size particle of the spread material.

d. Placement within the catchment

Recent work in Scandinavia has shown that liming may have little effect on drainage water quality unless the source areas in the catchment are treated. These are the areas where the water table is at or very close to the surface and which act as direct sources to the stream; they are usually wet flushes, damp areas with wet-loving vegetation or bogs. Agricultural liming often avoids these areas. The source areas will, however, vary in extent over time; during storms, and flood events, the source areas will comprise a larger proportion of the catchment than during low flows. The source areas can be identified by field observation of topography and vegetation, by remote sensing or by using mathematical models, such as TOPMODEL to predict and delimit them. Warfvinge and Sverdrup[33] have shown the importance of liming a large proportion of the source zones in determining not only the stream pH reached in the early years after liming but also in controlling acidity during floods. They suggested that more than 50% of the discharge areas active at high flow should be treated in severely acidified watersheds.

2. Burning

Until the last 20–30 years, burning was important in the management of the semi-natural vegetation of upland Wales. *Molinia* and heather moorland, in particular, were burnt at regular intervals. When vegetation is burnt, nutrients in the vegetation are deposited in the ash formed or are volatilised and lost to the atmosphere.[2] Most of the base cations remain in the ash, as oxides which are gradually converted into soluble salts. A large proportion of these base cations is rapidly leached into the underlying soil and can produce an increase in soil pH.[3, 13]

Increases in soil pH and base cation concentrations have also been reported following forest fires or burning of felling trimmings.[11, 12] Increases in potassium concentrations,[21] alkalinity and concentrations of base cations[31] have been noted after fires but such effects are not universal.[19]

Because of the reported influence of fires on drainage water pH, Brown[10] suggested that the recent acidification of lakes in parts of Canada may be an indirect result of the reduction of the number of forest fires. Bayley and Schindler[5] dispute this suggestion; they found not only increased concentrations of calcium and magnesium in streams following a forest fire but also increased sulphate concentrations. However, the latter resulted from oxidation of organic sulphur compounds during a drought which occurred after the fire.

While there is clearly insufficient data to suggest that burning of vegetation ameliorates drainage water acidity in Wales, the existing information suggests that further research should be undertaken.

3. Drainage

Liming and burning will increase the buffering, or neutralisation capacity of soil and hence may reduce drainage water acidity. Drainage of peats and peaty soils can, however, increase acidity. Wet, anaerobic peatlands can act as sinks for sulphate and nitrogen compounds from the atmosphere or in water moving into and through the peats.[4, 32] Plant uptake is presumably the main mechanism for retention of nitrogen compounds while sulphate is reduced to sulphide and then incorporated into organic molecules.[8, 9] If the water table is lowered in these bogs, by drainage, or ploughing, the peat may become aerobic and cease to function as a sink for incoming sulphate. The sulphate would then be available to 'drive' acidification. In this respect, the very wet peats alongside and at the head of streams, which act as source zones, may play an important role in retention of sulphate and nitrate. It would seem worthwhile leaving such wet peat undrained to exploit its potential as a sink for atmospheric inputs. Drying of peats and organic surface horizons as a result of drainage, or of forest ploughing, can also increase the rate of mineralisation of organic sulphur and nitrogen accumulated in peats in the past. The organic material may then become a net source of sulphate and nitrate which can result in mobilisation of protons and/or aluminium. The sulphur retained in peats as sulphides and organic sulphur, but which was originally derived from atmospheric inputs, has been referred to as an ecological 'time bomb' which can be 'exploited' by lowering the water table.[5] Possible modifications to drainage practice, which would ameliorate the impact of drainage on soil chemistry, are discussed later in this chapter.

Modifications to catchment hydrology

Forestry ploughing and drainage increase the rate of run-off especially of storm waters,[26] and the proportion of water from acid, near surface soil horizons in streams. Together these changes may produce an increase in drainage water acidity.[22] Drainage of peat and peaty surface soils for agricultural purposes will have a similar, but less pronounced effect. The effect is most marked in the early years after ploughing and/or drainage, when the furrows and drains are operating most efficiently. Ploughing and drainage could be discouraged in sensitive catchments or the practices modified.

Possible modifications to ploughing and draining practices are now being explored with the aim of increasing residence time, delaying storm run-off and ensuring penetration of waters to deep soil or drift layers where buffering may take place in some areas but not others. The alternative methods of site preparation being examined include:

i. no ploughing, screef ploughing, turf planting;
ii. contour ploughing;
iii. stopping drains short of permanent watercourses;
iv. deep ripping and excavation of sumps.

Screef ploughing involves suppression of ground flora. In turf planting the young trees are planted on overturned turfs which have been excavated by hand or machine. Neither of these techniques produces the continuous plough furrow which acts as a route for rapid transfer of near-surface water to drains and streams. Screef ploughing and turf planting can, however, only be used on relatively well drained soils.

Contour ploughing uses conventional forestry ploughs but aims to plough close to, but not on the contour. The furrows have a gentle slope; water movement is, therefore slower and the transfer to drains and streams delayed. This approach has disadvantages, however, as the trees may be less stable than with conventional ploughing, drainage being less efficient.

The current Forestry Commission guidelines suggest that drains are stopped short of permanent watercourses. This delays transfer of the drainwater to streams, especially during storms, thus increasing residence time and allowing further buffering. Plough furrows may also be stopped short of main drains. Deep ripping and excavation of sumps aim to ensure penetration of drainage waters to deeper soil horizons which contain carbonates or weatherable silicates or have a higher base saturation.

These alternative approaches to ploughing and drainage could play an important part in reducing the impact of these basic forest management practices on streamwater chemistry and are becoming part of normal planning of forestry schemes. Whilst modification to drainage in existing forests is difficult, there may be scope, particularly following harvesting, for blocking drains which currently run directly into streams and for excavation of sumps on existing drains.

Current experiments in Wales

To assess the value of different methods of catchment liming in maintaining or restoring stream quality, two treatments involving standard agricultural improvement and surface dressing of fine limestone within the catchments of Group IIIA streams have been completed. The effectiveness of limited application of fine-limestone to water source areas within a similar moorland catchment is also being evaluated.

Current practice at the ground preparation stage of afforestation schemes is for minimum ploughing following the contours of slopes and with drainage ditches cut short of watercourses, in recognition of the problems caused by the former practice of ploughing and drainage following the fall line of slopes. It has been demonstrated that this former technique leads to reduced stream pH, together with elevated levels of sulphate, nitrate, ammonia and aluminium.[34] Experimental contour ploughing of an area within a Group IIIA moorland has therefore been undertaken to provide information on the value of these modified practices, and to separate the effects of ground preparation from those brought about by the trees.

Standard agricultural land improvement

An area of 20 hectares (equivalent to 24% of the catchment area) within the

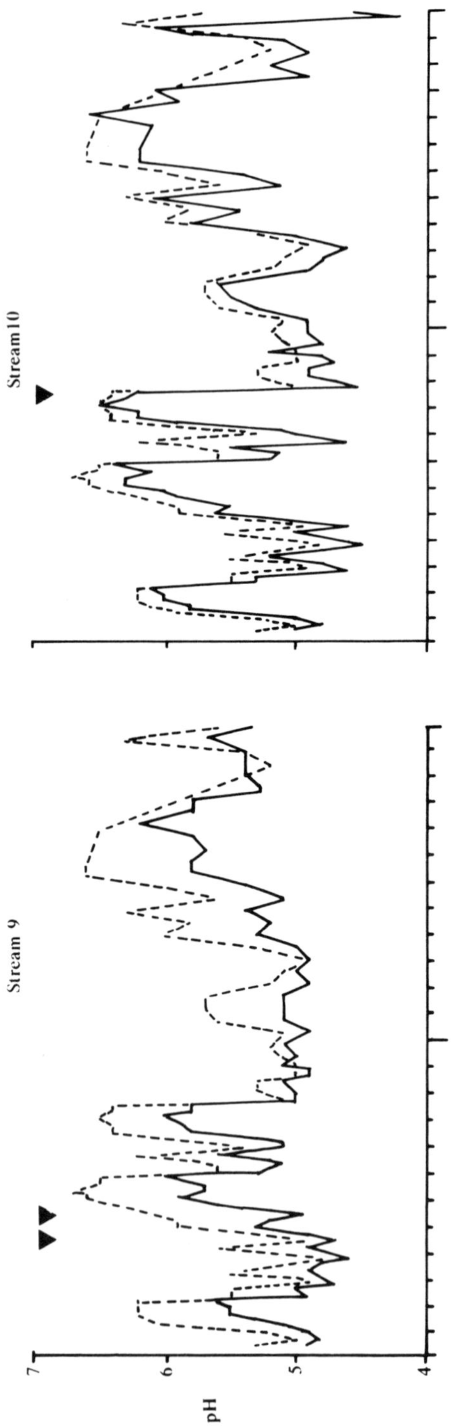
Stream 9
Stream10
pH
7
6
5
4

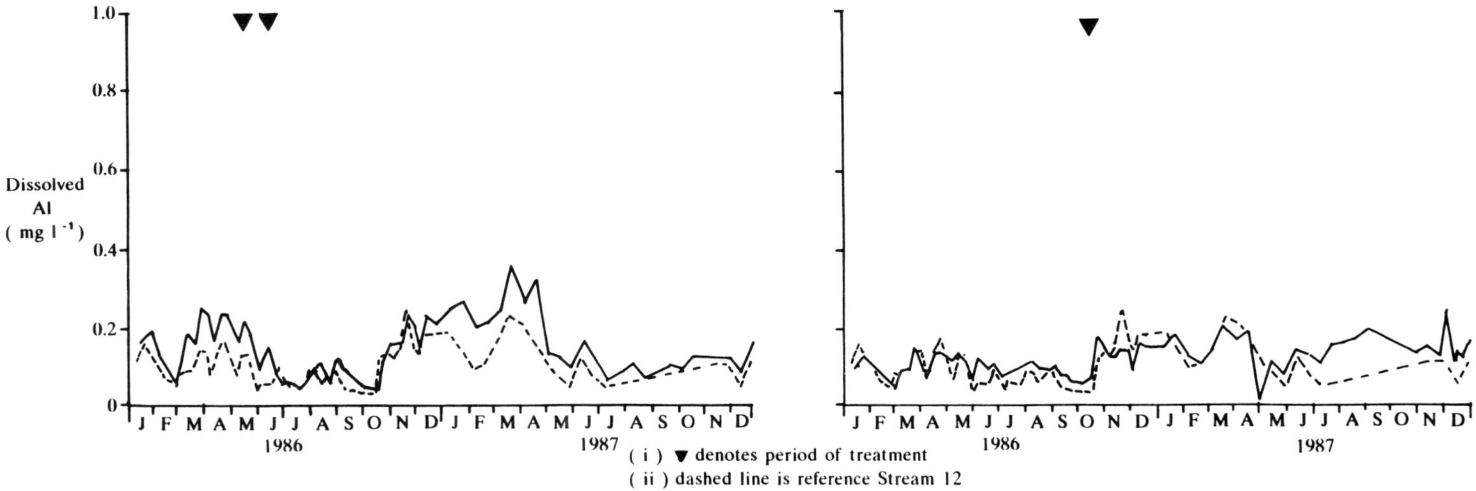

Fig. 19.1. pH and aluminium in streams 9 and 10 (1986–7) in relation to land treatment.

catchment of Stream 9 (Group IIIA) was treated following current guidelines for pasture improvement in upland Wales. The area was ploughed, limed with agricultural grade (4mm to dust) magnesium limestone (10 t ha^{-1}) and reseeded following application of NPK fertiliser during May and June 1986. Despite the addition of 200 tonnes of limestone to the catchment, the stream remains acidic and episodes with pH<5 accompanied by dissolved aluminium of up to 0.35 mgl^{-1} have continued to occur since completion of the treatment (Fig. 19.1). Soil waters have, however, responded to the treatment with an increase in pH and calcium concentrations. It appears that this area does not play an important role in controlling catchment hydrology and stream chemistry. The scheme has therefore been unsuccessful in achieving the required improvements in stream quality.

Surface liming

The complete catchment (34 ha) of Stream 11 (Group IIIA) was treated with finely powdered calcium carbonate limestone (50% < 10μm, 98% (<50 μm) applied at a rate of approximately 9 t ha^{-1}. The limestone was spread throughout September 1987, following which a dramatic improvement in stream pH (mean pH increased from 5.2 to 6.4) and reduction in dissolved aluminium (mean reduced from 0.16 mgl^{-1} to 0.07 mgl^{-1}) has resulted (Fig. 19.2). The improvement has been sustained, despite numerous storm events during the following winter which would otherwise have resulted in concentrations of dissolved aluminium as high as 0.5 mgl^{-1} together with pH <5. To date, however, there has been little impact on the chemistry of waters from stagnopodzols and brown podzolic soils in the catchment. The cultivation in the land improvement site resulted in rapid incorporation of the added lime while in this case the lime still remains at the surface. The effect on the drainage waters probably results from effects in the wet soils of the source areas.

The use of a fine grade of limestone has increased the availability of this material and, whilst the results are encouraging, it is too early to predict the longer term effects on the stream. However, on the basis of the currently observed calcium loss, the improvements in stream quality may be sustained for a number of years to come.

Forestry ploughing

Contour ploughing and drainage of an area of peat moorland (8 ha) within the catchment of stream 10 (Group IIIA) was completed in October 1986. This treatment has failed to have a clearly demonstrable effect on the stream, particularly when compared to reference Stream 12. No change in the pH variation in the stream has been detected and, whilst there is a slightly larger increase (0.1 mgl^{-1}) in dissolved aluminium in Stream 10 compared to Stream 12 in August and September 1987, a further period of monitoring will be

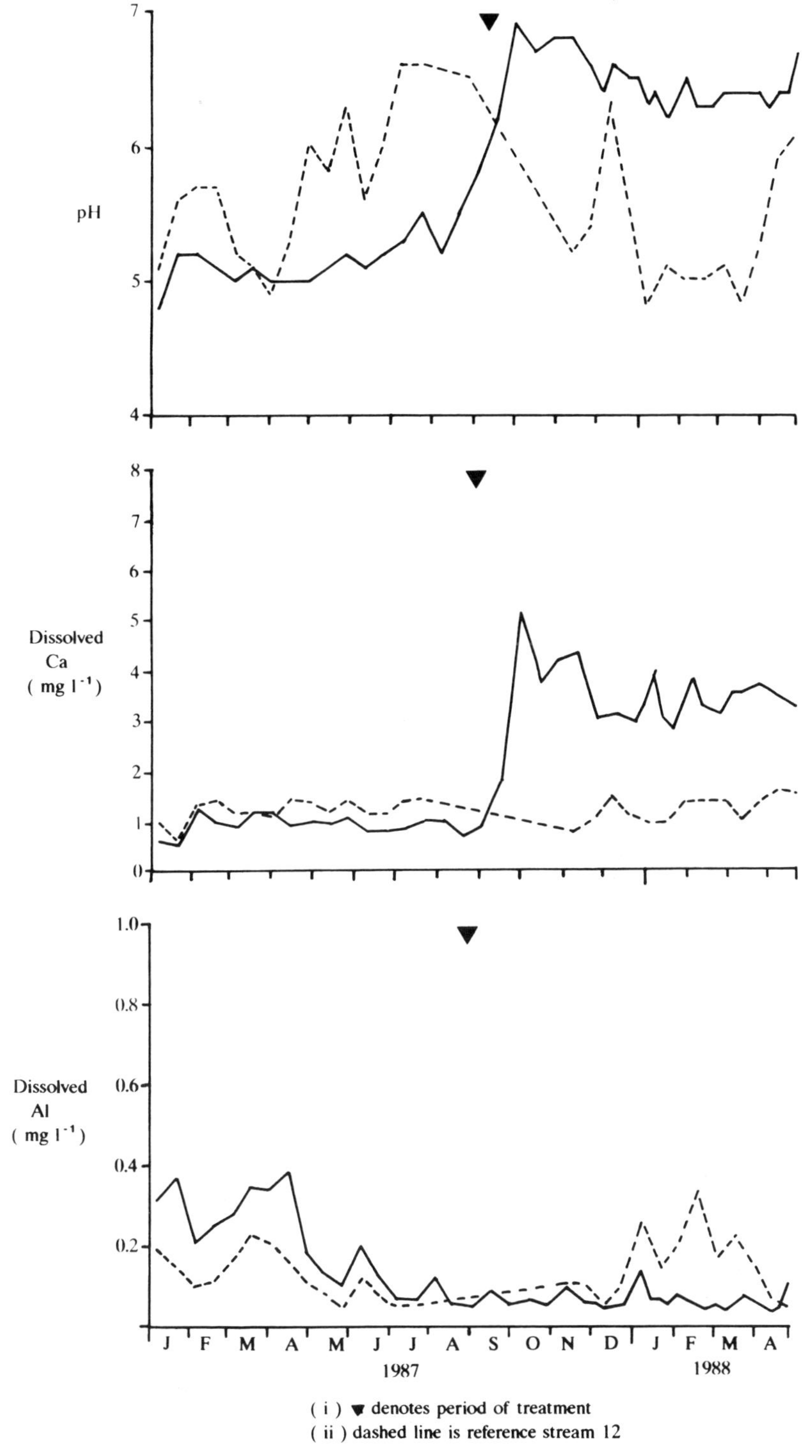

Fig. 19.2. pH, dissolved calcium and dissolved aluminium in stream 11 (1987–8) in relation to land treatment.

necessary to determine if this variation is attributable to the ploughing and drainage (Fig. 19.1). However, as the treated area comprises only 16% of the catchment any effects may be masked by drainage from other areas. Monitoring of new forestry schemes elsewhere in Wales should provide further information on the effects of ploughing greater proportions of catchment areas.

Conifer forest treatments

In 1982, an attempt was made to reduce the effects of acidic drainage from coniferous forest as part of studies by WWA in co-operation with the Forestry Commission. New coniferous planting schemes typically have provisions for unplanted and unploughed areas along stream banksides. It was found that felling of a 5 metre strip of trees along each bank of Stream 3 (Group IIB) to create a buffer zone was ineffective in reducing stream acidity.[29] A treatment liming of the cleared bankside zone was therefore formulated to assess if this modified technique may be appropriate for inclusion in future schemes.

Similarly, another treatment has involved intensive liming of an unplanted area (including the source area of the stream) in the headwaters of an afforested catchment to determine if limited liming of this type may be used to improve stream quality.

A further treatment to be undertaken involves the use of a pelletised limestone to assist in accurate targetting of the limestone to areas within the catchment where this will be most effective. In the case of all three forest treatments, the difficulties in spreading lime in areas of existing forest and the need to avoid possible adverse effects on the trees have been important considerations in devising possible management techniques.

Bankside clearance and liming

The banksides of Stream 2 (Group IIIB) were cleared to a minimum total width of 15m on the main stream and 10m on each of the principal tributaries during summer 1986. Magnesian limestone (60 tonnes; particle size 4mm to dust) was spread by helicopter throughout the cleared area (10 t ha^{-1}) during September 1986.

Prior to the liming treatment, Stream 2 was the most acidic of the Llyn Brianne study catchments with pH 0.2–0.3 units lower than reference Stream 1 (also Group IIIB), probably resulting from the slightly lower mean hardness and higher sulphate than in the reference stream, having high dissolved aluminium levels (mean 0.53 mgl^{-1}).

The initial response of the stream to the liming was rapid, and whilst, prior to treatment, measured pH had not exceeded 4.5 this rose to 6.0 during liming and remained at that level for two weeks, during which time dissolved aluminium levels decreased to less than 0.1 mgl^{-1} (Fig. 19.3). The treatment coincided with a period of low rainfall: a coincidental but lower increase in

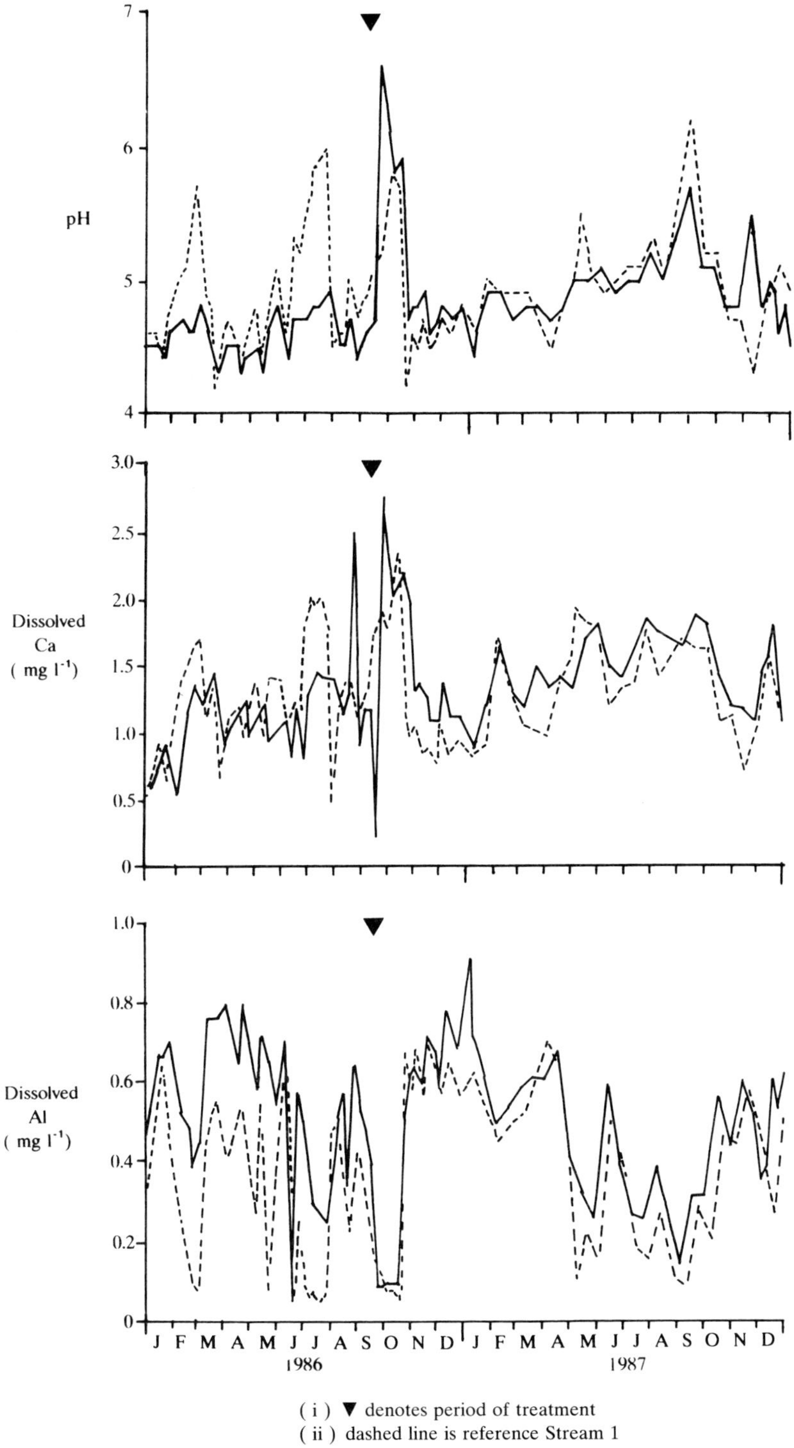

Fig. 19.3. pH, dissolved calcium and dissolved aluminium in stream 2 (1986–7) in relation to land treatment.

pH, together with a reduction in aluminium, was also observed in reference Stream 1. This factor alone cannot account for the magnitude of the changes in Stream 2 and it is clear that the treatment had a significant short term effect. This was almost certainly due to solution of limestone which fell into the stream and ditch channels. However, following a subsequent series of storm events during October 1986, pH and dissolved aluminium levels in Stream 2 returned to close to those recorded before treatment.

Despite the short duration of effect of the treatment, some longer term, marginal effects may still be discerned. Before liming, the levels of dissolved calcium in the reference Stream 1 were generally marginally higher than in Stream 2, although this situation has reversed following the treatment. Similarly, there have been small increases in pH and aluminium compared to the reference stream (Fig. 19.3).

Nevertheless, Stream 2 remains very acidic and, under high flows, pH and aluminium levels continue to reach the extremes (pH 4.0–4.5, dissolved aluminium up to 0.8 mgl^{-1}) that were observed under similar conditions before treatment. As with the moorland agricultural improvement, the failure of this technique appears to result from the limed area playing an insignificant role in the hydrology of the catchment, much of the run-off reaching the stream through well established drainage channels which by-pass the riparian zone within which the limestone was applied.

Intensive liming on an unplanted area within a forest

An uplanted area of predominantly peat (4 ha), including the stream source, within the upper catchment of Stream 4 (Group IIB) was treated between September and December 1987 using 120 tonnes of fine powdered limestone (50% <10 μm, 98% <50 μm), equivalent to a rate of 30 t ha^{-1}.

Prior to treatment, mean pH (5.0) and dissolved calcium (1.9 mgl^{-1}) in Stream 4 were similar to those of Stream 3 (mean pH 5.3, mean dissolved calcium 1.9 mgl^{-1}). Mean dissolved aluminium in Stream 4 (0.18 mgl^{-1}) was however lower than in Stream 3 (0.26 mgl^{-1}).

The liming was completed in two stages, 100 tonnes over the complete treatment area during early October and a further 20 tonnes into a ditch forming the source of the stream in December 1987. The liming in October led to a marginal increase in dissolved calcium in Stream 4 during October and November whilst levels in reference Stream 3 were declining (Fig. 19.4). Similarly, pH levels in Stream 4 remained around 5.3 in samples taken during this period whilst in Stream 3 a seasonal decline in pH occurred. On completion of the liming, a dramatic increase in dissolved calcium (to around 6.0 mgl^{-1}) and pH (to around 6.5) occurred in Stream 4 and further more gradual increase continued throughout the winter. By April 1988 dissolved calcium was variable between 6.0 and 7.5 mgl^{-1} and pH between 6.8 and 7.0. However, during dry weather in Spring 1988, the pH of the stream gradually declined to about 5.5, rising to around 7.0 again following heavy rain. This

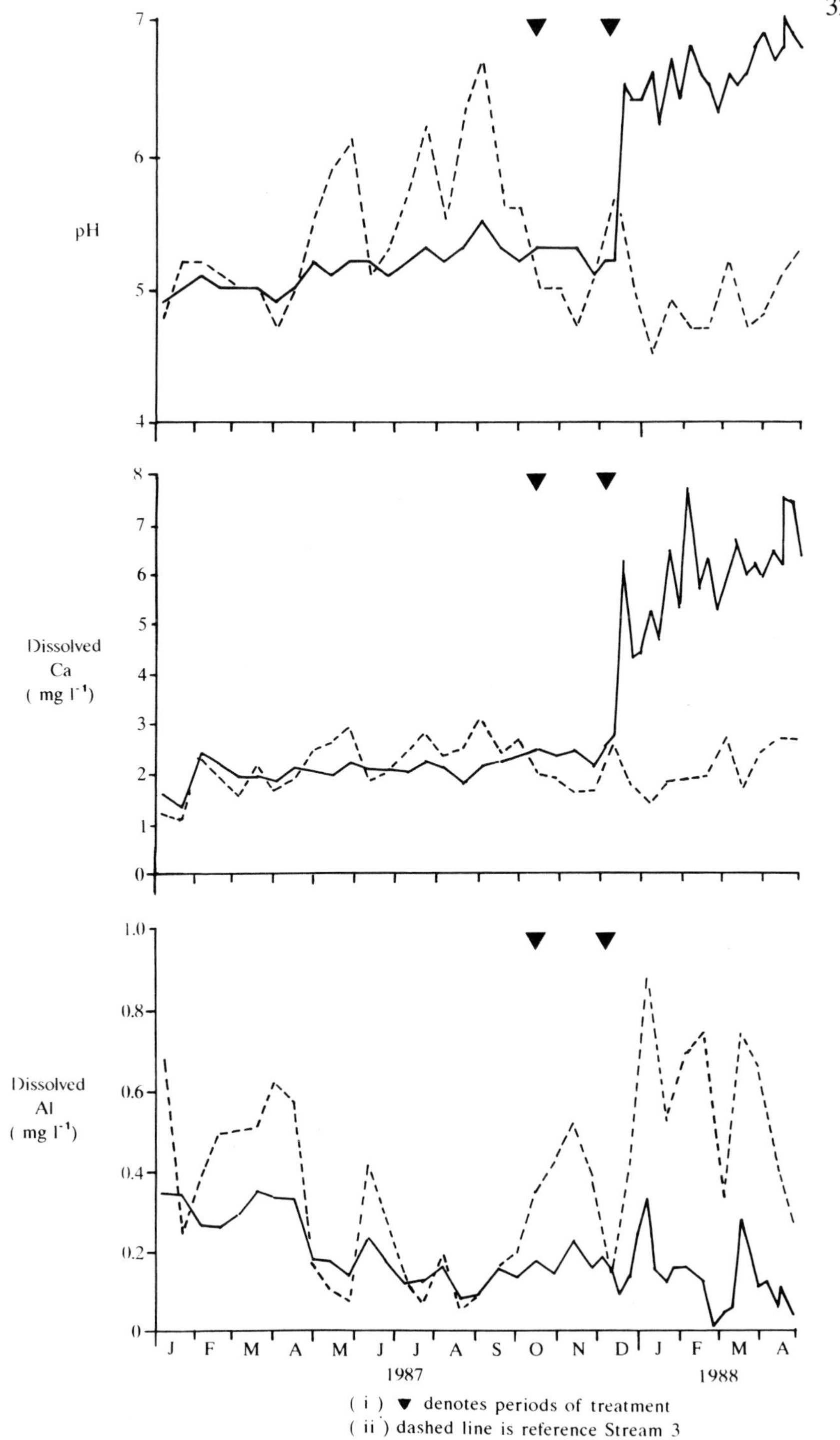

Fig. 19.4. pH, dissolved calcium and dissolved aluminium in stream 4 (1987–8) in relation to land treatment.

suggests that the dominant input to the stream in the dry period was either acid groundwater or water from deeper layers of the source which are unaffected by the surface liming.

Whilst some reduction in dissolved aluminium has also occurred in Stream 4, and the winter peaks during early 1988 were less than would normally be expected in relation to the corresponding period in 1987 and to Stream 3, peaks of up to 0.3 mgl^{-1} (40% labile) still occur, despite the increased pH. Further monitoring and investigations are therefore required to assess the ecological significance of these aluminium levels and to allow more accurate prediction of the longer term effectiveness of this treatment method.

Conclusions

It is clear that liming of stream catchments can lead to a reduction in surface water acidity. Finely ground materials should be used to obtain a rapid response; calcium carbonate limestone is preferable to magnesium limestone because of its higher solubility at normal surface temperatures. Surface liming produces an effective method of application. Treatment of a whole catchment, by surface liming, produces a reduction in acidity but the most cost-effective treatment might well be liming of source areas, which may form only a small part of the catchment, at high rates, e.g. 25–30 t ha^{-1}.

Warfvinge and Sverdrup[33] suggest that >50% of the source areas operative at high flows should be limed in sensitive catchments and that sufficient lime should be added to neutralise the exchange acidity in the top 20 cm. It must be stressed, however, that liming of acid soils and acidophilous vegetation may drastically modify the soils and vegetation. The changes in soils and vegetation which are produced may prove contentious on conservation grounds.

Experiments in Scotland and Scandinavia have shown that liming of forested catchments can be achieved using pelletised material or a slurry of limestone. The pelletised material is currently too expensive to be considered a realistic method for treatment in the UK. Liming of forests can cause a reduction in tree growth, and this could discourage the liming of existing forests. Treatment of small source areas, shown to be effective in reducing stream acidity, would have an insignificant impact on forest production.

Small areas of forests on source areas could be felled to facilitate liming by wheeled or tracked vehicles. Large-scale felling, to reduce stream acidity is not practical and may, on sensitive sites, cause an initial increase in acidity and/or aluminium concentrations. Decisions on the advisability of afforestation on sensitive sites should be made before planting, or after normal harvesting.

Modified ploughing and drainage techniques, and modified design of drainage schemes in forestry plantations, reduce the impact of site preparation on surface water acidity. Ripping and digging of sumps to ensure deep penetration, and increased buffering of drainage waters are unlikely to be effective at most upland Welsh sites.

References

1. Adamson, J.K., Hornung, M., Pyatt, D.G. and Anderson, A.R. (1987). Changes in solute chemistry of drainage waters following the clearfelling of a Sitka spruce plantation. *Forestry*, 60: 165–177.
2. Ahlgren, I.F. and Ahlgren, C.E. (1960). Ecological effects of forest fires. *Bot. Review*, 26: 483–533.
3. Allen, S.E. (1964). Chemical aspects of heather burning. *J. Appl. Ecol.*, 1: 347–367.
4. Bayley, S.E., Behr, R.S. and Kelly, C.A. (in press). Retention and release of sulphur from a freshwater wetland. *Water Air & Soil Pollution*.
5. Bayley, S.E. and Schindler, D.W. (1987). Sources of alkalinity in Precambrian shield watersheds under natural conditions and after fire or acidification. In: Effects of Atmospheric Pollutants on Forests, Wetlands and Agricultural Ecosystems, edited by T.C. Hutchinson & K.M. Meema. Berlin: Springer-Verlag.
6. Bell, R.M. and Hockin, D.C. (1986). The Loch Fleet Study: Land Intervention Options. In: The Loch Fleet Project. A Report of the Pre-Intervention Phase (1) 1984–1986, edited by G. Howells. Leatherhead: CERL.
7. Brown, D.J.A., Howells, G.D. and Paterson, K. (1987). The Loch Fleet Project. In: Acidification and Water Pathways. Oslo, Norway: Norwegian National Committee for Hydrology.
8. Brown, K.A. (1985). Sulphur distribution and metabolism in waterlogged peat. *Soil Biol. Biochem*, 17: 39–45.
9. Brown, K.A. (1986). Formation of organic sulphur in anaerobic peat. *Soil Biol. Biochem.*, 18: 131–140.
10. Brown, W. (1984). Maybe acid rain isn't the villain. *Fortune*, 109: 170–174.
11. Cleve, K. Van and Dyrness, C.T. (1983). Effects of forest-floor disturbance on soil-solution nutrient composition in a black spruce ecosystem. *Can. J. For. Res.*, 13: 894–902.
12. Dyck, W.J., Webber, B.D. and Barton, P.G. (1981). Soil water nutrient concentrations after clearfelling and burning of *Pinus radiata. New Zealand Journal of Forestry*, 11: 128–144.
13. Grier, C.C. (1975). Wildlife effects on nutrient distribution and leaching in a coniferous ecosystem. *Can. J. For. Res.*, 5: 599–607.
14. Hornung, M., Reynolds, B. and Hatton, A.A. (1985). Land management, geological and soil effects on streamwater chemistry in upland mid-Wales. *App. Geog.*, 5: 71–80.
15. Hornung, M. (1987). The effects of land management on acidification of aquatic ecosystems and the implications for the development of ameliorative measures. CEC Air Pollution and Ecosystems, 452. Proceedings of an International Symposium held in Greenoble, France, 18–22 May 1987.
16. Hornung, M., Reynolds, B., Stevens, P.A. and Neal, C. (1987). Increased acidity and aluminium concentrations in streams following afforestation: causative mechanisms and processes. In: Acidification and water pathways. Oslo, Norway: Norwegian National Committee for Hydrology.
17. Hornung, M., Adamson, J.K., Reynolds, B and Stevens, P.A. (In press). Impacts of forest management practices in plantation forests. In: Proceedings of CEC Cost 612 workshop, UWIST, Cardiff.
18. Hultberg, J. (1985). Budgets of base cations, chloride, nitrogen and sulphur in the acid Lake Gardsjon catchment in SW Sweden. *Ecol. bull.* (Stockholm), 37: 133–157.
19. Johnson, C.M. and Neeham, P.R. (1966). Ionic composition of Sagchen Creek, California following and adjacent fire. *Ecology*, 47: 636–639.
20. Lawrence, G.B., Fuller, R.D. and Driscoll, C.T. (1987). Release of aluminium following

whole-tree harvesting at the Hubbard Brook Experimental Forest, New Hampshire. *J. Environmental Quality*, 16: 383–390.
21. Lotspeich, F.B., Meuller, E.W. and Frey, P.J. (1970). Effects of large scale forest fires on water quality in interior Alaska. *US Dept. Inter. Fed. Water Poll. Control Adm.* Alaska Water Laboratory, College, Alaska.
22. Miller., H.G. (1985). The possible role of forests in streamwater acidification. *Soil Use & Management*, 1: 28–29.
23. Newbould, P. (1985). Improvement of native grassland in the uplands. *Soil Use & Management*, 1: 43–49.
24. Nye, P.H. and Ameloko, A.Y. (1987). Predicting the rate of discolouration of lime in soil. *Journal of Soil Science*, 38: 641–650.
25. Ormerod, S.J. and Edwards, R.W. (1985). Stream acidity in some areas of Wales in relation to historical trends in afforestation and the usage of agricultural limestone. *Journal of Environmental Management*, 20: 189–197.
25. Reynolds, B., Neal, C., Hornung, M. and Stevens, P.A. (1986). Baseflow buffering of streamwater acidity in five mid-Wales catchments. *J. Hydrol.*, 87: 167–185.
26. Robinson, M. and Newson, M.D. (1986). Comparison of forest and moor-land hydrology in an upland area with peat soils, *International Peat Journal*, 1: 49–68.
27. Smith, R.T. and Taylor, J.A. (1969). The post-glacial development of soils and vegetation in northern Cardiganshire. *Trans. Inst. Brit. Georgs.*, 48: 75–96.
28. Stoner, J.H., Gee, A.S. and Wade, K.R. (1984). The effects of acidification on the ecology of streams in the upper Tywi catchment in West Wales. *Environmental Pollution (series A)*, 35: 125–127.
29. Stoner, J.H. and Gee, A.S. (1985). The effects of forestry on water quality and fish in Welsh rivers and lakes. *J. Inst. Water Engineers & Scientists*, 39: 27–454.
30. Sverdrup, H. and Bjerle, I. (1982). Dissolution of calcite and other related minerals in acid aqueous solution in a pH-stat. *Vatten* 39: 59–73.
31. Tiedemann, A.R., Helvey, J.D. and Anderson, T.D. (1978). Stream chemistry and watershed nutrient economy following wildfire and fertilisation in eastern Washington. *J. Environ. Qual.*, 7: 580–588.
32. Urban, N.R. and Bayley, S.E. (1986). The acid-base balance of peatlands: a short-term perspective. *Water, Air & Soil Pollution*, 30: 791–800.
33. Warfvinge, P. and Sverdrup, H. (1988). Soil liming as a measure to mitigate acid run-off. Water Resources Research.
34. Welsh Water (1987). Acidification of run off following forestry drainage of a peaty podzol soil. Welsh Water Scientific Services, South Western District Report No. SW/87/1.

Conversion table (equivalents and weights)

The relationship between the units of concentration of the principal ions quoted in this book.

H^+	1 mg/l	=	992.10 μeq/l
Na^+	1 mg/l		43.50 μeq/l
K^+	1 mg/l		25.57 μeq/l
Ca^{++}	1 mg/l		49.90 μeq/l
Mg^{++}	1 mg/l		82.26 μeq/l
Al^{+++}	1 mg/l		111.2 μeq/l
Zn^{++}	1 mg/l		30.60 μeq/l
Mn^{++}	1 mg/l		72.81 μeq/l
Cu^{++}	1 mg/l		31.48 μeq/l
Pb^{++}	1 mg/l		9.65 μeq/l
NH_4^+	1 mg/l		55.44 μeq/l
SO_4^{--}	1 mg/l		20.83 μeq/l
NO_3^-	1 mg/l		16.13 μeq/l
CL^-	1 mg/l		28.21 μeq/l
F^-	1 mg/l		52.64 μeq/l
Si (as SiO_3^{--})	1 mg/l		26.29 μeq/l
1 mg Al^{+++}	= 37.06 μMOL		
1 mg Fe^{+++}	= 17.9 μMOL		

R.W. Edwards et al. (eds) Acid Waters in Wales. 329.

List of authors' addresses

P.G. Appleby, Department of Applied Mathematics and Theoretical Physics, University of Liverpool, P.O. Box 147, Liverpool L69 3BX

R.W. Battarbee, Palaeoecology Research Unit, Department of Geography, University College, 26 Bedford Way, London. WC1H OAP

S.C. Bird, NRA Welsh Region, St. Mellons Business Park, St. Mellons, Cardiff. CF3 OEG

R. Boon, Department of Geography, St. David's University College, Lampeter, Dyfed. SA48 7ED

N. Brown, ITE, Penrhos Road, Bangor, Gwynedd. LL57 2LQ

S.J. Brown, NRA Welsh Region, Highfield, Caernarfon, Gwynedd. LL55 1HR

C.B. Buckley, C/o Welsh Water PLC, Brecon, Powys. LD3 7HP

B.J. Cosby, Department of Environmental Science, University of Virginia, Charlottesville, Virginia, U.S.A.

R.G. Derwent, Modelling and Assessment Group, Environmental and Medical Sciences Division, Harwell Laboratory, Oxon. OX11 ORA

A.P. Donald, NRA Welsh Region, St. Mellons Business Park, St. Mellons, Cardiff. CF3 OEG

R.W. Edwards, School of Pure and Applied Biology, UWCC, P.O. Box 13, Cardiff. CF1 3XF

S.C. Fritz, Limnological Research Center, University of Minnesota 310 Pillsbury Drive S.E., Minneapolis, MN 55455, U.S.A.

A.S. Gee, NRA Welsh Region, St. Mellons Business Park, St. Mellons Cardiff CF3 OEG

X. Goenaga, Chemical Engineering Department, University College, Swansea SA2 8PP

M. Hornung, ITE Merlewood Research Station, Grange-over-Sands, Cumbria LA11 6JU

S. Hughes, ITE, Penrhos Road, Bangor, Gwynedd LL57 2LQ

D. Kay, Department of Geography, St. David's University College, Lampeter, Dyfed SA48 7ED

L. Keil, Welsh Water PLC, The Kingsway, Swansea

R.W. Edwards et al. (eds) Acid Waters in Wales. 331–332.

A.M. Kreiser, Palaeoecology Research Unit, Department of Geography, University College, 26 Bedford Way, London WC1H OAP
S. Le Grice, ADAS, Block C, Government Building, Brooklands Avenue, Cambridge CB3 2DR
I.G. Littlewood, IH, Wallingford, Oxon OX10 8BB
C.P. McCahon, School of Pure and Applied Biology, UWCC, P.O. Box 13, Cardiff CF1 3XF
S.E. Metcalfe, Department of Environmental Science, University of Stirling Stirling FK9 4LA
N. Milner, NRA Welsh Region, Highfield, Caernarfon, Gwynedd LL55 1HR
T.J. Musgrove, IH, Wallingford, Oxon OX10 8BB
J. Natkanski, Palaeoecology Research Unit, Department of Geography, University College, 26 Bedford Way, London WC1H OAP
C. Neal, IH, Wallingford, Oxon OX10 8BB
D. Norris, ITE, Penrhos Road, Bangor, Gwynedd LL57 2LQ
F. Oldfield, Department of Geography, University of Liverpool, Roxby Building, P.O. Box 147, Liverpool L69 3BX
S.J. Ormerod, Acid Waters Unit, School of Pure and Applied Biology, UWCC, c/o NRA Welsh Region, Penyfai House, Llanelli, Dyfed SA15
D. Pascoe, School of Pure and Applied Biology, UWCC, P.O. Box 13, Cardiff CF1 3XF
S.T. Patrick, Palaoecology Research Unit, Department of Geography University College, 26 Bedford Way, London WC1H OAP
A. Ranson, NRA Welsh Region, Penyfai House, Llanelli, Dyfed SA15 4EL
B. Reynolds, ITE, Penrhos Road, Bangor, Gwynedd LL57 2LQ
B. Rippey, Freshwater Laboratory, University of Ulster, Drumenagh, Magerafelt, Northern Ireland
P.A. Stevens, ITE, Penrhos Road, Bangor, Gwynedd LL57 2LQ
A.C. Stevenson, Department of Geography, University of Newcastle, Newcastle-upon-Tyne NE1 7RU
J.H. Stoner, NRA Welsh Region, St. Mellons Business Park, St. Mellons, Cardiff CF3 OEG
P.V. Varallo, NRA Welsh Region, St. Mellons Business Park, St. Mellons, Cardiff CF3 OEG
E. Vaughan, IH, Wallingford, Oxon OX10 8BB
K.R. Wade, Welsh Water PLC, Tremains House, Bridgend, Mid Glamorgan
R.P.D. Walsh, Department of Geography, University College Swansea SA2 8PP
N.S. Weatherley, Acid Waters Unit, School of Pure and Applied Biology, UWCC, c/o NRA Welsh Region, Penyfai House, Llanelli, Dyfed SA15 4EL
P.G. Whitehead, IH, Wallingford, Oxon OX10 8BB
D.J.A. Williams, Chemical Engineering Department, University College Swansea SA2 8PP

Index

Acid depositon (*see also* Atmospheric deposition) 1, 4, 11, 12, 13, 16, 22, 27, 31, 33, 36, 43, 45, 46, 51, 56, 59, 67, 75, 81, 93, 135, 173, 216, 269, 274, 292, 299, 303
Acidification 1, 2, 4, 11, 14–21, 27, 33, 36, 52, 76, 81–89, 96, 99, 110, 115, 121, 125, 140, 145, 154, 156, 174, 238, 255, 260, 269, 274, 276, 280
  artificial (experimental) 95, 104, 409, **160**, 295
  natural 27, 82
  susceptibility to 33, 56, 57, 67, 74, 76, 78, 81, 82, 110, 114, 121, 130, 135, 241, 269, 276
Adirondack Mountains 281
Adsorption 59, 95, 230
Aerosols 40, 223, 224, 263, 303
Afforestation – *see* Forestry
Agricultural improvement 2, 8, 67,68, 73, 75, 89, 239, 259, 312, 317–320
Agricultural policy 68–70, 73, 77, 313
Algae (*see also* Diatoms) 93, 95, **96**, 97–99, 100, 101, 103, 109
Alkalinity 11, 16, 93, 104, 146, 147, 157, 267, 268, 271, **274**, 280, 315
Aluminium (general references numerous and not included)
  adsorbed 174, 175, 189, 190, 196
  colloidal 191, 196
  complexation with Fluoride 174, 197
  in drinking water 149–150
  inorganic 173 *et seq*
  labile 130, 150, 161, 168, 197, 282, 326
  minerals 55, 174, 182, 183, 190, 293
  mobilization 228, 230, 235, 253, 280, 293
  monomeric 198
  organically complexed 160, 161, 166, 189, 196, 197
  response to liming 320–326
  smelting 42
  speciation 130, 161, 189
  toxicity to fish 128, 132, 140, 165, 168, 173, 266, 282
Ammonium 42, 43, 47, 48, 49, 52, 75, 99, 224, 227, 251, 264, 266, 268, 303, 305, 307
Amphibians 1, 105, **111**
Anglesey 42
Arkosic rocks 56, 64
Arctic Char 28, 121
Atmospheric deposition (*see also* Acid deposition) **39**, 40, 47, 52, 81, 89, 269, 280, 312
  dry 40, 208, 223, 263, 267, 280, 294, 303, 304, 307
  effects of reducing 266–268, 274, 276, 287, 289–291, 296–308
  occult 40, 43, 208, 223, 224, 263, 267, 280, 294
  wet 40, 208, 303, 304

Back-trajectories (of air masses) 45, 299
Baetis rhodani 17, 107, 109, 160, 163, 166
Bacteria 103, 104, 105, 109
Basic slag 313
Bankside clearance 322
Base saturation 59, 60, 229, 317
Beddgelert **2**, 77, 173, 175, 176, 178, 223, 224, 228, 312
Bicarbonate 63, 175, 177, 182, 233, 239, 280
Birds 1, 18, 93, 105, **112–113**, 115
  Dipper 18, 112–113
  Grey wagtail 113
BIRKENES model 256
Bracken 227
Brook Char 20, 110, 160
Brown Trout 19, 28, 111, 121, 124, 125, 135, 137, 139, 160, 161, 163, 165
Bryophytes 95, 99
Buffering capacity 1, 36, 55, 56, **57**, 59, 60, 62, 75, 121, 246, 251, 253, 269, 276, 311
Bulk deposition 3, 40, 224

Bullheads 111, 161
Burning of moorland 27, 28, 32, 81, 82, 85, **315**

Cadmium 50, 148, 152, 153
Caesium-137 28
Calcite 3, 55, 63, 238, 268
Calcium 1, 4, 16, 31, 41, 42, 43, 52, 55, 58, 74, 93, 95, 96, 98, 99, 101, 107, 110, 113, 122, 123, 125, 130, 140, 160, 168, 193, 229, 233, 234, 238, 239, 242, 243, 246, 249, 251, 267, 271, 285, 290, 295, 312, 313, 320, 324
Calcium carbonate 99, 134, 314, 320, 326
Calluna 32, 82, 84
Cambrian sediments 28, 269
Camddwr 4, 114
Canopy leaching 227
Carbon
  inorganic 94, 232
  organic 95, 104, 154, 232
Carbonates 55, 59, 177, 229, 238, 239
Carbon dioxide 175, 182, 232, 239, 261, 280, 294
Carboniferous Limestone 56
Cation exchange capacity 59, 261
Chloride 31, 41, 42, 122, 123, 168, 224, 228, 229, 232, 233, 242, 280, 312
Chlorine 148, 153, 154
Chlorophyl 'a' 97, 99, 104, 105
Circumneutral lakes 110
Circumneutral streams 93, 96, 99, 104, 105, 106, 107, 109, 113, 114, 115, 160, 283
Colour 146, 147
Conductivity 31, 41, 56, 63, 126, 146, 218, 251
Conifers – *see* Forestry
Conwy 14, 137, 140
Copper 31, 133, 148, 151, 152, 193
Cothi 135
'Cusum' 14

Daphnia 169
Deciduous woodland 8, 84, 98, 104, 107, 111, 224, 249, 283
DECORANA 99, 110, 284
Denitrification 104
Deposition – *see* Atmospheric deposition
Devonian sediments 56
Diatoms 11, 16, 22, 27, 33, 40, 84, 89, 96, 97, 98, 99, 100, 109, 294
  acidobiontic taxa 31, 33, 34, 35, 96
  acidophilous taxa 29, 33, 34, 35
  alkaliphilous taxa 34, 35
  circumneutral taxa 29, 31, 33, 34, 35
Documentary sources 81, 84, 85
Dolerite 2, 63, 75
Drainage 27, 36, 81, 85, 89, 259, 266, 315, **316**, 320
Drinking water – *see* Potable supplies
Driscoll technique 176
Dyfi 135
Dysyni 137

East Midlands 39, 45
EC Directives 68, 76, 145, 146, 150, 154–157, 308
Eels 110
Elan 114
Emissions 1, 40, 42, 47, 261, 300, 304
  relation to deposition 299, 301
Episodes of acidity 6, 45, 52, 95, 111, 128, 145, 150, 159, 169, 170, 205, 218, 251, 295, 320
Evapotranspiration 213, 264, 280, 294, 311

Feldspar 239
Fertiliser 74, 75, 233, 320
Fish
  alevins 124
  catch statistics 18, 19, 23, 121, 131, 138
  density 98, 114, 124, 127, 133, 134, 135, 137, 280, 289, 291
  eggs 124, 140
  fry 124, 140
  gill mucus 162, 166, 168
  ion regulation 122, 123, 165, 168
  mortalities 122, 124, 128, 140
  populations 18, 19, 128, 135, **136**, 280
  spawning habitat 22, 121
  ventilation frequency 163, 164
Fisheries 1, 11, 18, 21, 22, 74, 121, 128, 130, 134, 139
  economic losses to 137–139
  rod catches 121, 131, 138
Fishery decline 4, 22, 75, 82, 85, 85, 130, 135, 138
Fishery status 6, 82, 89, 135, 137
Flue Gas Desulphurization 306
Fluoride 174, 176, 192, 280
Fontinalis squamosa 100, 103

Forestry 1, 8, 12, 14, 18, 19, 22, 24, 28, 31, 33, 36, 67, 70–75, 82, 89, 97, 99, 100, 105, 109, 113, 133, 135, 137, 140, 173, 213, 233, 259, 263, 264, 288, 311, 312, 316
canopy closure 15, 130, 249, 264, 267
clearfelling (harvesting) 2, 3, 177, 312
impact on fisheries 19, 75, 134, 135
impact on hydrology 210, 251, 253, 259, 287
impact on stream biology 97, 113, 294
impact on water chemistry 12, 14, 22, 36, 67, 75, 133, 135, 173, 233, 259, 263, 264, 311, 312
Fossil fuel combustion 27, 31, 51, 52, 285
Frogs 112
Fungi 103

Galloway 27, 33, 81, 104, 270
Gammarus pulex 102, 104, 107, 113, 160, 161
Geology 5, 27, 55, 56, 60, 67, 215, 276, 313
drift 55, 59, 67, 175, 208, 238, 261
Glaslyn 137
Granite 27, 58
Grassland 177, 181, 228
Grazing
invertebrates 94, 96, 103, 105, 107, 109, 169
livestock 27, 28, 31, 33, 68, 82, 85
Groundflora 27, 312
Gwynydd (*see also* Snowdonia) 124, 125, 128, 137

Hafren 176, 177
Hardness 4, 6, 15, 64, 99, 127, 134, 135, 242, 243, 249, 283, 285
Hendriksen's acidification index 16
Hore 176, 177
Hydrogen ion 13, 40, 43, 44, 47, 48, 49, 52, 55, 93, 166, 167, 168, 175, 178, 179, 180, 182, 184, 230, 238, 243, 249, 259, 260
Hydrology
run-off processes 210, 212
modelling 209
stream hydrographs 213, 216, 251
effects on acidity 215, 259
Hydrolysis 59, 197, 229
Hysteresis 216

Indicator systems 18, 22, 115
Invertebrates 1, 17, 93, 94, 95, 96, 99, 102, 103, 104, 105, 106, 107, 109, 110, 111, 113, 115, 160, 166, 280, 285
caddisflies 105, 106, 107, 113, 160
crustaceans 105, 106, 107
density 166
drift 166
herbivores 94, 95
mayflies 17, 95, 105, 106, 107, 109, 113
molluscs 17, 105, 106, 107, 113
shredders 103, 105, 107
stoneflies 95, 106, 109, 113, 161, 283
taxon richness 105, 108, 109, 110
Ion exchange 59, 179, 180, 218, 229, 233, 235, 260
Irfon 18, 112, 113, 114
Iron 133, 146, 147, 148, **149**, 193

Juncus spp 96, 99

Lake sediments 27, 33, 39, 40, 81, 84, 311
Land management 2, 27, 33, 68, 78, 81, 85, 89, 241, 311, 317–326
Land use 2, 5, 24, 31, 62, 63, 67, 73, 76, 77, 81, 84, 85, 89, 99, 130, 180, 214, 233, 239, 241, 259, 287
Larch 224, 312
Leaching 58, 59, 93, 95, 228, 230, 269, 280
Lead 31, 46, 50, 51, 127, 133, 151, 193
Lead-210 dating 28, 33, 40
Leeches 112
Lichens 42
Liming 3, 8, 63, 74, 82, 89, 111, 122, 170, 233, 239, 276, 313–314
catchments 101, 176–7, 314, 315, 320, 326
lakes 83, 96, 122, 154
methods of application 314
Llyn Berwyn 82, 83, 85, 96, 103, 110
Llyn Blaenmelindwr 128, 130
Llyn Brianne 2, **4**, 12, 18, 63, 77, 93, 96, 98, 99, 105, 107, 133, 140, 173, 184, 190, 212, 223, 241, 256, 260, 263, 279, 285, 296, 305, 306
Llyn Conwy 82, 83, 85, 89, 122, 123, 128, 130
Llyn Dulyn 28, 36, 40, 41, 50, 51, 52, 83, 84, 89
Llyn Eiddew Bach 28, 36
Llyn Gamallt 83, 85, 89, 130, 140

Llyn Gynon 13, 28, 36, 83, 84, 89, 99
Llyn Hir 13, 28, 36, 82, 83, 84, 85, 89, 96, 99, 103
Llyn Pendam 96, 128, 130
Llyn Syfydrin 128
Llyn y Bi 28, 30, 31, 32, 33, 36, 83, 84, 85
Loch Dee 256, 260
Lower Palaeozoic sediments 56, 63, 175

Macrophytes 93, 95, 96, 99
MAGIC model **259**, 260–274, 280, 285, 288, 292–294
Magnesium 16, 31, 41, 133, 193, 224, 229, 232, 234, 238, 239, 242, 246, 267, 271, 290, 312, 313
Magnetic analysis 28, 40, 50, **51**
Mammals 1, 93
Mangese 33, 125, 126, 127, 133, 146, 147, 148–9, 154, 190, 193
Mawddach 114, 135, 139
Modelling
  hydrochemical 5, 255, 260–274, 292–294
  hydrological 209, 256–9
  biological 5, 141, 279 *et seq*
  emission-deposition relationships 299
Molinia 3, 224, 229, 285, 312
Monte Carlo Analysis 270, 287
Moorland 3, 6, 7, 8, 11, 12, 14, 15, 19, 22, 27, 28, 33, 68, 75, 82, 84, 97, 100, 104, 105, 109, 111, 185, 213, 233, 238, 243, 251, 253, 261, 313
  reclamation 69, 75

Nardia compressa 100, 103, 105
Nardus-Festuca grassland 3, 84, 85, 104, 175, 224, 285
Neutralizing capacity – *see* Buffering capacity
Newts 112
Nitrate 42, 43, 45, 47, 48, 49, 52, 74, 224, 232, 242, 243, 246, 251, 261, 280, 303, 305, 312
Nitrification 75, 99
Nitrogen dioxide (also NO_X) 42, 224, 301, 303, 304–308
Nutrient cycling 93, 311

Oak woodland – *see* Deciduous woodland
Oil combustion 51
Ordovician sediments (*see also* Lower Palaeozoic) 2, 3, 4, 56, 175, 238, 269
Organic acids 59, 95, 147, 148, 230, 234, 238
Otters 112, 113, 114

Palaeoecology 11, 14, 36, 40, 274
Palaeolimnology 27, 36
Perch 110, 121, 154
Pesticides 113
pH – references numerous and not included
Phosphorus 94, 95
Photosynthesis 95
Pine 4
Ploughing 27, 28, 33, 259, 294, **316**, 320, 326
Plynlimon **2**, 4, 12, 16, 40, 47, 77, 173, 175, 176, 178, 223, 224, 228, 256, 260, 313
Pollen analysis 28, 81, 84
Potable supplies 1, 75, 78, 145 *et seq*
  private supplies 145, 154, 156
Potassium 16, 31, 168, 224, 229, 232, 242, 249, 315
Power stations 42, 51, 306
Primary production 93, 94, 95, 99, 103, 105

Quartz 174

Rainfall
  quality – *see* Atmospheric deposition
  quantity 205
RAIN project 260
Respiration 95
Rheidol 14, 135

Salmon 20, 22, 121, 133, 135, 137, 138, 160, 161
Salmonids 1, 19, 74, 75, 78, 110, 121, 124, 133, 135, 137, 139, 294, 295
Scandanavia 1, 13, 230, 260
Scapania 18, 100, 104, 109
Scavenging processes 89, 261, 276
Scotland (*see also* Galloway) 21, 39
Sea salts 40, 41, 42, 45, 52, 216, 218, 232, 267, 276
Severn 2, 3, 16, 114, 176, 205
Sheep 31, 33, 67, 68, 69, 70, 73, 85
Shredding 103, 105, 107
Silicates 55, 59, 178, 229, 238, 313
Silurian sediments (*see also* Lower

Palaeozoic) 3, 4, 28, 56, 238, 261, 269
Sitka spruce 2, 4, 104, 176, 224–227, 228, 285, 312
Smelting 42, 51
Snowdonia 2, 28
Snowfall 205
Snowmelt 95, 159, 196, 251, **255**, 295
Sodium 16, 31, 41, 42, 43, 122, 123, 160, 165, 166, 168, 169, 218, 224, 229, 232, 239, 242, 243, 246, 312
Soils 55, 58–60, 67, 175, 209, 215, 261, 276, 313
  brown earths 3, 175
  brown podzolic soils 60, 62, 212, 234, 312, 320
  peats 3, 59, 60, 62, 130, 147, 210, 316
  podzols 60, 178, 210
  stagnogleys 3, 175
  stagnopodzols 2, 3, 62, 175, 212, **229**, 234, 238, 312, 320
Soil pipes 212
Soil waters 178, 180, 185, **229–235**, 312, 320
Stemflow 3, 12, 223, **228**
Sulphate 12, 16, 17, 27, 31, 41, 42, 45, 47, 48, 49, 52, 176, 197, 224, 228, 229, 233, 242, 243, 246, 249, 251, 253, 261, 263, 267, 271, 274, 280, 301, 303, 305, 312
Sulphur dioxide 42, 47, 51, 224, 301, 303
Suspended Solids 190, 199

Teifi 21
Throughfall 3, 12, 223, **224–226**
Till 58
TOPMODEL 256, 315
Toxicity 159, 160, 167
  fish 122, 128, 130, 133, 137, 140, 161, 280, 288
  invertebrates 93, 107, 109, 161
Trihalomethanes 148, 154
Trophic levels 93, 94, 107, 140
TWINSPAN 99, 101, 102, 110, 280, 283
Tywi 4, 14, 16, 17, 18, 21, 114, 135, 284

Vegetation type 3, 5, 323
Vehicle exhausts 42

Weathering 179, 180, 183, 229, 233, 238, 260, 261, 266, 274, 280, 294
Wetlands 11
Wye 2, 3, 14, 15, 20, 104, 112, 135, 139, 175, 205, 284

Ystwyth 14, 15, 114, 135

Zinc 31, 46, 50, 125, 127, 133, 135, 151, 152, 153, 282